Graduate Texts in Mathematics

38

H. Grauert

K. Fritzsche

Several Complex Variables

Springer-Verlag

New York Heidelberg Berlin

H. Grauert
Mathematischen Institut der Universität
Bunsenstrasse 3–5
34 Göttingen
Federal Republic of Germany

K. Fritzsche
Mathematischen Institut der Universität
Bunsenstrasse 3–5
34 Göttingen
Federal Republic of Germany

AMS Subject Classifications: 32-01, 32A05, 32A07, 32A10, 32A20, 32B10, 32C10, 32C35, 32D05, 32D10, 32E10

Library of Congress Cataloging in Publication Data

Grauert, Hans, 1930–
Several complex variables.
(Graduate texts in mathematics; 38)
Translation of Einführung in die Funktionentheorie mehrerer Veränderlicher.
Bibliography: p. 201
Includes index.
1. Functions of several complex variables.
I. Fritzsche, Klaus, joint author. II. Title. III. Series.
QA331.G69 515'.94 75-46503

Softcover reprint of the hardcover 1st edition 1976

ISBN-13: 978-1-4612-9876-2 e-ISBN-13: 978-1-4612-9874-8
DOI: 10.1007/978-1-4612-9874-8

Preface

The present book grew out of introductory lectures on the theory of functions of several variables. Its intent is to make the reader familiar, by the discussion of examples and special cases, with the most important branches and methods of this theory, among them, e.g., the problems of holomorphic continuation, the algebraic treatment of power series, sheaf and cohomology theory, and the real methods which stem from elliptic partial differential equations.

In the first chapter we begin with the definition of holomorphic functions of several variables, their representation by the Cauchy integral, and their power series expansion on Reinhardt domains. It turns out that, in contrast to the theory of a single variable, for $n \geqslant 2$ there exist domains $G, \hat{G} \subset \mathbb{C}^n$ with $G \subset \hat{G}$ and $G \neq \hat{G}$ such that each function holomorphic in G has a continuation on $\hat{G}$. Domains G for which such a $\hat{G}$ does not exist are called *domains of holomorphy*. In Chapter 2 we give several characterizations of these domains of holomorphy (theorem of Cartan–Thullen, Levi's problem). We finally construct the holomorphic hull $H(G)$ for each domain G, that is the largest (not necessarily schlicht) domain over $\mathbb{C}^n$ into which each function holomorphic on G can be continued.

The third chapter presents the Weierstrass formula and the Weierstrass preparation theorem with applications to the ring of convergent power series. It is shown that this ring is a factorization, a Noetherian, and a Hensel ring. Furthermore we indicate how the obtained algebraic theorems can be applied to the local investigation of analytic sets. One achieves deep results in this connection by using sheaf theory, the basic concepts of which are discussed in the fourth chapter. In Chapter V we introduce complex manifolds and give several examples. We also examine the different closures of $\mathbb{C}^n$ and the effects of modifications on complex manifolds.

Cohomology theory with values in analytic sheaves connects sheaf theory with the theory of functions on complex manifolds. It is treated and applied in Chapter VI in order to express the main results for domains of holomorphy and Stein manifolds (for example, the solvability of the Cousin problems).

The seventh chapter is entirely devoted to the analysis of real differentiability in complex notation, partial differentiation with respect to z, $\bar{z}$, and complex functional matrices, topics already mentioned in the first chapter. We define tangential vectors, differential forms, and the operators d, d', d''. The theorems of Dolbeault and de Rham yield the connection with cohomology theory.

The authors develop the theory in full detail and with the help of numerous figures. They refer to the literature for theorems whose proofs exceed the scope of the book. Presupposed are only a basic knowledge of differential and integral calculus and the theory of functions of one variable, as well as a few elements from vector analysis, algebra, and general topology. The book is written as an introduction and should be of interest to the specialist and the nonspecialist alike.

Göttingen, Spring 1976

H. Grauert
K. Fritzsche

Contents

CHAPTER I
Holomorphic Functions

Preliminaries

Let $\mathbb{C}$ be the field of complex numbers. If n is a natural number we call the set of ordered n-tuples of complex numbers the *n-dimensional complex number space*:

$$\mathbb{C}^n := \{\mathfrak{z} = (z_1, \ldots, z_n) : z_\nu \in \mathbb{C} \text{ for } 1 \leqslant \nu \leqslant n\}.$$

Each component of a point $\mathfrak{z} \in \mathbb{C}^n$ can be decomposed uniquely into real and imaginary parts: $z_\nu = x_\nu + iy_\nu$. This gives a unique 1—1 correspondence between the elements $(z_1, \ldots, z_n)$ of $\mathbb{C}^n$ and the elements $(x_1, \ldots, x_n, y_1, \ldots, y_n)$ of $\mathbb{R}^{2n}$, the $2n$-dimensional space of real numbers.

$\mathbb{C}^n$ is a vector space: addition of two elements as well as the multiplication of an element of $\mathbb{C}^n$ by a (real or complex) scalar is defined componentwise. As a complex vector space $\mathbb{C}^n$ is n-dimensional; as a real vector space it is $2n$-dimensional. It is clear that the $\mathbb{R}$ vector space isomorphism between $\mathbb{C}^n$ and $\mathbb{R}^{2n}$ leads to a topology on $\mathbb{C}^n$: For $\mathfrak{z} = (z_1, \ldots, z_n) = (x_1 + iy_1, \ldots, x_n + iy_n) \in \mathbb{C}^n$ let

$$\|\mathfrak{z}\| := \left(\sum_{k=1}^{n} z_k \bar{z}_k\right)^{1/2} = \left(\sum_{k=1}^{n} (x_k^2 + y_k^2)\right)^{1/2},$$
$$\|\mathfrak{z}\|^* := \max_{k=1,\ldots,n} (|x_k|, |y_k|).$$

Norms are defined on $\mathbb{C}^n$ by $\mathfrak{z} \mapsto \|\mathfrak{z}\|$ and $\mathfrak{z} \mapsto \|\mathfrak{z}\|^*$, with corresponding metrics given by

$$\operatorname{dist}(\mathfrak{z}_1, \mathfrak{z}_2) := \|\mathfrak{z}_1 - \mathfrak{z}_2\|,$$
$$\operatorname{dist}^*(\mathfrak{z}_1, \mathfrak{z}_2) := \|\mathfrak{z}_1 - \mathfrak{z}_2\|^*.$$

In each case we obtain a topology on $\mathbb{C}^n$ which agrees with the usual topology for $\mathbb{R}^{2n}$. Another metric on $\mathbb{C}^n$, defined by $|\mathfrak{z}| := \max_{k=1,\dots,n} |z_k|$ and $\text{dist}'(\mathfrak{z}_1, \mathfrak{z}_2) := |\mathfrak{z}_1 - \mathfrak{z}_2|$, induces the usual topology too.

A *region* $B \subset \mathbb{C}^n$ is an open set (with the usual topology) and a *domain* an open, connected set. An open set $G \subset \mathbb{C}^n$ is called *connected* if one of the following two equivalent conditions is satisfied:

a. For every two points $\mathfrak{z}_1, \mathfrak{z}_2 \in G$ there is a continuous mapping $\varphi:[0, 1] \to \mathbb{C}^n$ with $\varphi(0) = \mathfrak{z}_1$, $\varphi(1) = \mathfrak{z}_2$, and $\varphi([0, 1]) \subset G$.
b. If $B_1, B_2 \subset G$ are open sets with $B_1 \cup B_2 = G$, $B_1 \cap B_2 = \varnothing$ and $B_1 \neq \varnothing$, then $B_2 = \varnothing$.

Definition. Let $B \subset \mathbb{C}^n$ be a region, $\mathfrak{z}_0 \in B$ a point. The set $C_B(\mathfrak{z}_0) := \{\mathfrak{z} \in B : \mathfrak{z}$ and $\mathfrak{z}_0$ can be joined by a path in $B\}$ is called the *component of* $\mathfrak{z}_0$ *in B.*

Remark. Let $B \subset \mathbb{C}^n$ be an open set. Then:

a. For each $\mathfrak{z} \in B$, $C_B(\mathfrak{z})$ and $B - C_B(\mathfrak{z})$ are open sets.
b. For each $\mathfrak{z} \in B$, $C_B(\mathfrak{z})$ is connected.
c. From $C_B(\mathfrak{z}_1) \cap C_B(\mathfrak{z}_2) \neq \varnothing$ it follows that $C_B(\mathfrak{z}_1) = C_B(\mathfrak{z}_2)$.
d. $B = \bigcup_{\mathfrak{z} \in B} C_B(\mathfrak{z})$
e. If G is a domain with $\mathfrak{z} \in G \subset B$, it follows that $G \subset C_B(\mathfrak{z})$.
f. B has at most countably many components.

The proof is trivial.

Finally for $\mathfrak{z}_0 \in \mathbb{C}^n$ we define:

$$U_\varepsilon(\mathfrak{z}_0) := \{\mathfrak{z} \in \mathbb{C}^n : \text{dist}(\mathfrak{z}, \mathfrak{z}_0) < \varepsilon\},$$
$$U_\varepsilon^*(\mathfrak{z}_0) := \{\mathfrak{z} \in \mathbb{C}^n : \text{dist}^*(\mathfrak{z}, \mathfrak{z}_0) < \varepsilon\},$$
$$U_\varepsilon'(\mathfrak{z}_0) := \{\mathfrak{z} \in \mathbb{C}^n : \text{dist}'(\mathfrak{z}, \mathfrak{z}_0) < \varepsilon\}.$$

1. Power Series

Let M be a subset of $\mathbb{C}^n$. A mapping f from M to $\mathbb{C}$ is called a complex function on M. The polynomials

$$p(\mathfrak{z}) = \sum_{\nu_1,\dots,\nu_n=0}^{m_1,\dots,m_n} a_{\nu_1,\dots,\nu_n} z_1^{\nu_1} \cdot \ldots \cdot z_n^{\nu_n}, \qquad a_{\nu_1,\dots,\nu_n} \in \mathbb{C},$$

are particularly simple examples, defined on all of $\mathbb{C}^n$. In order to simplify notation we introduce multi-indices: let ν_i, $1 \leqslant i \leqslant n$, be non-negative integers and let $\mathfrak{z} = (z_1, \dots, z_n)$ be a point of $\mathbb{C}^n$. Then we define:

$$\nu := (\nu_1, \dots, \nu_n), \qquad |\nu| := \sum_{i=1}^n \nu_i, \qquad \mathfrak{z}^\nu := \prod_{i=1}^n z_i^{\nu_i}.$$

With this notation a polynomial has the form $p(\mathfrak{z}) = \sum_{\nu=0}^m a_\nu \mathfrak{z}^\nu$.

Def. 1.1. Let $\mathfrak{z}_0 \in \mathbb{C}^n$ be a point and for $|\nu| \geqslant 0$, a_ν be a complex number. Then the expression

$$\sum_{\nu=0}^{\infty} a_\nu(\mathfrak{z} - \mathfrak{z}_0)^\nu$$

is called a *formal power series* about $\mathfrak{z}_0$.

Now such an expression has, as the name says, only a formal meaning. For a particular $\mathfrak{z}$ it does not necessarily represent a complex number. Since the multi-indices can be ordered in several ways it is not clear how the summation is to be performed. Therefore we must introduce a suitable notion of convergence.

Def. 1.2. Let $\mathfrak{I} := \{\nu = (\nu_1, \ldots, \nu_n) : \nu_i \geqslant 0 \text{ for } 1 \leqslant i \leqslant n\}$, and $\mathfrak{z}_1 \in \mathbb{C}^n$ be fixed. We say that $\sum_{\nu=0}^{\infty} a_\nu(\mathfrak{z}_1 - \mathfrak{z}_0)^\nu$ *converges to the complex number* c if for each $\varepsilon > 0$ there exists a finite set $I_0 \subset \mathfrak{I}$ such that for any finite set I with $I_0 \subset I \subset \mathfrak{I}$

$$\left| \sum_{\nu \in I} a_\nu(\mathfrak{z}_1 - \mathfrak{z}_0)^\nu - c \right| < \varepsilon.$$

One then writes $\sum_{\nu=0}^{\infty} a_\nu(\mathfrak{z} - \mathfrak{z}_0)^\nu = c$.
Convergence in this sense is synonymous with absolute convergence.

Def. 1.3. Let M be a subset of $\mathbb{C}^n$, $\mathfrak{z}_0 \in M$, f a complex function on M. One says that the power series $\sum_{\nu=0}^{\infty} a_\nu(\mathfrak{z} - \mathfrak{z}_0)^\nu$ *converges uniformly* on M to $f(\mathfrak{z})$ if for each $\varepsilon > 0$ there is a finite set $I_0 \subset \mathfrak{I}$ such that

$$\left| \sum_{\nu \in I} a_\nu(\mathfrak{z} - \mathfrak{z}_0)^\nu - f(\mathfrak{z}) \right| < \varepsilon$$

for each finite I with $I_0 \subset I \subset \mathfrak{I}$ and each $\mathfrak{z} \in M$.

$\sum_{\nu=0}^{\infty} a_\nu(\mathfrak{z} - \mathfrak{z}_0)^\nu$ converges uniformly in the interior of a region B if the series converges uniformly in each compact subset of B.

Def. 1.4. Let $B \subset \mathbb{C}^n$ be a region and f be a complex function on B. f is called *holomorphic* in B if for each $\mathfrak{z}_0 \in B$ there is a neighborhood $U = U(\mathfrak{z}_0)$ in B and a power series $\sum_{\nu=0}^{\infty} a_\nu(\mathfrak{z} - \mathfrak{z}_0)^\nu$ which converges on U to $f(\mathfrak{z})$.

Note that uniform convergence on U is not required. We show now why pointwise convergence suffices.

Def. 1.5. The point set $V = \{\mathfrak{r} = (r_1, \dots, r_n) \in \mathbb{R}^n : r_\nu \geqslant 0 \text{ for } 1 \leqslant \nu \leqslant n\}$ will be called *absolute space*. $\tau : \mathbb{C}^n \to V$ with $\tau(\mathfrak{z}) := (|z_1|, \dots, |z_n|)$ is the *natural projection* of $\mathbb{C}^n$ onto V.

V is a subset of $\mathbb{R}^n$ and as such inherits the topology induced from $\mathbb{R}^n$ to V (relative topology). Then $\tau : \mathbb{C}^n \to V$ is a continuous surjective mapping. If $B \subset V$ is open, then $\tau^{-1}(B) \subset \mathbb{C}^n$ is also open.

Def. 1.6. Let $\mathfrak{r} \in V_+ := \{\mathfrak{r} = (r_1, \dots, r_n) \in \mathbb{R}^n : r_k > 0\}$, $\mathfrak{z}_0 \in \mathbb{C}^n$. Then $P_\mathfrak{r}(\mathfrak{z}_0) := \{\mathfrak{z} \in \mathbb{C}^n : |z_k - z_k^{(0)}| < r_k \text{ for } 1 \leqslant k \leqslant n\}$ is called the *polycylinder about* $\mathfrak{z}_0$ with *(poly-)radius* $\mathfrak{r}$. $T = T(P) := \{\mathfrak{z} \in \mathbb{C}^n : |z_k - z_k^{(0)}| = r_k\}$ is called the *distinguished boundary of P* (see Fig. 1).

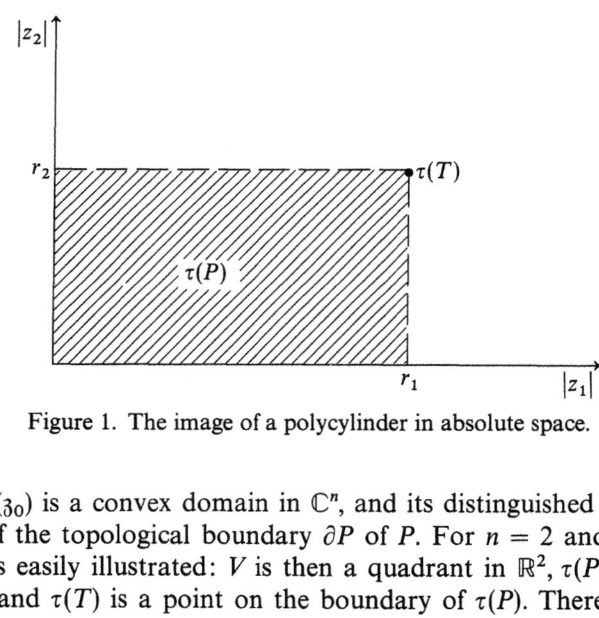

Figure 1. The image of a polycylinder in absolute space.

$P = P_\mathfrak{r}(\mathfrak{z}_0)$ is a convex domain in $\mathbb{C}^n$, and its distinguished boundary is a subset of the topological boundary ∂P of P. For $n = 2$ and $\mathfrak{z}_0 = 0$ the situation is easily illustrated: V is then a quadrant in $\mathbb{R}^2$, $\tau(P)$ is an open rectangle, and $\tau(T)$ is a point on the boundary of $\tau(P)$. Therefore

$$\begin{aligned} T &= \{\mathfrak{z} \in \mathbb{C}^2 : |z_1| = r_1, |z_2| = r_2\} \\ &= \{\mathfrak{z} = (r_1 \cdot e^{i\theta_1}, r_2 \cdot e^{i\theta_2}) \in \mathbb{C}^2 : 0 \leqslant \theta_1 < 2\pi, 0 \leqslant \theta_2 < 2\pi\} \end{aligned}$$

is a 2-dimensional torus. Similarly in the n-dimensional case we get an n-dimensional torus (the cartesian product of n circles).

If $\mathfrak{z}_1 \in \mathring{\mathbb{C}}^n := \{\mathfrak{z} = (z_1, \dots, z_n) \in \mathbb{C}^n : z_k \neq 0 \text{ for } 1 \leqslant k \leqslant n\}$, then $P_{\mathfrak{z}_1} := \{\mathfrak{z} \in \mathbb{C}^n : |z_k| < |z_k^{(1)}| = r_k \text{ for } 1 \leqslant k \leqslant n\}$ is a polycylinder about 0 with radius $\mathfrak{r} = (r_1, \dots, r_n)$.

Theorem 1.1. *Let* $\mathfrak{z}_1 \in \mathring{\mathbb{C}}^n$. *If the power series* $\sum_{\nu=0}^{\infty} a_\nu \mathfrak{z}^\nu$ *converges at* $\mathfrak{z}_1$, *then it converges uniformly in the interior of the polycylinder* $P_{\mathfrak{z}_1}$.

PROOF

1. Since the series converges at $\mathfrak{z}_1$, the set $\{a_\nu \mathfrak{z}_1^\nu : |\nu| \geqslant 0\}$ is bounded. Let $M \in \mathbb{R}$ be chosen so that $|a_\nu \mathfrak{z}_1^\nu| < M$ for all ν. If $\mathfrak{z}_1 \in \mathring{\mathbb{C}}^n$ and $0 < q < 1$ then $q \cdot \mathfrak{z}_1 \in \mathring{\mathbb{C}}^n$. Let $P^*: = P_{q \cdot \mathfrak{z}_1}$. For $\mathfrak{z} \in P^*$, $|\mathfrak{z}^\nu| = |z_1|^{\nu_1} \cdot \ldots \cdot |z_n|^{\nu_n} < |q \cdot z_1^{(1)}|^{\nu_1} \cdot \ldots \cdot |q \cdot z_n^{(1)}|^{\nu_n} = q^{\nu_1 + \cdots + \nu_n} \cdot |z_1^{(1)}|^{\nu_1} \cdot \ldots \cdot |z_n^{(1)}|^{\nu_n} = q^{|\nu|} \cdot |\mathfrak{z}_1^\nu|$, that is, $\sum_{\nu=0}^{\infty} |a_\nu| \cdot |\mathfrak{z}_1^\nu| \cdot q^{|\nu|}$ is a majorant of $\sum_{\nu=0}^{\infty} a_\nu \mathfrak{z}^\nu$ and therefore

$$M \cdot \sum_{\nu=0}^{\infty} q^{\nu_1 + \cdots + \nu_n} = M \cdot \left(\sum_{\nu_1=0}^{\infty} q^{\nu_1} \right) \cdot \ldots \cdot \left(\sum_{\nu_n=0}^{\infty} q^{\nu_n} \right) = M \cdot \left(\frac{1}{1-q} \right)^n.$$

The set $\mathfrak{J}$ of multi-indices is countable, so there exists a bijection $\Phi: \mathbb{N}_0 \to \mathfrak{J}$. Let $b_n(\mathfrak{z}): = a_{\Phi(n)} \cdot \mathfrak{z}^{\Phi(n)}$. Then $\sum_{n=0}^{\infty} b_n(\mathfrak{z})$ is absolutely and uniformly convergent on P^*. Given $\varepsilon > 0$ there is an $n_0 \in \mathbb{N}$ such that $\sum_{n=n_0+1}^{\infty} |b_n(\mathfrak{z})| < \varepsilon$ on P^*. Let $I_0: = \Phi(\{0, 1, 2, \ldots, n_0\})$. If I is a finite set with $I_0 \subset I \subset \mathfrak{J}$, then $\{0, 1, \ldots, n_0\} \subset \Phi^{-1}(I)$, so

$$\left| \sum_{n=0}^{\infty} b_n(\mathfrak{z}) - \sum_{\nu \in I} a_\nu \mathfrak{z}^\nu \right| = \left| \sum_{n=0}^{\infty} b_n(\mathfrak{z}) - \sum_{n \in \Phi^{-1}(I)} b_n(\mathfrak{z}) \right|$$
$$= \left| \sum_{n \notin \Phi^{-1}(I)} b_n(\mathfrak{z}) \right| \leqslant \sum_{n=n_0+1}^{\infty} |b_n(\mathfrak{z})| < \varepsilon \text{ for } \mathfrak{z} \in P^*.$$

But then $\sum_{\nu=0}^{\infty} a_\nu \mathfrak{z}^\nu$ is uniformly convergent in P^*.

2. Let $K \subset P_{\mathfrak{z}_1}$ be compact. $\{P_{q \cdot \mathfrak{z}_1}: 0 < q < 1\}$ is an open covering of $P_{\mathfrak{z}_1}$, and thus of K. But then there is a finite subcovering $\{P_{q_1 \cdot \mathfrak{z}_1}, \ldots, P_{q_\ell \cdot \mathfrak{z}_1}\}$. If we set $q: = \max(q_1, \ldots, q_\ell)$, then $K \subset P_{q \cdot \mathfrak{z}_1}$, and $P_{q \cdot \mathfrak{z}_1}$ is a P^* such as in 1). Therefore $\sum_{\nu=0}^{\infty} a_\nu \mathfrak{z}^\nu$ is uniformly convergent on K, which was to be shown. □

Next we shall examine on what sets power series converge. In order to be brief we choose $\mathfrak{z}_0 = 0$ as our point of expansion. The corresponding statements always hold in the general case.

Def. 1.7. An open set $B \subset \mathbb{C}^n$ is called a *Reinhardt domain* if $\mathfrak{z}_1 \in B \Rightarrow T_{\mathfrak{z}_1}: = \tau^{-1}\tau(\mathfrak{z}_1) \subset B$.

Comments. $T_{\mathfrak{z}_1}$ is the torus $\{\mathfrak{z} \in \mathbb{C}^n: |z_k| = |z_k^{(1)}|\}$. The conditions of definition 1.7 mean that $\tau^{-1}\tau(B) = B$; a Reinhardt domain is characterized by its image $\tau(B)$ in absolute space.

Theorem 1.2. *An open set $B \subset \mathbb{C}^n$ is a Reinhardt domain if and only if there exists an open set $W \subset V$ with $B = \tau^{-1}(W)$.*

PROOF

1. Let $B = \tau^{-1}(W)$, $W \subset V$ open. For $\mathfrak{z} \in B$, $\tau(\mathfrak{z}) \in W$; therefore $\tau^{-1}\tau(\mathfrak{z}) \subset \tau^{-1}(W) = B$.

2. Let B be a Reinhardt domain. Then $B = \tau^{-1}\tau(B)$ and it suffices to show that $\tau(B)$ is open in V. Assume that $\tau(B)$ is not open. Then there is a point $\mathfrak{r}_0 \in \tau(B)$ which is not an interior point of $\tau(B)$ and therefore is a cluster point of $V - \tau(B)$. Let $(\mathfrak{r}_j)$ be a sequence in $V - \tau(B)$ which converges to $\mathfrak{r}_0$. There are points $\mathfrak{z}_j \in \mathbb{C}^n$ with $\mathfrak{r}_j = \tau(\mathfrak{z}_j)$, so that $|z_p^{(j)}| = r_p^{(j)}$ for all j and $1 \leqslant p \leqslant n$. Since $(\mathfrak{r}_j)$ is convergent there is an $M \in \mathbb{R}$ such that $|r_p^{(j)}| < M$ for all j and p. Hence the sequence $(\mathfrak{z}_j)$ is also bounded. It must have a cluster point $\mathfrak{z}_0$, and a subsequence $(\mathfrak{z}_{j_\nu})$ with $\lim_{\nu \to \infty} \mathfrak{z}_{j_\nu} = \mathfrak{z}_0$. Since τ is continuous $\tau(\mathfrak{z}_0) = \lim_{\nu \to \infty} \tau(\mathfrak{z}_{j_\nu}) = \lim_{\nu \to \infty} \mathfrak{r}_{j_\nu} = \mathfrak{r}_0$. B is a Reinhardt domain; it follows that $\mathfrak{z}_0 \in \tau^{-1}(\mathfrak{r}_0) \subset \tau^{-1}\tau(B) = B$. B is an open neighborhood of $\mathfrak{z}_0$; therefore almost all $\mathfrak{z}_{j_\nu}$ must lie in B, and then almost all $\mathfrak{r}_{j_\nu} = \tau(\mathfrak{z}_{j_\nu})$ must lie in $\tau(B)$. This is a contradiction, and therefore $\tau(B)$ is open. □

The image of a Reinhardt domain in absolute space is always an open set (of arbitrary form), and the inverse image of this set is again the domain.

Def. 1.8. Let $G \subset \mathbb{C}^n$ be a Reinhardt domain.

1. G is called *proper* if

   a. G is connected, and
   b. $0 \in G$.

2. G is called *complete* if

$$\mathfrak{z}_1 \in G \cap \mathring{\mathbb{C}}^n \Rightarrow P_{\mathfrak{z}_1} \subset G.$$

Figure 2 illustrates Def. 1.8. for the case $n = 2$ in absolute space.

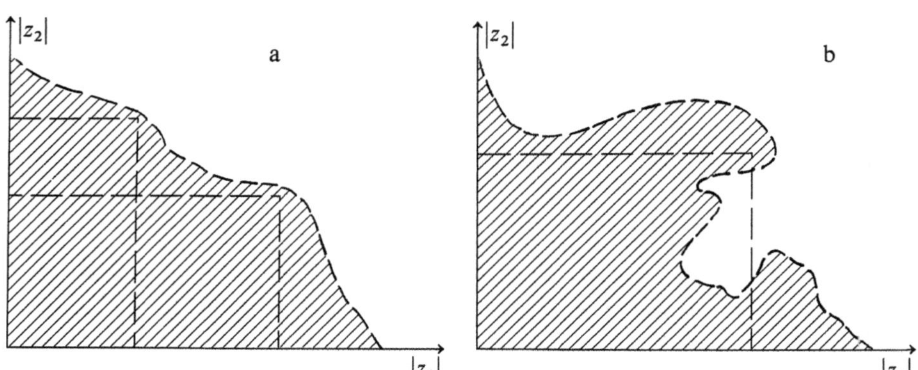

Figure 2. (a) Complete Reinhardt domain; (b) Proper Reinhardt domain.

For $n = 1$ Reinhardt domains are the unions of open annuli. There is no difference between complete and proper Reinhardt domains in this case; we are dealing with open circular discs.

Clearly for $n > 1$ the polycylinders and balls $K = \{\mathfrak{z}: |z_1|^2 + \cdots + |z_n|^2 < R^2\}$ are proper and complete Reinhardt domains. In general:

Theorem 1.3. *Every complete Reinhardt domain is proper.*

PROOF. Let G be a complete Reinhardt domain. There exists a point $\mathfrak{z}_1 \in G \cap \mathring{\mathbb{C}}^n$, and by definition $0 \in P_{\mathfrak{z}_1} \subset G$. It remains to show that G is connected.

a. Let $\mathfrak{z}_1 \in G$ be a point in a general position (i.e., $\mathfrak{z}_1 \in G \cap \mathring{\mathbb{C}}^n$). Then the connecting line segment between $\mathfrak{z}_1$ and 0 lies entirely within $P_{\mathfrak{z}_1}$ and hence within G.

b. $\mathfrak{z}_1$ lies on one of the "axes." Since G is open there exists a neighborhood $U_\varepsilon(\mathfrak{z}_1) \subset G$, and we can find a point $\mathfrak{z}_2 \in U_\varepsilon(\mathfrak{z}_1) \cap \mathring{\mathbb{C}}^n$. Hence there is a path in U_ε which connects $\mathfrak{z}_1$ and $\mathfrak{z}_2$, and a path in G which connects $\mathfrak{z}_2$ and 0. Together they give a path in G which joins $\mathfrak{z}_1$ and 0.

From (a) and (b) it follows that G is connected. □

Let $\mathfrak{P}(\mathfrak{z}) = \sum_{\nu=0}^{\infty} a_\nu \mathfrak{z}^\nu$ be a power series about zero. The set $M \subset \mathbb{C}^n$ on which $\mathfrak{P}(\mathfrak{z})$ converges is called the *convergence set* of $\mathfrak{P}(\mathfrak{z})$. $\mathfrak{P}(\mathfrak{z})$ always converges in $\mathring{M}$ and diverges outside $\bar{M}$. $B(\mathfrak{P}(\mathfrak{z})): = \mathring{M}$ is called the *region of convergence* of the power series $\mathfrak{P}(\mathfrak{z})$.

Theorem 1.4. *Let $\mathfrak{P}(\mathfrak{z}) = \sum_{\nu=0}^{\infty} a_\nu \mathfrak{z}^\nu$ be a formal power series in $\mathbb{C}^n$. Then the region of convergence $B = B(\mathfrak{P}(\mathfrak{z}))$ is a complete Reinhardt domain. $\mathfrak{P}(\mathfrak{z})$ converges uniformly in the interior of B.*

PROOF

1. Let $\mathfrak{z}_1 \in B$. Then $U'_\varepsilon(\mathfrak{z}_1) = \{\mathfrak{z} \in \mathbb{C}^n: |\mathfrak{z} - \mathfrak{z}_1| < \varepsilon\} = U_\varepsilon(z_1^{(1)}) \times \cdots \times U_\varepsilon(z_n^{(1)})$ is a polycylinder about $\mathfrak{z}_1$ with radius $(\varepsilon, \ldots, \varepsilon)$. For a sufficiently small ε, $U'_\varepsilon(\mathfrak{z}_1)$ lies in B. For $k = 1, \ldots, n$ we can find a $z_k^{(2)} \in U_\varepsilon(z_k^{(1)})$ such that $|z_k^{(2)}| > |z_k^{(1)}|$. Let $\mathfrak{z}_2: = (z_1^{(2)}, \ldots, z_n^{(2)})$. Then $\mathfrak{z}_2 \in B$ and $\mathfrak{z}_1 \in P_{\mathfrak{z}_2}$. For each point $\mathfrak{z}_1 \in B$ choose such a fixed point $\mathfrak{z}_2$.

2. If $\mathfrak{z}_1 \in B$, then there is a $\mathfrak{z}_2 \in B$ with $\mathfrak{z}_1 \in P_{\mathfrak{z}_2}$. $\mathfrak{P}(\mathfrak{z})$ converges at $\mathfrak{z}_2$, therefore in $P_{\mathfrak{z}_2}$ (from Theorem 1.1). Hence $P_{\mathfrak{z}_2} \subset B$. Since $P_{\mathfrak{z}_1} \subset P_{\mathfrak{z}_2}$ and $T_{\mathfrak{z}_1} \subset P_{\mathfrak{z}_2}$, it follows that B is a complete Reinhardt domain.

3. Let $P^*_{\mathfrak{z}_1}: = P_{\mathfrak{z}_2}$ where $\mathfrak{z}_2$ is chosen for $\mathfrak{z}_1$ as in 1). Clearly $B = \bigcup_{\mathfrak{z}_1 \in B} P^*_{\mathfrak{z}_1}$. Now for each $\mathfrak{z}_2$ select a q with $0 < q < 1$ and such that $\mathfrak{z}_3: = (1/q)\mathfrak{z}_2$ lies in B. This is possible and it follows that for each $\mathfrak{z}_1 \in B$ $\mathfrak{P}(\mathfrak{z})$ is uniformly convergent in $P^*_{\mathfrak{z}_1}$. If $K \subset B$ is compact, then K can be covered by a finite number of sets $P^*_{\mathfrak{z}_1}$. Therefore $\mathfrak{P}(\mathfrak{z})$ converges uniformly on K. □

The question arises whether every complete Reinhardt domain is the region of convergence for some power series. This is not true; additional properties are necessary. However, we shall not pursue this matter here.

Since each complete Reinhardt domain is connected, we can speak of the *domain of convergence* of a power series. We now return to the notion of holomorphy.

Let f be a holomorphic function on a region B, $\mathfrak{z}_0$ a point in B. Let the power series $\sum_{\nu=0}^{\infty} a_\nu(\mathfrak{z} - \mathfrak{z}_0)^\nu$ converge to $f(\mathfrak{z})$ in a neighborhood U of $\mathfrak{z}_0$. Then there is a $\mathfrak{z}_1 \in U$ with $z_\nu^{(1)} \neq z_\nu^{(0)}$ for $1 \leqslant \nu \leqslant n$ and $P_{\mathfrak{r}(\mathfrak{z}_1 - \mathfrak{z}_0)}(\mathfrak{z}_0) \subset U$. Now let $0 < \varepsilon < \min_{\nu=1,\ldots,n} (|z_\nu^{(1)} - z_\nu^{(0)}|)$. From Theorem 1.1 the series converges uniformly on $U'_\varepsilon(\mathfrak{z}_0)$. For each $\nu \in \mathfrak{J}$ one can regard $a_\nu(\mathfrak{z} - \mathfrak{z}_0)^\nu$ as a complex-valued function on $\mathbb{R}^{2n}$. This function is clearly continuous at $\mathfrak{z}_0$ and consequently the limit function is continuous at $\mathfrak{z}_0$. We have:

Theorem 1.5. *Let $B \subset \mathbb{C}^n$ be a region, and f a function holomorphic on B. Then f is continuous on B.*

2. Complex Differentiable Functions

Def. 2.1. Let $B \subset \mathbb{C}^n$ be a region, $f: B \to \mathbb{C}$ a complex function. f is called *complex differentiable at* $\mathfrak{z}_0 \in B$ if there exist complex functions $\Delta_1, \ldots, \Delta_n$ on B which are all continuous at $\mathfrak{z}_0$ and which satisfy the equality

$$f(\mathfrak{z}) = f(\mathfrak{z}_0) + \sum_{\nu=1}^{n} (z_\nu - z_\nu^{(0)})\, \Delta_\nu(\mathfrak{z}) \text{ in } B.$$

Differentiability is a local property. If there exists a neighborhood $U = U(\mathfrak{z}_0) \subset B$ such that $f|U$ is complex differentiable at $\mathfrak{z}_0$, then $f|B$ is complex differentiable at $\mathfrak{z}_0$ since the functions $\Delta_\nu(\mathfrak{z})$ can be continued outside U in such a way that the desired equation holds.

At $\mathfrak{z}_0$ the following is true:

Theorem 2.1. *Let $B \subset \mathbb{C}^n$ be a region and $f: B \to \mathbb{C}$ complex differentiable at $\mathfrak{z}_0 \in B$. Then the values of the functions $\Delta_1, \ldots, \Delta_n$ at $\mathfrak{z}_0$ are uniquely determined.*

PROOF. $E_\nu := \{\mathfrak{z} \in \mathbb{C}^n : z_\lambda = z_\lambda^{(0)} \text{ for } \lambda \neq \nu\}$ is a complex one-dimensional plane. Let $B_\nu := \{\zeta \in \mathbb{C} : (z_1^{(0)}, \ldots, z_{\nu-1}^{(0)}, \zeta, z_{\nu+1}^{(0)}, \ldots, z_n^{(0)}) \in E_\nu \cap B\}$. $f_\nu^*(z_\nu) := f(z_1^{(0)}, \ldots, z_{\nu-1}^{(0)}, z_\nu, z_{\nu+1}^{(0)}, \ldots, z_n^{(0)})$ defines a complex function on B_ν. Since f is differentiable at $\mathfrak{z}_0$, we have on B_ν

$$\begin{aligned} f_\nu^*(z_\nu) &= f(z_1^{(0)}, \ldots, z_{\nu-1}^{(0)}, z_\nu, z_{\nu+1}^{(0)}, \ldots, z_n^{(0)}) \\ &= f(\mathfrak{z}_0) + (z_\nu - z_\nu^{(0)}) \cdot \Delta_\nu(z_1^{(0)}, \ldots, z_\nu, \ldots, z_n^{(0)}) \\ &= f_\nu^*(z_\nu^{(0)}) + (z_\nu - z_\nu^{(0)}) \cdot \Delta_\nu^*(z_\nu). \end{aligned}$$

Thus $\Delta_\nu^*(z_\nu) := \Delta_\nu(z_1^{(0)}, \dots, z_{\nu-1}^{(0)}, z_\nu, z_{\nu+1}^{(0)}, \dots, z_n^{(0)})$ is continuous at $z_\nu^{(0)}$. Therefore $f_\nu^*(z_\nu)$ is complex differentiable at $z_\nu^{(0)} \in \mathbb{C}^n$, and $\Delta_\nu^*(z_\nu^{(0)}) = \Delta_\nu(\mathfrak{z}_0)$ is uniquely determined. This holds for each ν. □

Def. 2.2. Let the complex function f defined on the region $B \subset \mathbb{C}^n$ be complex differentiable at $\mathfrak{z}_0 \in B$. If $f(\mathfrak{z}) = f(\mathfrak{z}_0) + \sum_{\nu=1}^{n} (z_\nu - z_\nu^{(0)}) \Delta_\nu(\mathfrak{z})$, then we call $\Delta_\nu(\mathfrak{z}_0)$ the *partial derivative of f with respect to z_ν at $\mathfrak{z}_0$*, and write $\Delta_\nu(\mathfrak{z}_0) = \dfrac{\partial f}{\partial z_\nu}(\mathfrak{z}_0) = f_{z_\nu}(\mathfrak{z}_0) = f_{,\nu}(\mathfrak{z}_0)$.

Theorem 2.2. *Let $B \subset \mathbb{C}^n$ be a region and f complex differentiable at $\mathfrak{z}_0 \in B$. Then f is continuous at $\mathfrak{z}_0$.*

PROOF. We have $f(\mathfrak{z}) = f(\mathfrak{z}_0) + \sum_{\nu=1}^{n} (z_\nu - z_\nu^{(0)}) \Delta_\nu(\mathfrak{z})$; the right side of this equation is clearly continuous at $\mathfrak{z}_0$. □

Let $B \subset \mathbb{C}^n$ be a region. f is called complex differentiable on B if f is complex differentiable at each point of B.

Sums, products, and quotients (with nonvanishing denominators) of complex differentiable functions are again complex differentiable. The proof is analogous to the real case, and we do not present it here.

Theorem 2.3. *Let $B \subset \mathbb{C}^n$ be a region, f holomorphic in B. Then f is complex differentiable in B.*

PROOF. Let $\mathfrak{z}_0 \in B$. Then there is a neighborhood $U = U(\mathfrak{z}_0)$ and a power series $\sum_{\nu=0}^{\infty} a_\nu(\mathfrak{z} - \mathfrak{z}_0)^\nu$ which in U converges uniformly to $f(\mathfrak{z})$. Without loss of generality let $\mathfrak{z}_0 = 0$. Then

$$\sum_{\nu=0}^{\infty} a_\nu \mathfrak{z}^\nu = a_{0\cdots 0} + z_1 \cdot \sum_{\nu_1 \geq 1} a_{\nu_1 \cdots \nu_n} z_1^{\nu_1 - 1} \cdot z_2^{\nu_2} \cdots z_n^{\nu_n}$$
$$+ z_2 \cdot \sum_{\nu_2 \geq 1} a_{0, \nu_2 \cdots \nu_n} z_2^{\nu_2 - 1} \cdot z_3^{\nu_3} \cdots z_n^{\nu_n} + \cdots + z_n \cdot \sum_{\nu_n \geq 1} a_{0 \cdots 0, \nu_n} z_n^{\nu_n - 1}.$$

For now, this decomposition has only formal meaning. Choose a polycylinder of the form $P = U_\varepsilon(0) \times \cdots \times U_\varepsilon(0) \subset U(0)$ and a point $\mathfrak{z}_1 \in T = \{\mathfrak{z} \in \mathbb{C}^n : |z_k| = \varepsilon\}$. Then $P_{\mathfrak{z}_1} = P$ and $\mathfrak{z}_1 \in U$ (if ε is chosen sufficiently small). $\sum_{\nu=0}^{\infty} a_\nu \mathfrak{z}_1^\nu$ converges, therefore $\sum_{\nu=0}^{\infty} |a_\nu \mathfrak{z}_1^\nu|$ also converges. Since $\mathfrak{z}_1 \in \mathring{\mathbb{C}}^n$, $|z_k^{(1)}| \neq 0$ for all k. Therefore every subseries in the above representation at $\mathfrak{z}_1$ also converges absolutely and uniformly in the interior of $P_{\mathfrak{z}_1}$. The limit functions are continuous and are denoted by $\Delta_1, \dots, \Delta_n$. Since $f(\mathfrak{z}) = f(\mathfrak{z}_0) + z_1 \cdot \Delta_1(\mathfrak{z}) + \cdots + z_n \cdot \Delta_n(\mathfrak{z})$, it follows that f is complex differentiable at $\mathfrak{z}_0$. □

From this proof we obtain the values of the partial derivatives at a point $\mathfrak{z}_0$. For

$$f(\mathfrak{z}) = \sum_{\nu_1,\ldots,\nu_n=0}^{\infty} a_{\nu_1\cdots\nu_n}(z_1 - z_1^{(0)})^{\nu_1}\cdots(z_n - z_n^{(0)})^{\nu_n}$$

We obtain

$$f_{z_1}(\mathfrak{z}_0) = a_{1,0,\ldots,0},$$
$$\vdots$$
$$f_{z_n}(\mathfrak{z}_0) = a_{0,\ldots,0,1}.$$

3. The Cauchy Integral

In this section we shall seek additional characterizations of holomorphic functions.

Let $\mathfrak{r} = (r_1, \ldots, r_n)$ be a point in absolute space with $r_\nu \neq 0$ for all ν. Then $P = \{\mathfrak{z} \in \mathbb{C}^n : |z_\nu| < r_\nu \text{ for all } \nu\}$ is a nondegenerate polycylinder about the origin and $T = \{\mathfrak{z} \in \mathbb{C}^n : \tau(\mathfrak{z}) = \mathfrak{r}\}$ is the corresponding distinguished boundary. It will turn out that the values of an arbitrary holomorphic function on P are determined by its values on T.

First of all we must generalize the notion of a complex line integral. Let $K = \{z \in \mathbb{C} : z = re^{i\theta}, r > 0 \text{ fixed}, 0 \leqslant \theta \leqslant 2\pi\}$ be a circle in the complex plane, f a function continuous on K. As usual one writes

$$\int_K f(z)\,dz = \int_0^{2\pi} f(re^{i\theta})\cdot rie^{i\theta}\,d\theta.$$

The expression on the right is reduced to real integrals by

$$\int_a^b \varphi(t)\,dt: = \int_a^b \operatorname{Re}\varphi(t)\,dt + i\cdot\int_a^b \operatorname{Im}\varphi(t)\,dt.$$

Now let $f = f(\xi)$ be continuous on the n-dimensional torus $T = \{\xi \in \mathbb{C}^n : \tau(\xi) = \mathfrak{r}\}$. Then $h: P \times T \to \mathbb{C}$ with

$$h(\mathfrak{z}, \xi): = \frac{f(\xi)}{(\xi_1 - z_1)\cdots(\xi_n - z_n)}$$

is also continuous. We define

$$\begin{aligned}
F(\mathfrak{z}): &= \left(\frac{1}{2\pi i}\right)^n \cdot \int_T h(\mathfrak{z}, \xi)\,d\xi_1\cdots d\xi_n:\\
&= \left(\frac{1}{2\pi i}\right)^n \cdot \int_{|\xi_1|=r_1}\frac{d\xi_1}{\xi_1 - z_1}\int_{|\xi_2|=r_2}\frac{d\xi_2}{\xi_2 - z_2}\cdots\int_{|\xi_n|=r_n}\frac{d\xi_n}{\xi_n - z_n} f(\xi_1,\ldots,\xi_n)\\
&= \left(\frac{1}{2\pi}\right)^n \cdot \int_0^{2\pi}\int_0^{2\pi}\cdots\int_0^{2\pi}\frac{f(r_1e^{i\theta_1},\ldots,r_ne^{i\theta_n})}{(r_1e^{i\theta_1} - z_1)\cdots(r_ne^{i\theta_n} - z_n)}\\
&\quad\times r_1\cdots r_n e^{i(\theta_1+\cdots+\theta_n)}\,d\theta_1\cdots d\theta_n.
\end{aligned}$$

For each $\mathfrak{z} \in P$, F is well defined and even continuous on P.

Def. 3.1. Let P be a polycylinder and T the corresponding n-dimensional torus. Let f be a continuous function on T. Then the continuous function $\mathrm{ch}(f):P \to \mathbb{C}$ defined by

$$\mathrm{ch}(f)(\mathfrak{z}): = \left(\frac{1}{2\pi i}\right)^n \cdot \int_T \frac{f(\xi)\,d\xi}{(\xi_1 - z_1)\cdots(\xi_n - z_n)}$$

is called the *Cauchy integral of f over T.*

Theorem 3.1. *Let $B \subset \mathbb{C}^n$ be a region, P a polycylinder with $\bar{P} \subset B$ and T the n-dimensional torus belonging to P. If f is complex differentiable in B then $f|P = \mathrm{ch}(f|T)$.*

PROOF. This theorem is a generalization of the familar 1-dimensional Cauchy integral formula.

The function f_n^* with $f_n^*(z_n): = f(\xi_1, \ldots, \xi_{n-1}, z_n)$ is complex differentiable for fixed $(\xi_1, \ldots, \xi_{n-1}) \in \mathbb{C}^{n-1}$ in $B_n: = \{z_n \in \mathbb{C}:(\xi_1, \ldots, \xi_{n-1}, z_n) \in E_n \cap B\}$, where E_n is the plane $\{\mathfrak{z} \in \mathbb{C}^n: z_\lambda = \xi_\lambda \text{ for } \lambda \neq n\}$. But then f_n^* is holomorphic in B_n. B_n is an open set in $\mathbb{C}$. $K_n: = \{\xi_n \in \mathbb{C}:|\xi_n| = r_n\}$ is contained in B_n, and the Cauchy integral formula for a single variable says

$$f_n^*(z_n) = \frac{1}{2\pi i}\int_{|\xi_n| = r_n} \frac{f_n^*(\xi_n)}{\xi_n - z_n}\,d\xi_n.$$

Therefore

$$f(\xi_1, \ldots, \xi_{n-1}, z_n) = \frac{1}{2\pi i}\int_{K_n} \frac{f(\xi_1, \ldots, \xi_n)}{\xi_n - z_n}\,d\xi_n.$$

Similarly for the penultimate variable we obtain

$$\begin{aligned} f(\xi_1, \ldots, \xi_{n-2}, z_{n-1}, z_n) &= \frac{1}{2\pi i}\int_{K_{n-1}} \frac{f(\xi_1, \ldots, \xi_{n-1}, z_n)}{\xi_{n-1} - z_{n-1}}\,d\xi_{n-1} \\ &= \frac{1}{2\pi i}\int_{K_{n-1}} \frac{d\xi_{n-1}}{\xi_{n-1} - z_{n-1}}\left[\frac{1}{2\pi i}\int_{K_n} \frac{f(\xi_1, \ldots, \xi_n)}{\xi_n - z_n}\,d\xi_n\right] \end{aligned}$$

And, after n steps

$$\begin{aligned} f(z_1, \ldots, z_n) = \frac{1}{2\pi i}\int_{K_1} \frac{d\xi_1}{\xi_1 - z_1}\left[\frac{1}{2\pi i}\cdot\int_{K_2}\cdot\frac{d\xi_2}{\xi_2 - z_2}\right. \\ \left.\left[\cdots\left[\frac{1}{2\pi i}\int_{K_n} \frac{f(\xi_1, \ldots, \xi_n)}{\xi_n - z_n}\,d\xi_n\right]\cdots\right]\right] = \mathrm{ch}(f|T)(\mathfrak{z}). \quad \square \end{aligned}$$

Theorem 3.2. *Let $P \subset \mathbb{C}^n$ be a polycylinder, T the corresponding torus, and h a function continuous on T. Then $f: = \mathrm{ch}(h)$ can be expanded in a power series which converges in all of P.*

PROOF. For simplicity we consider only the case of two variables. Let $T = \{(\xi_1, \xi_2) \in \mathbb{C}^2:|\xi_1| = r_1, |\xi_2| = r_2\}$, with fixed $\mathfrak{z} = (z_1, z_2) \in P$. Then

$|z_1| < r_1, |z_2| < r_2$ and therefore $q_j := (|z_j|/r_j) < 1$ for $j = 1, 2$. Hence $\sum_{\nu_j=0}^{\infty} q_j^{\nu_j}$ dominates $\sum_{\nu_j=0}^{\infty} \left(\frac{z_j}{\xi_j}\right)^{\nu_j}$ for $j = 1, 2$ and

$$\frac{1}{(\xi_1 - z_1)(\xi_2 - z_2)} = \frac{1}{\xi_1 \cdot \xi_2} \cdot \frac{1}{\left(1 - \frac{z_1}{\xi_1}\right)\left(1 - \frac{z_2}{\xi_2}\right)}$$
$$= \frac{1}{\xi_1 \cdot \xi_2}\left(\sum_{\nu_1=0}^{\infty} \left(\frac{z_1}{\xi_1}\right)^{\nu_1} \cdot \sum_{\nu_2=0}^{\infty} \left(\frac{z_2}{\xi_2}\right)^{\nu_2}\right)$$

is absolutely and uniformly convergent for $(\xi_1, \xi_2) \in T$. In particular arbitrary substitutions are allowed, so,

$$\frac{1}{\xi_1\xi_2} \cdot \sum_{\nu_1, \nu_2=0}^{\infty} \left(\frac{z_1}{\xi_1}\right)^{\nu_1} \cdot \left(\frac{z_2}{\xi_2}\right)^{\nu_2}$$

also converges uniformly and absolutely on T. Since h is continuous on T and T is compact, h is uniformly bounded on T: $|h| \leqslant M$. Then, for fixed $(z_1, z_2) \in P$,

$$\frac{h(\xi_1, \xi_2)}{(\xi_1 - z_1)(\xi_2 - z_2)} = \frac{1}{\xi_1\xi_2} \cdot \sum_{\nu_1, \nu_2=0}^{\infty} h(\xi_1, \xi_2) \cdot \left(\frac{z_1}{\xi_1}\right)^{\nu_1}\left(\frac{z_2}{\xi_2}\right)^{\nu_2}$$

converges absolutely and uniformly on T, and we can interchange summation and integration:

$$f(\mathfrak{z}) = \left(\frac{1}{2\pi i}\right)^2 \cdot \int_T \frac{h(\xi_1, \xi_2)}{(\xi_1 - z_1)(\xi_2 - z_2)}\, d\xi_1\, d\xi_2$$
$$= \sum_{\nu_1, \nu_2=0}^{\infty} z_1^{\nu_1} z_2^{\nu_2} \cdot \left(\frac{1}{2\pi i}\right)^2 \cdot \int_T \frac{f(\xi_1, \xi_2)}{\xi_1^{\nu_1+1}\xi_2^{\nu_2+1}}\, d\xi_1\, d\xi_2 = \sum_{\nu_1, \nu_2=0}^{\infty} a_{\nu_1\nu_2} z_1^{\nu_1} z_2^{\nu_2}$$

with

$$a_{\nu_1\nu_2} := \left(\frac{1}{2\pi i}\right)^2 \cdot \int_T \frac{f(\xi_1, \xi_2)}{\xi_1^{\nu_1+1}\xi_2^{\nu_2+1}}\, d\xi_1\, d\xi_2$$

The series converges for each $\mathfrak{z} = (z_1, z_2) \in P$. □

Theorem 3.3. *Let $B \subset \mathbb{C}^n$ be a region, f complex differentiable in B. Then f is holomorphic in B.*

Proof. Let $\mathfrak{z}_0 \in B$. For the sake of simplicity we assume $\mathfrak{z}_0 = 0$. Then there exists a polycylinder P about $\mathfrak{z}_0$ such that $\bar{P} \subset B$. Let T be the distinguished boundary of P. From Theorem 3.1 $f|P = \operatorname{ch}(f|T)$. $f|T$ is continuous so from Theorem 3.2 f is holomorphic at $\mathfrak{z}_0$. □

Theorem 3.4. *Let $B \subset \mathbb{C}^n$ be a region, f holomorphic in B and $\mathfrak{z}_0$ a point in B. If $P \subset B$ is a polycylinder about $\mathfrak{z}_0$ with $\bar{P} \subset B$, then there is a power series $\mathfrak{P}(\mathfrak{z}) = \sum_{\nu=0}^{\infty} a_\nu(\mathfrak{z} - \mathfrak{z}_0)^\nu$ which converges to f on all of P.*

PROOF. If f is holomorphic in B, then $f|P = \operatorname{ch}(f|T)$, where the distinguished boundary of P is denoted by T. From Theorem 3.2 $f|P$ can be expanded as a power series in all of P. □

Theorem 3.5. *Let the sequence of functions (f_ν) converge uniformly to f on the region B with all f_ν holomorphic in B. Then f is holomorphic in B.*

PROOF. Let $\mathfrak{z}_0 \in B$. Again, we assume that $\mathfrak{z}_0 = 0$. Let P be a polycylinder about $\mathfrak{z}_0$ with $\bar{P} \subset B$. Let $\mathfrak{z} = (z_1, \ldots, z_n) \in P$. $N(\xi) := (\xi_1 - z_1) \cdot \ldots \cdot (\xi_n - z_n)$ is continuous and $\neq 0$ on T; therefore, $1/N(\xi)$ is also continuous on T and there exists an $M \in \mathbb{R}$ such that $|1/N(\xi)| < M$ on T. (f_ν) converges uniformly on T to f so for every $\varepsilon > 0$ there exists a $\nu_0 = \nu_0(\varepsilon)$ such that $|f_\nu - f| < \varepsilon/M$ on all of T for $\nu \geqslant \nu_0$. But then

$$\left|\frac{f_\nu}{N} - \frac{f}{N}\right| = \left|\frac{1}{N}\right| \cdot |f_\nu - f| < \varepsilon.$$

Hence f_ν/N converges uniformly on T to f/N and one can interchange the integral and the limit.

$$f|P = \lim_{\nu\to\infty} (f_\nu|P) = \lim_{\nu\to\infty} \operatorname{ch}(f_\nu|T) = \operatorname{ch}\left(\lim_{\nu\to\infty} (f_\nu|T)\right) = \operatorname{ch}(f|T).$$

f is continous on T since all the f_ν are continuous on T. From Theorem 3.2 it follows that f is holomorphic at the origin. □

Theorem 3.6. *Let $\mathfrak{P}(\mathfrak{z}) = \sum_{\nu=0}^{\infty} a_\nu \mathfrak{z}^\nu$ be a formal power series and G the domain of convergence for $\mathfrak{P}(\mathfrak{z})$. Then f with $f(\mathfrak{z}) := \mathfrak{P}(\mathfrak{z})$ is holomorphic in G.*

PROOF. Let $\mathfrak{I}$ be the set of all multi-indices $\nu = (\nu_1, \ldots, \nu_n)$, $I_0 \subset \mathfrak{I}$ a finite subset. Clearly the polynomial $\sum_{\nu \in I_0} a_\nu \mathfrak{z}^\nu$ is holomorphic on all of $\mathbb{C}^n$.

Let $\mathfrak{z}_0 \in G$ be a point, P a polycylinder about $\mathfrak{z}_0$ with $\bar{P} \subset G$. $\mathfrak{P}(\mathfrak{z})$ converges uniformly on P to the function $f(\mathfrak{z})$. If one sets $\varepsilon_k := 1/k$ for $k \in \mathbb{N}$ then in each case there is a finite set $I_k \subset \mathfrak{I}$ such that $\left|\sum_{\nu \in I} a_\nu \mathfrak{z}^\nu - f(\mathfrak{z})\right| < \varepsilon_k$ on all of P for any finite set I with $I_k \subset I \subset \mathfrak{I}$. For $f_k := \sum_{\nu \in I_k} a_\nu \mathfrak{z}^\nu$ we have f_k holomorphic and for each $k \in \mathbb{N}$, $|f_k - f| < 1/k$ on all of P. Therefore (f_k) converges uniformly on P to f. From Theorem 3.5 f is holomorphic in P and in particular at $\mathfrak{z}_0$. □

Theorem 3.7. *Let f be holomorphic on the region B. Then all the partial derivatives f_{z_μ}, $1 \leqslant \mu \leqslant n$, are also holomorphic in B. If $P \subset B$ is a polycyclinder with center at the origin and $f(\mathfrak{z}) = \sum_{\nu=0}^{\infty} a_\nu \mathfrak{z}^\nu$ on P, then*

$$f_{z_\mu}(\mathfrak{z}) = \sum_{\nu=0}^{\infty} a_\nu \cdot \nu_\mu \cdot z_1^{\nu_1} \cdots z_\mu^{\nu_\mu - 1} \cdots z_n^{\nu_n}$$

on P.

PROOF

1. Let $P \subset B$, $\mathfrak{z}_1 \in P \cap \mathring{\mathbb{C}}^n$. Then there is an $M \in \mathbb{R}$ such that $|a_\nu \mathfrak{z}_1^\nu| < M$ for all ν, where $\sum_{\nu=0}^{\infty} a_\nu \mathfrak{z}$ is the power series expansion of f in P. If $0 < q < 1$ and $\mathfrak{z}_2 := q \cdot \mathfrak{z}_1$, then $\sum_{\nu=0}^{\infty} a_\nu \mathfrak{z}_2^\nu$ is dominated by $M \cdot \sum_{\nu=0}^{\infty} q^{|\nu|}$. Now $\mathfrak{z}_2 = (z_1, \ldots, z_n)$ with $|z_k| \neq 0$ for $k = 1, \ldots, n$. It follows that

$$|a_\nu \cdot \nu_j \cdot z_1^{\nu_1} \cdots z_j^{\nu_j - 1} \cdots z_n^{\nu_n}| = \frac{\nu_j}{|z_j|} \cdot |a_\nu \mathfrak{z}_2^\nu| \leqslant \frac{\nu_j}{|z_j|} M \cdot q^{|\nu|}.$$

Formally

$$\sum_{\nu=0}^{\infty} \nu_j \cdot q^{|\nu|} = \left(\sum_{\nu_1=0}^{\infty} q^{\nu_1}\right) \cdots \left(\sum_{\nu_j=0}^{\infty} \nu_j q^{\nu_j}\right) \cdots \left(\sum_{\nu_n=0}^{\infty} q^{\nu_n}\right).$$

For $\mu \neq j$, $\sum_{\nu_\mu=0}^{\infty} q^{\nu_\mu}$ is a geometric series and therefore convergent. For $\mu = j$ the convergence of $\sum_{\nu_j=0}^{\infty} \nu_j q^{\nu_j}$ follows from the ratio test:

$$\lim_{\nu_j \to \infty} \frac{(\nu_j + 1) q^{\nu_j + 1}}{\nu_j \cdot q^{\nu_j}} = q \cdot \lim_{\nu_j \to \infty} \frac{\nu_j + 1}{\nu_j} = q < 1.$$

Hence the series

$$\sum_{\nu=0}^{\infty} \frac{\nu_j}{|z_j|} \cdot M \cdot q^{|\nu|} = \frac{M}{|z_j|} \cdot \sum_{\nu=0}^{\infty} \nu_j \cdot q^{|\nu|}$$

converges. By the comparison test the series $\sum_{\nu=0}^{\infty} a_\nu \nu_j z_1^{\nu_1} \cdots z_j^{\nu_j - 1} \cdots z_n^{\nu_n}$ is also convergent at the point $\mathfrak{z}_2$, and therefore in $P_{\mathfrak{z}_2}$. Since P is the union of all the $P_{\mathfrak{z}_2}$ the series converges in all of P to a holomorphic function g_j.

2. Let

$$f^*(\mathfrak{z}) := \int_0^{z_j} g_j(z_1, \ldots, z_{j-1}, \xi, z_{j+1}, \ldots, z_n)\, d\xi + f(z_1, \ldots, 0, \ldots, z_n).$$

The path of integration can be chosen in such a way that it consists of the line segments connecting 0 to z_j in the z_j-plane. Thus f^* is defined on P. For $h_\nu(\mathfrak{z}) := a_\nu \mathfrak{z}^\nu$, we have $f(\mathfrak{z}) = \sum_{\nu=0}^{\infty} h_\nu(\mathfrak{z})$ and $g_j(\mathfrak{z}) = \sum_{\nu=0}^{\infty} (h_\nu)_{z_j}(\mathfrak{z})$. The path of integration is a compact subset of P and the series converges uniformly there. Hence one may interchange summation and integration and obtains

$$f^*(\mathfrak{z}) = \sum_{\nu=0}^{\infty}\left(\int_0^{z_j} (h_\nu)_{z_j}(z_1, \ldots, z_{j-1}, \xi, z_{j+1}, \ldots, z_n)\, d\xi + h_\nu(z_1, \ldots, 0, \ldots, z_n)\right)$$

$$= \sum_{\nu=0}^{\infty} h_\nu(\mathfrak{z})$$

$$= f(\mathfrak{z})$$

Hence $f_{z_j}(\mathfrak{z}) = f^*_{z_j}(\mathfrak{z}) = g_j(\mathfrak{z})$. □

We conclude this section with a summary of our results.

Theorem 3.8. *Let $B \subset \mathbb{C}^n$ be a region and f a complex function on B. The following statements about f are equivalent:*

a. *f is complex differentiable in B*
b. *f is arbitrarily often complex differentiable in B*
c. *f is holomorphic in B. For every $\mathfrak{z}_0 \in B$ there is a neighborhood U such that $f(\mathfrak{z}) = \sum_{\nu=0}^{\infty} a_\nu(\mathfrak{z} - \mathfrak{z}_0)^\nu$ in U. Here the a_ν are the "coefficients of the Taylor series expansion":*

$$a_{\nu_1 \ldots \nu_n} = \frac{1}{\nu_1! \cdots \nu_n!} \cdot \frac{\partial^{\nu_1 + \cdots + \nu_n} f}{\partial z_1^{\nu_1} \cdots \partial z_n^{\nu_n}}(\mathfrak{z}_0)$$

d. *For each polycylinder P with $\bar{P} \subset B$, $f|T$ is continuous and $f|T =$ ch$(f|T)$.*

PROOF. Nearly everything has already been proved, but we must still calculate the coefficients a_ν. For simplicity let $\mathfrak{z}_0 = 0$ and $n = 2$. In the proof of Theorem 3.2 we obtained:

$$a_{\nu_1\nu_2} = \frac{1}{(2\pi i)^2}\int_T \frac{f(z_1, z_2)}{z_1^{\nu_1+1} \cdot z_2^{\nu_2+1}}\, dz_1\, dz_2.$$

From the Cauchy integral formula for one variable it now follows that

$$a_{\nu_1\nu_2} = \frac{1}{2\pi i}\int_{K_1} \frac{1}{z_1^{\nu_1+1}}\left[\frac{1}{2\pi i}\int_{K_2} \frac{f(z_1, z_2)}{z_2^{\nu_2+1}}\, dz_2\right] dz_1$$

$$= \frac{1}{\nu_2!}\frac{1}{2\pi i}\int_{K_1} \frac{\partial^{\nu_2} f}{\partial z_2^{\nu_2}}(z_1, 0)\frac{dz_1}{z_1^{\nu_1+1}} = \frac{1}{\nu_1!\nu_2!} \cdot \frac{\partial^{\nu_1+\nu_2} f}{\partial z_1^{\nu_1}\, \partial z_2^{\nu_2}}(0, 0). \qquad \square$$

4. Identity Theorems

Different from the theory of one complex variable, the following theorem does not hold in $\mathbb{C}^n$: "Let G be a domain, $M \subset G$ have a cluster point in G and f_1, f_2 be holomorphic on G with $f_1 = f_2$ on M. Then $f_1 = f_2$ in G."

There is already a counter-example for $n = 2$. Let $G: = \mathbb{C}^2$, $M: = \{(z_1, z_2) \in G: z_2 = 0\}$, $f_1(z_1, z_2): = z_2 \cdot g(z_1, z_2)$, $f_2(z_1, z_2): = z_2 \cdot h(z_1, z_2)$ with g and h holomorphic on all of $\mathbb{C}^2$. Then $f_1|M = f_2|M$, but $f_1 \neq f_2$ for $g \neq h$.

Theorem 4.1 (Identity theorem for holomorphic functions). *Let $G \subset \mathbb{C}^n$ be a domain and f_1, f_2 be holomorphic in G. Let $B \subset G$ be a nonempty region with $f_1|B = f_2|B$. Then $f_1|G = f_2|G$.*

PROOF. Let B_0 be the interior of the set $\{\mathfrak{z} \in G: f_1(\mathfrak{z}) = f_2(\mathfrak{z})\}$ and $W_0: = G - B_0$. Because $B \subset B_0$, $B_0 \neq \emptyset$. Since G is connected it suffices to show that W_0 is open, for then $B_0 = G$ follows. Let us assume W_0 contains a point $\mathfrak{z}_0$ which is not an interior point. Then for every polycylinder P about $\mathfrak{z}_0$ with $\bar{P} \subset G$, $P \cap B_0 \neq \emptyset$. Let $r \in \mathbb{R}$ and $P: = \{\mathfrak{z}: |z_j - z_j^0| < r\} = \{\mathfrak{z}: \text{dist}'(\mathfrak{z}, \mathfrak{z}_0) < r\}$ be such a polycylinder. Let

$$P': = \{\mathfrak{z}: \text{dist}'(\mathfrak{z}, \mathfrak{z}_0) < r/2\} \subset P.$$

Then also $P' \cap B_0 \neq \emptyset$. Choose an arbitrary point $\mathfrak{z}_1 \in P' \cap B_0$ and set $P^*: = \{\mathfrak{z}: \text{dist}'(\mathfrak{z}, \mathfrak{z}_1) < r/2\}$. Clearly $\mathfrak{z}_0 \in P^*$ and $P^* \subset P$ (triangle inequality). Therefore $\bar{P}^* \subset \bar{P} \subset G$. Let

$$f_1(\mathfrak{z}) = \sum_{\nu=0}^{\infty} a_\nu(\mathfrak{z} - \mathfrak{z}_1)^\nu \quad \text{and} \quad f_2(\mathfrak{z}) = \sum_{\nu=0}^{\infty} b_\nu(\mathfrak{z} - \mathfrak{z}_1)^\nu$$

be the Taylor series expansions of f_1 and f_2 in P^*. Since f_1 and f_2 coincide in the neighborhood of $\mathfrak{z}_1 \in B_0$, $a_\nu = b_\nu$ for all ν. (The coefficients are uniquely determined by the function; cf. Theorem 3.8.) Therefore $f_1|P^* = f_2|P^*$ and $P^* \subset B_0$. It follows that $\mathfrak{z}_0 \in B_0$, a contradiction. □

Theorem 4.2 (Identity theorem for power series). *Let $G \subset \mathbb{C}^n$ be a domain with $0 \in G$, and $\sum_{\nu=0}^{\infty} a_\nu \mathfrak{z}^\nu$, $\sum_{\nu=0}^{\infty} b_\nu \mathfrak{z}^\nu$ two power series convergent in G. If there is an $\varepsilon > 0$ such that $\sum_{\nu=0}^{\infty} a_\nu \mathfrak{z}^\nu = \sum_{\nu=0}^{\infty} b_\nu \mathfrak{z}^\nu$ in $U_\varepsilon(0) \subset G$, then $a_\nu = b_\nu$ for all ν.*

PROOF. Let $f(\mathfrak{z}): = \sum_{\nu=0}^{\infty} a_\nu \mathfrak{z}^\nu$, $g(\mathfrak{z}): = \sum_{\nu=0}^{\infty} b_\nu \mathfrak{z}^\nu$ for $\mathfrak{z} \in G$. By Theorem 3.6 f and g are holomorphic in G, and differentiation gives:

$$\nu_1! \cdots \nu_n! \cdot a_\nu = \frac{\partial^{\nu_1 + \cdots + \nu_n} f}{\partial z_1^{\nu_1} \cdots \partial z_n^{\nu_n}}(0) = \frac{\partial^{\nu_1 + \cdots + \nu_n} g}{\partial z_1^{\nu_1} \cdots \partial z_n^{\nu_n}}(0) = \nu_1! \cdots \nu_n! \cdot b_\nu.$$

Then $a_\nu = b_\nu$. □

5. Expansion in Reinhardt Domains

In this section we shall study the properties of certain domains in $\mathbb{C}^n$ in some detail.

Let r'_ν, r''_ν be real numbers with $0 < r'_\nu < r''_\nu$ for $1 \leqslant \nu \leqslant n$. Let $\mathfrak{r} = (r_1, \ldots, r_n) \in V$ be chosen so that $r'_\nu < r_\nu < r''_\nu$ for all ν. Then $T_\mathfrak{r}: = \{\mathfrak{z}: |z_\nu| = r_\nu \text{ for all } \nu\}$ is an n-dimensional torus. We define

$$H: = \{\mathfrak{z}: r'_\nu < |z_\nu| < r''_\nu \text{ for all } \nu\}$$
$$P: = \{\mathfrak{z}: |z_\nu| < r'_\nu \text{ for all } \nu\}.$$

Clearly H and P are Reinhardt domains.

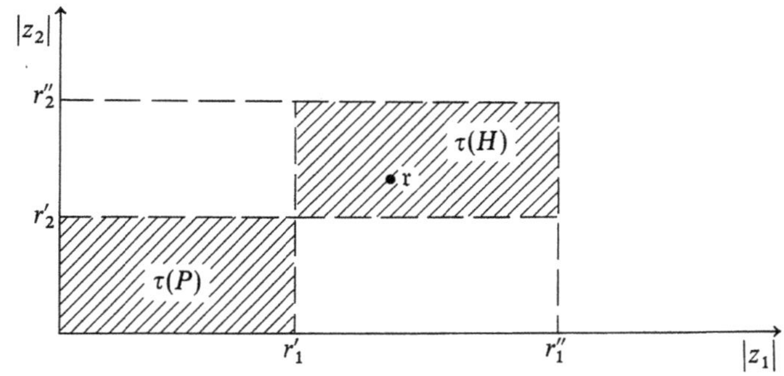

Figure 3. Expansion in Reinhardt domains.

Let f be a holomorphic function in H. Then for all $\mathfrak{r} \in \tau(H)$, $\text{ch}(f|T_\mathfrak{r})$ is a holomorphic function in $P_\mathfrak{r} = \{\mathfrak{z}: |z_\nu| < r_\nu \text{ for all } \nu\}$ (and therefore *a fortiori* in P).

Proposition. *$g: P \times \tau(H) \to \mathbb{C}$ with $g(\mathfrak{z}, \mathfrak{r}): = \text{ch}(f|T_\mathfrak{r})(\mathfrak{z})$ is independent of $\mathfrak{r}$.*

PROOF. We have

$$\text{ch}(f|T_\mathfrak{r})(\mathfrak{z}) = \left(\frac{1}{2\pi i}\right)^n \cdot \int_{|\xi_n| = r_n} \frac{d\xi_n}{\xi_n - z_n} \int_{|\xi_{n-1}| = r_{n-1}} \frac{d\xi_{n-1}}{\xi_{n-1} - z_{n-1}} \cdots \int_{|\xi_1| = r_1} \frac{d\xi_1}{\xi_1 - z_1} f(\xi_1, \ldots, \xi_n).$$

For each j with $1 \leqslant j \leqslant n$ we have $|z_j| < r_j = |\xi_j|$; therefore $z_j \neq \xi_j$. Hence the integrand is holomorphic on the annulus $\{z_j: r'_j < |z_j| < r''_j\}$ and from the Cauchy integral formula for one variable it follows that if $\mathfrak{r} = (r_1, \ldots, r_n) \in \tau(H)$ and $\mathfrak{r}^* = (r_1^*, \ldots, r_n^*) \in \tau(H)$, then

$$\int_{|\xi_j| = r_j} \frac{f(\xi_1, \ldots, \xi_n)}{\xi_j - z_j} d\xi_j = \int_{|\xi_j| = r_j^*} \frac{f(\xi_1, \ldots, \xi_n)}{\xi_j - z_j} d\xi_j.$$

This yields the proposition. □

Theorem 5.1. *Let $G \subset \mathbb{C}^n$ be a domain and $E := \{\mathfrak{z} = (z_1, \ldots, z_n) \in \mathbb{C}^n$ with $z_1 = 0\}$. Then the set $G' := G - E$ is also a domain in $\mathbb{C}^n$.*

PROOF

1. E is closed, therefore $\mathbb{C}^n - E$ is open, and hence $G' = G \cap (\mathbb{C}^n - E)$ is also open. Moreover, E contains no interior points.

2. We write the points $\mathfrak{z} \in \mathbb{C}^n$ in the form $\mathfrak{z} = (z_1, \mathfrak{z}^*)$ with $\mathfrak{z}^* \in \mathbb{C}^{n-1}$. Now let $\mathfrak{z}_0 = (z_1^{(0)}, \mathfrak{z}^{*(0)}) \in G$ and let $U'_\varepsilon(\mathfrak{z}_0) = U_\varepsilon(z_1^{(0)}) \times U'_\varepsilon(\mathfrak{z}^{*(0)})$ be an ε-neighborhood of $\mathfrak{z}_0$. We show that $U'_\varepsilon - E$ is still connected. Let $\mathfrak{z}_1 = (z_1^{(1)}, \mathfrak{z}^{*(1)})$ and $\mathfrak{z}_2 = (z_1^{(2)}, \mathfrak{z}^{*(2)})$ be two arbitrary points in $U'_\varepsilon - E$. Then we define $\mathfrak{z}_3 := (z_1^{(2)}, \mathfrak{z}^{*(1)})$.

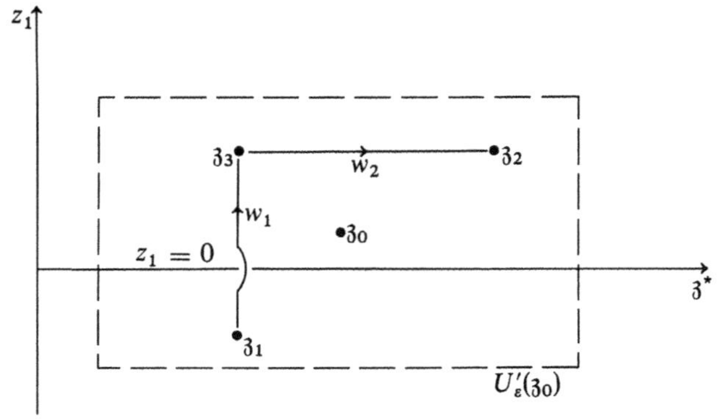

Figure 4. Proof of Theorem 5.1.

Clearly $\mathfrak{z}_3 \in U'_\varepsilon - E$. $U_\varepsilon(z_1^{(0)})$ is an open circular disk in the z_1-plane, and $U_\varepsilon(z_1^{(0)}) - \{0\}$ is still connected. Hence there is a path φ which joins $z_1^{(1)}$ and $z_1^{(2)}$ and lies entirely within $U_\varepsilon(z_1^{(0)}) - \{0\}$; naturally there is also a path ψ which joins $\mathfrak{z}^{*(1)}$ and $\mathfrak{z}^{*(2)}$ and which lies within $U'_\varepsilon(\mathfrak{z}^{*(0)})$.

We now define paths w_1, w_2 by $w_1(t) := (\varphi(t), \mathfrak{z}^{*(1)})$ and $w_2(t) := (z_1^{(2)}, \psi(t))$. Then w_1 joins $\mathfrak{z}_1$ and $\mathfrak{z}_3$, w_2 joins $\mathfrak{z}_3$ and $\mathfrak{z}_2$, and the composition joins $\mathfrak{z}_1$ and $\mathfrak{z}_2$ in $U'_\varepsilon - E$. Therefore $U'_\varepsilon - E$ is connected.

3. Let $\mathfrak{z}', \mathfrak{z}'' \in G - E$ and let φ be an arbitrary path which joins $\mathfrak{z}'$ and $\mathfrak{z}''$ in G. Since $\varphi(I)$ is compact, one can cover it with finitely many polycylinders $U_1, \ldots, U_\ell$ such that $U_\lambda \subset G$ for $\lambda = 1, \ldots, \ell$.

Lemma. *There is a $\delta > 0$ such that for all $t', t'' \in I$ with $|t' - t''| < \delta$, $\varphi(t')$, $\varphi(t'')$ lie in the same polycylinder U_k.*

PROOF. Let there be sequences $(t'_j), (t''_j) \in I$ with $|t'_j - t''_j| \to 0$ such that $\varphi(t'_j), \varphi(t''_j)$ do not lie in the same polycylinder U_k. There are convergent subsequences $(t'_{j_\nu}), (t''_{j_\mu})$ of $(t'_j), (t''_j)$. Let $t_0 := \lim_{\nu \to \infty} t'_{j_\nu} = \lim_{\mu \to \infty} t''_{j_\mu}$. If $\varphi(t_0) \in U_k$, then there is an open neighborhood $V = V(t_0) \subset I$ with $\varphi(V) \subset U_k$. Then for almost all $\nu \in \mathbb{N}$, $t'_{j_\nu} \in V$ and $t''_{j_\nu} \in V$, so that $\varphi(t'_{j_\nu}) \in U_k$ and $\varphi(t''_{j_\nu}) \in U_k$. This is a contradiction, which proves the lemma.

Now let δ be suitably chosen and $0 = t_0 < t_1 < \cdots < t_k = 1$ be a partition of I with $t_j - t_{j-1} < \delta$ for $j = 1, \ldots, k$. Let $\mathfrak{z}_j := \varphi(t_j)$ and V_j be the polycylinder which contains $\mathfrak{z}_j, \mathfrak{z}_{j-1}$ (it can happen that $V_{j_1} = V_{j_2}$ for $j_1 \neq j_2$). By construction $\mathfrak{z}_{j-1}$ lies in $V_j \cap V_{j-1}$, so $V_j \cap V_{j-1}$ is always a non-empty open set. Indeed, $V_j \cap V_{j-1} - E \neq \varnothing$ for $j = 1, \ldots, k$.

We join $\mathfrak{z}' = \mathfrak{z}_0 \in V_1 - E$ and a point $\mathfrak{z}_1^* \in V_1 \cap V_2 - E$ by a path φ_1 interior to $V_1 - E$. By (2) this is possible. Next we join $\mathfrak{z}_1^*$ with a point $\mathfrak{z}_2^* \in V_2 \cap V_3 - E$ by a path φ_2 interior to $V_2 - E$, and so on.

Finally, let φ_k be a path in $V_k - E$ which joins $\mathfrak{z}_{k-1}^*$ with $\mathfrak{z}_k = \mathfrak{z}'' \in V_k - E$. The composition of the paths $\varphi_1, \ldots, \varphi_k$ connects $\mathfrak{z}'$ and $\mathfrak{z}''$ in $G - E$. □

Theorem 5.2. *Let G be a domain in $\mathbb{C}^n$, $E_0 := \{\mathfrak{z} = (z_1, \ldots, z_n) \in \mathbb{C}^n : z_\nu = 0$ for at least one $\nu\}$. Then $G_0 := G - E_0$ is also a domain.*

PROOF. For each μ with $1 \leqslant \mu \leqslant n$, $G_\mu : G - E_\mu$ is connected, where $E_\mu := \{\mathfrak{z} = (z_1, \ldots, z_n) \in \mathbb{C}^n : z_\mu = 0\}$. This follows from Theorem 5.1 by a simple permutation of the coordinates.

Clearly $E_0 = \bigcup_{\mu=1}^{n} E_\mu$; therefore $G_0 = (((G - E_1) - E_2) \cdots) - E_n$. A trivial induction proof yields the proposition. □

Theorem 5.3. *Let $G \subset \mathbb{C}^n$ be a proper Reinhardt domain, f holomorphic on G, $\mathfrak{z}_0 \in G \cap \mathring{\mathbb{C}}^n$. Then $\operatorname{ch}(f|T_{\mathfrak{z}_0})$ coincides with f in a neighborhood of the origin.*

PROOF. We have $G_0 := \tau(G \cap \mathring{\mathbb{C}}^n) \subset \{\mathfrak{r} \in V : r_j \neq 0 \text{ for } j = 1, \ldots, n\}$.

1. G_0 is a domain:

a. $G \cap \mathring{\mathbb{C}}^n$ is a Reinhardt domain; therefore $G_0 = \tau(G \cap \mathring{\mathbb{C}}^n)$ is open by Theorem 1.2.

b. If $\mathfrak{r}_1, \mathfrak{r}_2$ are points in G_0, then there are points $\mathfrak{z}_p \in G \cap \mathring{\mathbb{C}}^n$ with $\tau(\mathfrak{z}_p) = \mathfrak{r}_p$ for $p = 1, 2$. As shown above, $G \cap \mathring{\mathbb{C}}^n$ is a domain, so that there is a path φ in $G \cap \mathbb{C}^n$ which joins $\mathfrak{z}_1$ and $\mathfrak{z}_2$. Then $\tau \circ \varphi$ is a path in G_0 which joins $\mathfrak{r}_1$ and $\mathfrak{r}_2$.

2. Let

$$B := \{\mathfrak{r} \in G_0 : \operatorname{ch}(f|T) \text{ coincides with } f \text{ in the vicinity of } 0\}.$$

a. B is open: If $\mathfrak{r}_0 \in B \subset G_0$, then there is a neighborhood $U'_\varepsilon(\mathfrak{r}_0) \subset G_0$ which can be written $\tau(H)$. This follows from the way we chose the set $\tau(H)$ at the beginning of this section. Let $P = P(0)$ be the corresponding polycylinder. Then for $\mathfrak{z} \in P$ and $\mathfrak{r} \in U'_\varepsilon(\mathfrak{r}_0)$ we have $\operatorname{ch}(f|T_\mathfrak{r})(\mathfrak{z}) = \operatorname{ch}(f|T_{\mathfrak{r}_0})(\mathfrak{z})$. Moreover $g(\mathfrak{z}) := \operatorname{ch}(f|T_{\mathfrak{r}_0})(\mathfrak{z})$ is a holomorphic function on P which coincides with f near the origin because $\mathfrak{r} \in B$. Therefore $U'_\varepsilon(\mathfrak{r}_0) \subset B$.

b. $W := G_0 - B$ is open: The proof goes as in (a).

c. $B \neq \varnothing$: There is a polycylinder $P_{\mathfrak{z}_0}$ about 0 with $\bar{P}_{\mathfrak{z}_0} \subset G$. Then $f|P_{\mathfrak{z}_0} = \operatorname{ch}(f|T_{\mathfrak{z}_0})$, and $\mathfrak{r}_0 := (|z_1^{(0)}|, \ldots, |z_n^{(0)}|)$ lies in B.

(1) and (2) imply $B = G_0$. □

Theorem 5.4. *Let $G \subset \mathbb{C}^n$ be a proper Reinhardt domain, f holomorphic in G. Then there is a power series $\mathfrak{P}(\mathfrak{z}) = \sum_{\nu=0}^{\infty} a_\nu \mathfrak{z}^\nu$ which converges in G with $f(\mathfrak{z}) = \mathfrak{P}(\mathfrak{z})$ for $\mathfrak{z} \in G$.*

PROOF. If $\mathfrak{z}_0 \in G$ then there is a $\mathfrak{z}_1 \in G$ with $|z_j^{(0)}| < |z_j^{(1)}|$ for $j = 1, \ldots, n$; therefore $\mathfrak{z}_0 \in P_{\mathfrak{z}_1}$. Let $\operatorname{ch}(f|T_{\mathfrak{z}_1})(\mathfrak{z}) = \sum_{\nu=0}^{\infty} a_\nu \mathfrak{z}^\nu$ for $\mathfrak{z} \in P_{\mathfrak{z}_1}$. The coefficients a_ν are those of the Taylor series about 0; they do not depend on $\mathfrak{z}_1$. Since $\mathfrak{z}_0$ was arbitrary it follows that the Taylor series of f about 0 converges in all of G. It defines a holomorphic function g, which coincides with f near the origin. By the uniqueness theorem, $f = g$ on G. □

Def. 5.1. If $G \subset \mathbb{C}^n$ is a proper Reinhardt domain, then $\hat{G} := \bigcup_{\mathfrak{z} \in G \cap \mathring{\mathbb{C}}^n} P_\mathfrak{z}$ is called the *complete hull of* G.

Remarks

1. $\hat{G}$ is open.
2. $G \subset \hat{G}$. If $\mathfrak{z}_0 \in G$, then there is a $\mathfrak{z}_1 \in G \cap \mathring{\mathbb{C}}^n$ with $\mathfrak{z}_0 \in P_{\mathfrak{z}_1} \subset \hat{G}$.
3. $\hat{G}$ is a Reinhardt domain. Let $\mathfrak{z}_0 \in \hat{G}$, $\mathfrak{z}_1 \in G \cap \mathring{\mathbb{C}}^n$ with $\mathfrak{z}_0 \in P_{\mathfrak{z}_1}$. Then $T_{\mathfrak{z}_0} \subset P_{\mathfrak{z}_1} \subset \hat{G}$.
4. $\hat{G}$ is complete. Let $\mathfrak{z}_0 \in \hat{G} \cap \mathring{\mathbb{C}}^n$, $\mathfrak{z}_1 \in G \cap \mathring{\mathbb{C}}^n$ with $\mathfrak{z}_0 \in P_{\mathfrak{z}_1}$. Then $P_{\mathfrak{z}_0} \subset P_{\mathfrak{z}_1} \subset \hat{G}$.
5. $\hat{G}$ is minimal for the properties (1) through (4). Let $G \subset G_1$, G_1 a complete Reinhardt domain. If $\mathfrak{z} \in G \cap \mathring{\mathbb{C}}^n$, then $P_\mathfrak{z} \subset G_1$. Therefore $\hat{G} \subset G_1$.

$\hat{G}$ is the smallest complete domain which contains G and we have the following important theorem.

Theorem 5.5. *Let G be a proper Reinhardt domain, f holomorphic in G. Then there is exactly one holomorphic function F in $\hat{G}$ with $F|G = f$.*

PROOF. By Theorem 5.4 we can write in G

$$f(\mathfrak{z}) = \sum_{\nu=0}^{\infty} a_\nu \mathfrak{z}^\nu.$$

The series is still convergent on $\hat{G}$, and actually converges to a holomorphic function F. Clearly $F|G = f$. The uniqueness of the continuation follows from the identity theorem. □

For $n \geqslant 2$ we can choose sets G and $\hat{G}$ in $\mathbb{C}^n$ so that $G \neq \hat{G}$. This constitutes a vital difference from the theory of functions of a single complex variable, where for each domain G there exists a function holomorphic on G which cannot be continued to any proper superdomain.

We conclude this section with an important example of such a pair of sets $(\hat{G}, G)$ with $\hat{G} \neq G$ for $n = 2$.

Let $P: = \{\mathfrak{z} \in \mathbb{C}^2 : |\mathfrak{z}| < 1\}$ be the unit polycylinder about the origin and $D: = \{\mathfrak{z} \in \mathbb{C}^2 : q_1 \leqslant |z_1| < 1, |z_2| \leqslant q\}$ with $0 < q_1 < 1$ and $0 < q < 1$. Then $H: = P - D$ is a proper Reinhardt domain, and $\hat{H} = \bigcup_{\mathfrak{z} \in H_0} P_{\mathfrak{z}} = P$.

The pair (P, H) is called a *Euclidean Hartogs figure*. Their image in absolute space appears in Fig. 5.

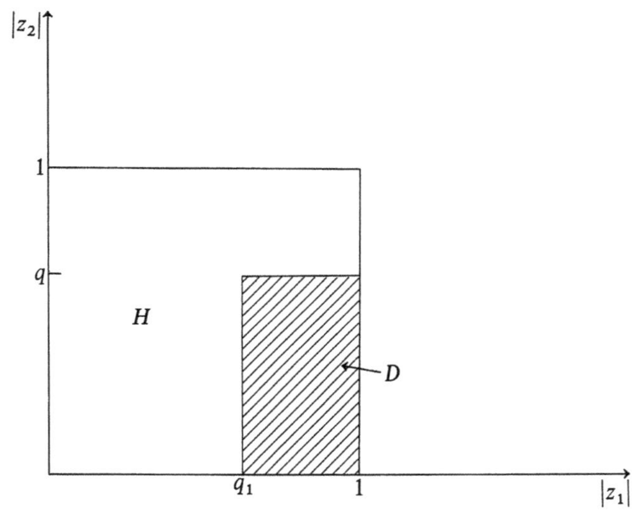

Figure 5. Euclidean Hartogs figure in $\mathbb{C}^2$.

The basis for the difference here between the theories of one and several variables is that such a Hartogs figure does not exist in $\mathbb{C}$. We already noted that Reinhardt domains in $\mathbb{C}$ are open disks and annuli. Therefore a proper Reinhardt domain in $\mathbb{C}$ is an open disk, i.e., a complete Reinhardt domain. Hence $\hat{G}$ is not a proper superset of G.

6. Real and Complex Differentiability

Let $M \subset \mathbb{C}^n$ be a set, f a complex function on M. At each point $\mathfrak{z}_0 \in M$ there is a unique representation $f(\mathfrak{z}_0) = \operatorname{Re} f(\mathfrak{z}_0) + i \operatorname{Im} f(\mathfrak{z}_0)$.
Therefore one can define real functions g and h on M by

$$g(\mathfrak{x}, \mathfrak{y}) = \operatorname{Re} f(\mathfrak{z})$$
$$h(\mathfrak{x}, \mathfrak{y}) = \operatorname{Im} f(\mathfrak{z})$$

where $\mathfrak{z} = \mathfrak{x} + i\mathfrak{y}$. We then write:

$$f = g + ih.$$

Def. 6.1. Let $B \subset \mathbb{C}^n$ be a region, $f = g + ih$ a complex function on B, $\mathfrak{z}_0$ a point of B. f is called *real differentiable at* $\mathfrak{z}_0$ if g and h are totally (real) differentiable.

What does real differentiability mean? If g and h are differentiable, then

$$(1)\quad \begin{aligned} g(\mathfrak{x}, \mathfrak{y}) &= g(\mathfrak{x}_0, \mathfrak{y}_0) + \sum_{\nu=1}^{n} (x_\nu - x_\nu^{(0)})\alpha_\nu^*(\mathfrak{x}, \mathfrak{y}) + \sum_{\nu=1}^{n} (y_\nu - y_\nu^{(0)})\alpha_\nu^{**}(\mathfrak{x}, \mathfrak{y}),\\ h(\mathfrak{x}, \mathfrak{y}) &= h(\mathfrak{x}_0, \mathfrak{y}_0) + \sum_{\nu=1}^{n} (x_\nu - x_\nu^{(0)})\beta_\nu^*(\mathfrak{x}, \mathfrak{y}) + \sum_{\nu=1}^{n} (y_\nu - y_\nu^{(0)})\beta_\nu^{**}(\mathfrak{x}, \mathfrak{y}); \end{aligned}$$

where $\alpha_\nu^*, \alpha_\nu^{**}, \beta_\nu^*, \beta_\nu^{**}$ are real functions on B which are continuous at $(\mathfrak{x}_0, \mathfrak{y}_0)$ and for which

$$\begin{aligned} \alpha_\nu^*(\mathfrak{x}_0, \mathfrak{y}_0) &= g_{x_\nu}(\mathfrak{x}_0, \mathfrak{y}_0)\\ \alpha_\nu^{**}(\mathfrak{x}_0, \mathfrak{y}_0) &= g_{y_\nu}(\mathfrak{x}_0, \mathfrak{y}_0)\\ \beta_\nu^*(\mathfrak{x}_0, \mathfrak{y}_0) &= h_{x_\nu}(\mathfrak{x}_0, \mathfrak{y}_0)\\ \beta_\nu^{**}(\mathfrak{x}_0, \mathfrak{y}_0) &= h_{y_\nu}(\mathfrak{x}_0, \mathfrak{y}_0). \end{aligned}$$

We combine the equations:

$$(2)\quad f(\mathfrak{z}) = f(\mathfrak{z}_0) + \sum_{\nu=1}^{n} (x_\nu - x_\nu^{(0)})\, \Delta_\nu^*(\mathfrak{z}) + \sum_{\nu=1}^{n} (y_\nu - y_\nu^{(0)})\, \Delta_\nu^{**}(\mathfrak{z}),$$

where $\Delta_\nu^* = \alpha_\nu^* + i\beta_\nu^*$ and $\Delta_\nu^{**} = \alpha_\nu^{**} + i\beta_\nu^{**}$ are continuous at $\mathfrak{z}_0$ and where

$$\begin{aligned} \Delta_\nu^*(\mathfrak{z}_0) &= g_{x_\nu}(\mathfrak{z}_0) + ih_{x_\nu}(\mathfrak{z}_0) = : f_{x_\nu}(\mathfrak{z}_0)\\ \Delta_\nu^{**}(\mathfrak{z}_0) &= g_{y_\nu}(\mathfrak{z}_0) + ih_{y_\nu}(\mathfrak{z}_0) = : f_{y_\nu}(\mathfrak{z}_0). \end{aligned}$$

Theorem 6.1. *Let $B \subset \mathbb{C}^n$ be a region, $\mathfrak{z}_0 \in B$ a point, f a complex function on B. f is real differentiable at $\mathfrak{z}_0$ if and only if there are functions $\Delta'_\nu, \Delta''_\nu$ on B which are continuous at $\mathfrak{z}_0$ and satisfy in B the following equation*:

$$(3)\quad f(\mathfrak{z}) = f(\mathfrak{z}_0) + \sum_{\nu=1}^{n} (z_\nu - z_\nu^{(0)})\, \Delta'_\nu(\mathfrak{z}) + \sum_{\nu=1}^{n} (\overline{z}_\nu - \overline{z}_\nu^{(0)})\, \Delta''_\nu(\mathfrak{z}).$$

PROOF

1. Let f be real differentiable at $\mathfrak{z}_0$. We use the equations

$$x_\nu - x_\nu^{(0)} = \tfrac{1}{2}[(z_\nu - z_\nu^{(0)}) + (\overline{z}_\nu - \overline{z}_\nu^{(0)})]$$

and

$$y_\nu - y_\nu^{(0)} = \frac{1}{2i}[(z_\nu - z_\nu^{(0)}) - (\overline{z}_\nu - \overline{z}_\nu^{(0)})].$$

Then

$$f(\mathfrak{z}) = f(\mathfrak{z}_0) + \sum_{\nu=1}^{n} (z_\nu - z_\nu^{(0)}) \frac{\Delta_\nu^*(\mathfrak{z}) - i\Delta_\nu^{**}(\mathfrak{z})}{2} + \sum_{\nu=1}^{n} (\overline{z}_\nu - \overline{z}_\nu^{(0)}) \frac{\Delta_\nu^*(\mathfrak{z}) + i\Delta_\nu^{**}(\mathfrak{z})}{2}.$$

If we define

$$\Delta'_\nu : = \frac{\Delta^*_\nu - i\Delta^{**}_\nu}{2} \quad \text{and} \quad \Delta''_\nu : = \frac{\Delta^*_\nu + i\Delta^{**}_\nu}{2}$$

then (3) is satisfied.

2. Let $f(\mathfrak{z}) = f(\mathfrak{z}_0) + \sum_{\nu=1}^{n} (z_\nu - z_\nu^{(0)}) \Delta'_\nu(\mathfrak{z}) + \sum_{\nu=1}^{n} (\bar{z}_\nu - \bar{z}_\nu^{(0)}) \Delta''_\nu(\mathfrak{z})$, Δ'_ν, Δ''_ν continuous at $\mathfrak{z}_0$. The equations $\Delta'_\nu = (\Delta^*_\nu - i\Delta^{**}_\nu)/2$, $\Delta''_\nu = (\Delta^*_\nu + i\Delta^{**}_\nu)/2$ appear in matrix form as

$$\begin{pmatrix} \Delta'_\nu \\ \Delta''_\nu \end{pmatrix} = \frac{1}{2} \cdot \begin{pmatrix} 1 & -i \\ 1 & i \end{pmatrix} \circ \begin{pmatrix} \Delta^* \\ \Delta^{**}_\nu \end{pmatrix}$$

Let

$$A : = \begin{pmatrix} 1 & -i \\ 1 & i \end{pmatrix}.$$

Then $\det A = 2i \neq 0$. That means that the equations can be solved for Δ^*_ν and Δ^{**}_ν. The solution functions satisfy equation (2); (1) follows from decomposition into real and imaginary parts. Since the values of the functions α^*_ν, α^{**}_ν, β^*_ν, β^{**}_ν are uniquely determined at the point $\mathfrak{z}_0$, the same must be true of the functions Δ'_ν, Δ''_ν. □

We now write:

$$f_{z_\nu}(\mathfrak{z}_0) : = \Delta'_\nu(\mathfrak{z}_0) = \frac{1}{2}\left[f_{x_\nu}(\mathfrak{z}_0) - i f_{y_\nu}(\mathfrak{z}_0) \right],$$

$$f_{\bar{z}_\nu}(\mathfrak{z}_0) : = \Delta''_\nu(\mathfrak{z}_0) = \frac{1}{2}\left[f_{x_\nu}(\mathfrak{z}_0) + i f_{y_\nu}(\mathfrak{z}_0) \right].$$

Theorem 6.2. *Let $B \subset \mathbb{C}^n$ be a region $\mathfrak{z}_0 \in B$, f a complex function on B. f is complex differentiable at $\mathfrak{z}_0$ if and only if f is real differentiable at $\mathfrak{z}_0$ and $f_{\bar{z}_\nu}(\mathfrak{z}_0) = 0$ for $1 \leqslant \nu \leqslant n$. (This means that the Cauchy-Riemann differential equations must be satisfied:*

$$\left.\begin{aligned} g_{x_\nu} &= h_{y_\nu} \\ h_{x_\nu} &= -g_{y_\nu} \end{aligned}\right\} \quad \text{for } 1 \leqslant \nu \leqslant n.)$$

PROOF.

1. Let $f(\mathfrak{z}) = f(\mathfrak{z}_0) + \sum_{\nu=1}^{n} (z_\nu - z_\nu^{(0)}) \Delta_\nu(\mathfrak{z})$, $\Delta_\nu(\mathfrak{z})$ continuous at $\mathfrak{z}_0$. Then $f(\mathfrak{z}) = f(\mathfrak{z}_0) + \sum_{\nu=1}^{n} (z_\nu - z_\nu^{(0)}) \Delta'_\nu(\mathfrak{z}) + \sum_{\nu=1}^{n} (\bar{z}_\nu - \bar{z}_\nu^{(0)}) \Delta''_\nu(\mathfrak{z})$ with $\Delta'_\nu(\mathfrak{z}) = \Delta_\nu(\mathfrak{z})$ and $\Delta''_\nu(\mathfrak{z}) = 0$, so that $f_{\bar{z}_\nu}(\mathfrak{z}_0) = 0$ for $1 \leqslant \nu \leqslant n$.

2. Let f be real differentiable and $f_{\bar{z}_\nu}(\mathfrak{z}_0) = 0$ for $1 \leqslant \nu \leqslant n$. Then $f(\mathfrak{z}) = f(\mathfrak{z}_0) + \sum_{\nu=1}^{n} (z_\nu - z_\nu^{(0)}) \Delta'_\nu(\mathfrak{z}) + \sum_{\nu=1}^{n} (\bar{z}_\nu - \bar{z}_\nu^{(0)}) \Delta''_\nu(\mathfrak{z})$ with $\Delta''_\nu(\mathfrak{z}_0) = 0$ for $\nu = 1, \ldots, n$.

We define

$$\alpha_\nu(\mathfrak{z}) := \begin{cases} 0 & \text{if } z_\nu = z_\nu^{(0)} \\ \dfrac{\bar{z}_\nu - \bar{z}_\nu^{(0)}}{z_\nu - z_\nu^{(0)}} \cdot \Delta''_\nu(\mathfrak{z}) & \text{otherwise.} \end{cases}$$

Since

$$\frac{\bar{z}_\nu - \bar{z}_\nu^{(0)}}{z_\nu - z_\nu^{(0)}}$$

is bounded except at $z_\nu^{(0)}$ and $\lim_{\mathfrak{z} \to \mathfrak{z}_0} \Delta''_\nu(\mathfrak{z}) = 0$, it follows that α_ν is continuous at $\mathfrak{z}_0$. But then

$$\begin{aligned} f(\mathfrak{z}) &= f(\mathfrak{z}_0) + \sum_{\nu=1}^{n} (z_\nu - z_\nu^{(0)}) \Delta'_\nu(\mathfrak{z}) + \sum_{\nu=1}^{n} (\bar{z}_\nu - \bar{z}_\nu^{(0)}) \Delta''_\nu(\mathfrak{z}) \\ &= f(\mathfrak{z}_0) + \sum_{\nu=1}^{n} (z_\nu - z_\nu^{(0)})(\Delta'_\nu + \alpha_\nu)(\mathfrak{z}). \end{aligned}$$

Therefore f is complex differentiable at $\mathfrak{z}_0$. □

We mention another differentiation formula.

1. If f is real differentiable at $\mathfrak{z}_0$, we have at $\mathfrak{z}_0$

   a. $\bar{f}_{z_\mu} = (\bar{f})_{\bar{z}_\mu}$ for $1 \leqslant \mu \leqslant n$.
   b. $\bar{f}_{\bar{z}_\mu} = (\bar{f})_{z_\mu}$ for $1 \leqslant \mu \leqslant n$.

2. Let f be twice real differentiable in a neighborhood of $\mathfrak{z}_0$. Then at $\mathfrak{z}_0$

$$\left.\begin{aligned} &\text{a. } f_{z_\nu z_\mu} = f_{z_\mu z_\nu} \\ &\text{b. } f_{z_\nu \bar{z}_\mu} = f_{\bar{z}_\mu z_\nu} \\ &\text{c. } f_{\bar{z}_\nu \bar{z}_\mu} = f_{\bar{z}_\mu \bar{z}_\nu} \end{aligned}\right\} \quad \text{for all } \nu \text{ and } \mu.$$

Theorem 6.3 (Chain rule). *Let B_1, B_2 be regions in $\mathbb{C}^n$, respectively $\mathbb{C}^m$. $g = (g_1, \ldots, g_m): B_1 \to \mathbb{C}^m$ be a mapping with $g(B_1) \subset B_2$. Let $\mathfrak{z}_0 \in B_1$, $\mathfrak{w}_0 := g(\mathfrak{z}_0)$ and f a complex function on B_2. If all g_μ, $1 \leqslant \mu \leqslant m$, are real differentiable at $\mathfrak{z}_0$ and f is real differentiable at $\mathfrak{w}_0$, then $f \circ g$ is real differentiable at $\mathfrak{z}_0$ and*

$$(f \circ g)_{z_\nu}(\mathfrak{z}_0) = \sum_{\mu=1}^{m} (f_{w_\mu}(\mathfrak{w}_0)) \cdot ((g_\mu)_{z_\nu}(\mathfrak{z}_0)) + \sum_{\mu=1}^{m} (f_{\bar{w}_\mu}(\mathfrak{w}_0)) \cdot ((\bar{g}_\mu)_{z_\nu}(\mathfrak{z}_0)),$$

$$(f \circ g)_{\bar{z}_\nu}(\mathfrak{z}_0) = \sum_{\mu=1}^{m} (f_{w_\mu}(\mathfrak{w}_0)) \cdot ((g_\mu)_{\bar{z}_\nu}(\mathfrak{z}_0)) + \sum_{\mu=1}^{m} (f_{\bar{w}_\mu}(\mathfrak{w}_0)) \cdot ((\bar{g}_\mu)_{\bar{z}_\nu}(\mathfrak{z}_0)).$$

Proof. As in the real case, the proof follows from the definitions. □

Let $B \subset \mathbb{C}^n$ be a region, $f = (f_1, \ldots, f_n): B \to \mathbb{C}^n$ a real differentiable mapping. Then we can define the complex functional matrix of f:

$$J_f := \left(\begin{array}{c|c} (f_{\nu, z_\mu})_{\substack{\nu = 1, \ldots, n \\ \mu = 1, \ldots, n}} & (f_{\nu, \bar{z}_\mu})_{\substack{\nu = 1, \ldots, n \\ \mu = 1, \ldots, n}} \\ \hline (\bar{f}_{\nu, z_\mu})_{\substack{\nu = 1, \ldots, n \\ \mu = 1, \ldots, n}} & (\bar{f}_{\nu, \bar{z}_\mu})_{\substack{\nu = 1, \ldots, n \\ \mu = 1, \ldots, n}} \end{array}\right)$$

We assert that $\Delta_f := \det J_f$ agrees with the usual functional determinant as it is known for the real case. A series of row and column transformations is necessary for the proof: We have

$$f_{\nu, z_\mu} = \tfrac{1}{2}(f_{\nu, x_\mu} - i f_{\nu, y_\mu}),$$
$$f_{\nu, \bar{z}_\mu} = \tfrac{1}{2}(f_{\nu, x_\mu} + i f_{\nu, y_\mu}).$$

If we add the $(n + \mu)$-th to the μ-th column, we obtain

$$\Delta_f = \det \left(\begin{array}{c|c} (f_{\nu, x_\mu}) & (\tfrac{1}{2}(f_{\nu, x_\mu} + i f_{\nu, y_\mu})) \\ \hline (\bar{f}_{\nu, x_\mu}) & (\tfrac{1}{2}(\bar{f}_{\nu, x_\mu} + i \bar{f}_{\nu, y_\mu})) \end{array}\right),$$

therefore

$$\Delta_f = 2^{-n} \det \left(\begin{array}{c|c} (f_{\nu, x_\mu}) & (f_{\nu, x_\mu} + i f_{\nu, y_\mu}) \\ \hline (\bar{f}_{\nu, x_\mu}) & (\bar{f}_{\nu, x_\mu} + i \bar{f}_{\nu, y_\mu}) \end{array}\right).$$

Subtracting the μ-th from the $(n + \mu)$-th column yields

$$\Delta_f = 2^{-n} \det \left(\begin{array}{c|c} (f_{\nu, x_\mu}) & (i f_{\nu, y_\mu}) \\ \hline (\bar{f}_{\nu, x_\mu}) & (i \bar{f}_{\nu, y_\mu}) \end{array}\right),$$

therefore

$$\Delta_f = 2^{-n} i^n \det \left(\begin{array}{c|c} (f_{\nu, x_\mu}) & (f_{\nu, y_\mu}) \\ \hline (\bar{f}_{\nu, x_\mu}) & (\bar{f}_{\nu, y_\mu}) \end{array}\right).$$

Since $f_\nu = g_\nu + i h_\nu$,

$$f_{\nu, x_\mu} = g_{\nu, x_\mu} + i h_{\nu, x_\mu}, \qquad \bar{f}_{\nu, x_\mu} = g_{\nu, x_\mu} - i h_{\nu, x_\mu},$$
$$f_{\nu, y_\mu} = g_{\nu, y_\mu} + i h_{\nu, y_\mu}, \qquad \bar{f}_{\nu, y_\mu} = g_{\nu, y_\mu} - i h_{\nu, y_\mu}.$$

We add the $(n + \nu)$-th row to the ν-th row and obtain

$$\begin{aligned} \Delta_f &= 2^{-n} i^n \det \left(\begin{array}{c|c} (2 g_{\nu, x_\mu}) & (2 g_{\nu, y_\mu}) \\ \hline (g_{\nu, x_\mu} - i h_{\nu, x_\mu}) & (g_{\nu, y_\mu} - i h_{\nu, y_\mu}) \end{array}\right) \\ &= i^n \det \left(\begin{array}{c|c} (g_{\nu, x_\mu}) & (g_{\nu, y_\mu}) \\ \hline (g_{\nu, x_\mu} - i h_{\nu, x_\mu}) & (g_{\nu, y_\mu} - i h_{\nu, y_\mu}) \end{array}\right). \end{aligned}$$

Subtraction of the ν-th from the $(n + \nu)$-th row gives

$$\Delta_f = i^n \det \left(\begin{array}{c|c} (g_{\nu, x_\mu}) & (g_{\nu, y_\mu}) \\ \hline (-i h_{\nu, x_\mu}) & (-i h_{\nu, y_\mu}) \end{array}\right) = \det \left(\begin{array}{c|c} (g_{\nu, x_\mu}) & (g_{\nu, y_\mu}) \\ \hline (h_{\nu, x_\mu}) & (h_{\nu, y_\mu}) \end{array}\right).$$

This is precisely the functional determinant $\det J_F$ of the real mapping $F = (g_1, \ldots, g_n, h_1, \ldots, h_n)$.

7. Holomorphic Mappings

Def. 7.1. Let $B \subset \mathbb{C}^n$ be a region, $g_1, \ldots, g_m$ complex functions on B. $g = (g_1, \ldots, g_m): B \to \mathbb{C}^m$ is called a *holomorphic mapping* if all the component functions g_μ are holomorphic in B.

Theorem 7.1. *Let $B_1 \subset \mathbb{C}^n$, $B_2 \subset \mathbb{C}^m$ be regions, $g = (g_1, \ldots, g_m): B_1 \to B_2$ be a mapping. g is holomorphic if and only if for each holomorphic function f on B_2 $f \circ g$ is a holomorphic function on B_1.*

PROOF. Let g be a holomorphic mapping. Then all the component functions g_μ are holomorphic, that is, $(g_\mu)_{\bar{z}_\nu} = 0$ for all ν and μ. If f is holomorphic, then $f_{\bar{w}_\mu} = 0$ for all μ, $f \circ g$ is real differentiable, and from the chain rule it follows that

$$(f \circ g)_{\bar{z}_\nu} = \sum_{\mu=1}^{m} f_{w_\mu} \cdot (g_\mu)_{\bar{z}_\nu} + \sum_{\mu=1}^{m} f_{\bar{w}_\mu} \cdot (\bar{g}_\mu)_{\bar{z}_\nu} = 0 \qquad \text{for} \qquad \nu = 1, \ldots, n.$$

Conversely, set $f(\mathfrak{w}) \equiv w_\mu$, if the condition is satisfied. Then $f \circ g(\mathfrak{z}) \equiv g_\mu(\mathfrak{z})$. □

From this theorem it follows that $f \circ g: B_1 \to \mathbb{C}^l$ is a holomorphic mapping if $g: B_1 \to B_2$ is a holomorphic mapping and $f: B_2 \to \mathbb{C}^l$ is a holomorphic mapping.

Def. 7.2. Let $B \subset \mathbb{C}^n$ be a region, $g = (g_1, \ldots, g_m)$ a holomorphic mapping from B into $\mathbb{C}^m$. We call

$$\mathfrak{M}_g := \left((g_{\mu, z_\nu}) \begin{matrix} \mu = 1, \ldots, m \\ \nu = 1, \ldots, n \end{matrix} \right)$$

the holomorphic functional matrix of g.

Theorem 7.2. *Let $\mathfrak{z}_0 \in B$, $\mathfrak{w}_0 = g(\mathfrak{z}_0)$, f and g as above. Then*

$$\mathfrak{M}_{f \circ g}(\mathfrak{z}_0) = \mathfrak{M}_f(\mathfrak{w}_0) \circ \mathfrak{M}_g(\mathfrak{z}_0).$$

PROOF. $(\mathfrak{M}_{f \circ g})_{\nu\mu} = (f_\nu \circ g)_{z_\mu} = \sum_{\lambda=1}^{m} f_{\nu, w_\lambda} \cdot g_{\lambda, z_\mu} = (\mathfrak{M}_f \circ \mathfrak{M}_g)_{\nu\mu}$. □

Def. 7.3. Let $B \subset \mathbb{C}^n$ be a region, $g = (g_1, \ldots, g_n): B \to \mathbb{C}^n$ a holomorphic mapping. $M_g := \det \mathfrak{M}_g$ is called the *holomorphic functional determinant of g*.

Theorem 7.2 implies:

Theorem 7.3. *Let the notation be as above and let $m = n = 1$. Then $M_{f \circ g} = M_f \cdot M_g$.*

Complex functional determinants of a holomorphic mapping have the following form:

$$\begin{aligned}\Delta_g &= \det\begin{pmatrix}(g_{\nu,z_\mu}) & (g_{\nu,\bar z_\mu})\\ (\bar g_{\nu,z_\mu}) & (\bar g_{\nu,\bar z_\mu})\end{pmatrix} = \det\begin{pmatrix}(g_{\nu,z_\mu}) & 0\\ 0 & (\overline{g_{\nu,z_\mu}})\end{pmatrix}\\ &= \det((g_{\nu,z_\mu}))\cdot\det((\overline{g_{\nu,z_\mu}}))\\ &= \det((g_{\nu,z_\mu}))\cdot\overline{\det((g_{\nu,z_\mu}))} = |\det((g_{\nu,z_\mu}))|^2 = |M_g|^2,\end{aligned}$$

i.e., they are real and nonnegative. This means that holomorphic mappings are orientation-preserving.

Def. 7.4. Let B_1, B_2 be regions in $\mathbb{C}^n$. A mapping $g: B_1 \to B_2$ is called *biholomorphic* (resp. *invertably holomorphic*) if

a. g is bijective, and
b. g and g^{-1} are holomorphic.

Theorem 7.4. *Let $B \subset \mathbb{C}^n$ be a region, $g: B \to \mathbb{C}^n$ a holomorphic mapping. Let $\mathfrak{z}_0 \in B$ and $\mathfrak{w}_0 = g(\mathfrak{z}_0)$. There are open neighborhoods $U = U(\mathfrak{z}_0) \subset B$ and $V = V(\mathfrak{w}_0) \subset \mathbb{C}^n$ such that $g: U \to V$ is biholomorphic if and only if $M_g(\mathfrak{z}_0) \neq 0$.*

PROOF

1. There are open neighborhoods U, V such that $g: U \to V$ is biholomorphic. Then $1 = M_{\mathrm{id}_U}(\mathfrak{z}_0) = M_{g^{-1}}(\mathfrak{w}_0)\cdot M_g(\mathfrak{z}_0)$, hence $M_g(\mathfrak{z}_0) \neq 0$.

2. g is continuously differentiable, and the functional determinant M_g is continuous. If $M_g(\mathfrak{z}_0) \neq 0$, then there exists an open neighborhood $W = W(\mathfrak{z}_0) \subset B$ with $(M_g|W) \neq 0$. So $\Delta_g|W \neq 0$ and g is regular (in the real sense) at $\mathfrak{z}_0$.

There are open neighborhoods $U = U(\mathfrak{z}_0) \subset W$, $V = V(\mathfrak{w}_0)$ such that $g: U \to V$ is bijective and $g^{-1} = (\breve g_1, \ldots, \breve g_n)$ is continuously differentiable.

$g \circ g^{-1}|V = \mathrm{id}_V$ is a holomorphic mapping. It follows that

$$0 = (g_\nu \circ g^{-1})_{\bar w_\mu} = \sum_{\lambda=1}^{n} g_{\nu,z_\lambda}\cdot \breve g_{\lambda,\bar w_\mu} + \sum_{\lambda=1}^{n} g_{\nu,\bar z_\lambda}\cdot \overline{\breve g}_{\lambda,\bar w_\mu} = \sum_{\lambda=1}^{n} g_{\nu,z_\lambda}\cdot \breve g_{\lambda,\bar w_\mu}.$$

For each μ, $1 \leqslant \mu \leqslant n$, we obtain a system of linear equations:

$$0 = \mathfrak{M}_g \circ \begin{pmatrix}\breve g_{1,\bar w_\mu}\\ \cdot\\ \cdot\\ \cdot\\ \breve g_{n,\bar w_\mu}\end{pmatrix}$$

Since $\det \mathfrak{M}_g \neq 0$ there is only the trivial solution: $\breve g_{\lambda,\bar w_\mu} = 0$ for all λ and all μ. This holds in all of V. Therefore the Cauchy–Riemann differential equations are satisfied and g^{-1} is holomorphic. □

Theorem 7.5. *Let $B \subset \mathbb{C}^n$ be a region, $g = (g_1, \ldots, g_n)$ holomorphic and one-to-one in B. Then $M_g \neq 0$ throughout B.*

This theorem is wrong in the real case: for example $y = x^3$ is one-to-one, but the derivative $y' = 3x^2$ vanishes at the origin.

We shall not carry out the proof of Theorem 7.5 here. (It can be found as Theorem 5 of Chapter 5 in R. Narasimhan: *Several Complex Variables*, Chicago Lectures in Mathematics, 1971.)

Theorem 7.6. *Let $B_1 \subset \mathbb{C}^n$ be a region, $g: B_1 \to \mathbb{C}^n$ one-to-one and holomorphic. Then $B_2 := g(B_1)$ is also an open set and $g^{-1}: B_2 \to B_1$ is holomorphic.*

PROOF

1. Let $\mathfrak{w}_0 \in B_2$. Then there exists a $\mathfrak{z}_0 \in B_1$ with $g(\mathfrak{z}_0) = \mathfrak{w}_0$. From Theorem 7.5 $M_g \neq 0$ on B_1, and therefore there are open neighborhoods $U(\mathfrak{z}_0) \subset B_1$, $V(\mathfrak{w}_0) \subset \mathbb{C}^n$ such that $g: U \to V$ is biholomorphic. But then $V = g(U) \subset g(B_1) = B_2$; that is, $\mathfrak{w}_0$ is an interior point.

2. From (1) for each $\mathfrak{w}_0 \in B_2$ there exists an open neighborhood $V(\mathfrak{w}_0) \subset B_2$, such that $g^{-1}|V$ is holomorphic. □

CHAPTER II

Domains of Holomorphy

1. The Continuity Theorem

In this and the following sections we shall systematically treat the problem of analytic continuation of holomorphic functions.

Let $P = \{\mathfrak{z} \in \mathbb{C}^n : |\mathfrak{z}| < 1\}$ be the unit polycylinder, $q_1, \ldots, q_n$ with $0 < q_\nu < 1$ for $1 \leqslant \nu \leqslant n$ be real numbers. Then for $2 \leqslant \mu \leqslant n$ we define: $D_\mu := \{\mathfrak{z} \in P : |z_1| \leqslant q_1 \text{ and } q_\mu \leqslant |z_\mu| < 1\}$, $D := \bigcup_{\mu=2}^{n} D_\mu$ and $H := P - D = \bigcap_{\mu=2}^{n} (P - D_\mu)$. Then

$$\begin{aligned} H &= \{\mathfrak{z} \in P : |z_1| > q_1 \text{ or } |z_\mu| < q_\mu \text{ for } 2 \leqslant \mu \leqslant n\} \\ &= \{\mathfrak{z} \in P : q_1 < |z_1|\} \cup \{\mathfrak{z} \in P : |z_\mu| < q_\mu \text{ for } 2 \leqslant \mu \leqslant n\}. \end{aligned}$$

(P, H) is called a "Euclidean Hartogs figure in $\mathbb{C}^n$." H is a proper Reinhardt domain, $\hat{H} = P$ its complete hull.

Def. 1.1. Let (P, H) be a Euclidean Hartogs figure in $\mathbb{C}^n$, $g := (g_1, \ldots, g_n)$: $P \to \mathbb{C}^n$ be a biholomorphic mapping, and let $\tilde{P} := g(P)$, $\tilde{H} := g(H)$. Then $(\tilde{P}, \tilde{H})$ is called a *general* Hartogs *figure*.

We shall try to illustrate this concept intuitively for $n = 3$. The Euclidean Hartogs figure in absolute space appears in Fig. 6. In the future we shall use the following symbolic representation in $\mathbb{C}^n$. (Actually the situation is much more complicated.)

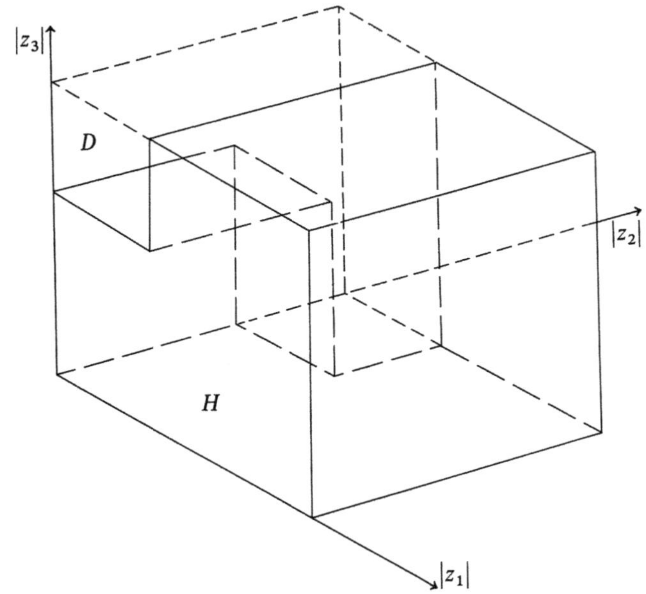

Figure 6. Euclidean Hartogs figure in $\mathbb{C}^3$.

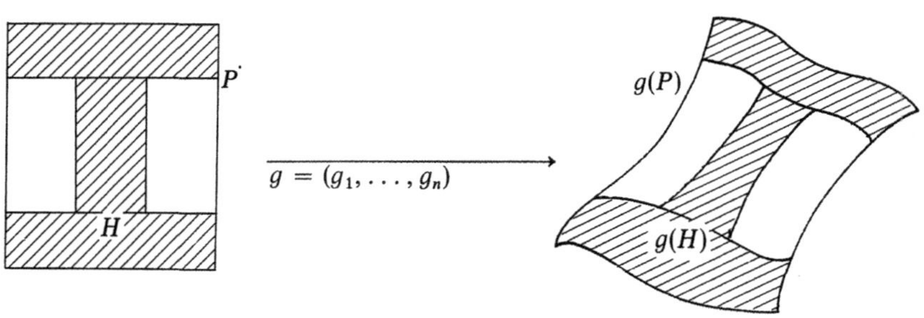

Figure 7. Symbolic representation of a general Hartogs figure.

Theorem 1.1. *Let $(\tilde{P}, \tilde{H})$ be a general Hartogs figure in $\mathbb{C}^n$, f holomorphic in $\tilde{H}$. Then there is exactly one holomorphic function F on $\tilde{P}$ with $F|\tilde{H} = f$.*

PROOF. Let $(\tilde{P}, \tilde{H}) = (g(P), g(H))$, $g: P \to \mathbb{C}^n$ be biholomorphic. Then $f \circ g$ is holomorphic in H and by Theorem 5.5 of Chapter I there is exactly one holomorphic function F^* on P with $F^*|H = f \circ g$. Let $F = F^* \circ g^{-1}$. Then F is holomorphic in $\tilde{P}$, $F|\tilde{H} = f$, and the uniqueness of the continuation follows from the uniqueness of F^*. □

Theorem 1.2 (Continuity theorem). *Let $B \subset \mathbb{C}^n$ be a region, $(\tilde{P}, \tilde{H})$ a general Hartogs figure with $\tilde{H} \subset B$, f a holomorphic function in B. If $\tilde{P} \cap B$ is connected, then f can be continued uniquely to $B \cup \tilde{P}$.*

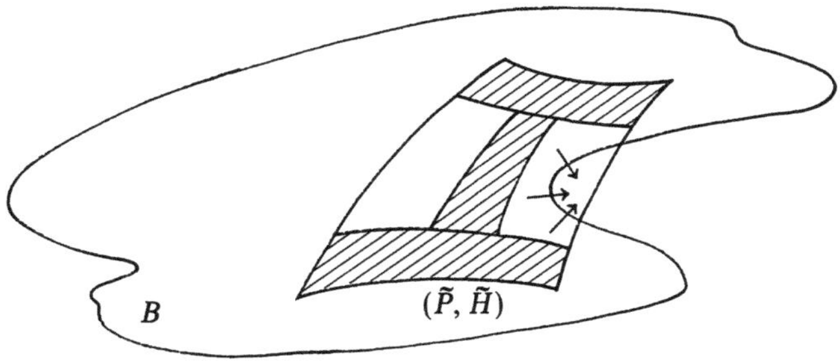

Figure 8. Illustration of the continuity theorem.

PROOF. $f_1 := f|\tilde{H}$ is holomorphic in $\tilde{H}$. Therefore there exists exactly one holomorphic function f_2 in $\tilde{P}$ with $f_2|\tilde{H} = f_1$.

Let
$$F(\mathfrak{z}) := \begin{cases} f(\mathfrak{z}) & \text{for } \mathfrak{z} \in B \\ f_2(\mathfrak{z}) & \mathfrak{z} \in \tilde{P}. \end{cases}$$

Since $B \cap \tilde{P}$ is a domain and $f|\tilde{H} = f_2|\tilde{H}$ it follows (from the identity theorem) that F is a well-defined holomorphic function on $B \cup \tilde{P}$. Clearly $F|B = f$. The uniqueness of the continuation is a further consequence of the identity theorem. □

The continuity theorem is fundamental to all further considerations.

Theorem 1.3. *Let $n \geqslant 2$, $P := \{\mathfrak{z} : |\mathfrak{z}| < 1\}$ be the unit polycylinder, $0 \leqslant r_\nu^0 < 1$ for $\nu = 1, \ldots, n$, $\bar{P}_{\mathfrak{r}_0} := \{\mathfrak{z} : |z_\nu| \leqslant r_\nu^0$ for all $\nu\}$ and $G := P - \bar{P}_{\mathfrak{r}_0}$. Then every holomorphic function f on G can be extended uniquely to a function holomorphic on P.*

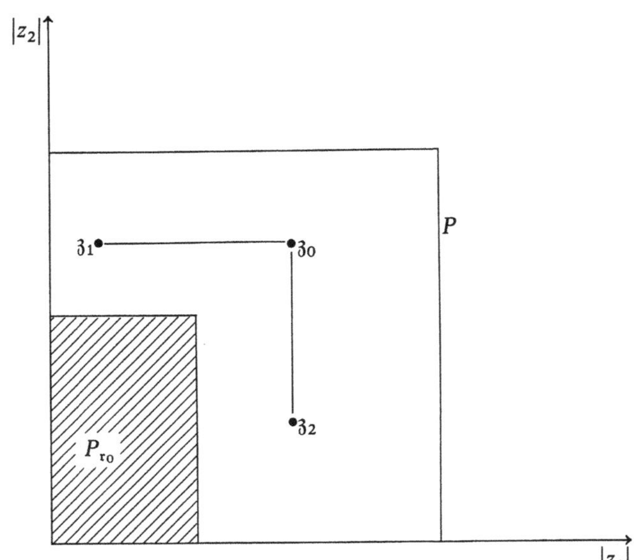

Figure 9. The proof of Theorem 1.3.

PROOF

1. Clearly G is a region. If $\mathfrak{z}_\lambda = (z_1^{(\lambda)}, \ldots, z_n^{(\lambda)})$, $\lambda = 1, 2$, are given, then the points $\tau(\mathfrak{z}_1)$, $\tau(\mathfrak{z}_2)$ also lie in G. For $\lambda = 1, 2$ we can connect $\mathfrak{z}_\lambda$ on the torus $T_{\mathfrak{z}_\lambda} \subset G$ with $\tau(\mathfrak{z}_\lambda)$. Define $\varphi_\lambda : I \to \mathbb{C}^n$ *by* $\varphi_\lambda(t) := (z_1^{(\lambda)}(t), \ldots, z_n^{(\lambda)}(t))$ with $z_\nu^{(\lambda)}(t) := |z_\nu^{(\lambda)}| + t \cdot (\max(|z_\nu^{(1)}|, |z_\nu^{(2)}|) - |z_\nu^{(\lambda)}|)$ for $\lambda = 1, 2$, $\nu = 1, \ldots, n$. Clearly $|z_\nu^{(\lambda)}(t)| \geqslant |z_\nu^{(\lambda)}| > r_\nu^0$ for $\nu = 1, \ldots, n$ so that $\varphi_\lambda(t) \in G$ for $t \in I$ and $\lambda = 1, 2$.

Let
$$\varphi(t) := \begin{cases} \varphi_1(2t) & 0 \leqslant t \leqslant \frac{1}{2} \\ \varphi_2(2 - 2t) & \frac{1}{2} \leqslant t \leqslant 1. \end{cases}$$

φ joins $\tau(\mathfrak{z}_1)$ with $\tau(\mathfrak{z}_2)$. Hence G is connected, and so is a domain.

2. For $\nu = 1, \ldots, n$ let $E_{(\nu)} := \{z_\nu \in \mathbb{C} : |z_\nu| < 1\}$. Choose $z_n^0 \in \mathbb{C}$ with $r_n^0 < |z_n^0| < 1$ and set

$$T(z_n) := \frac{z_n - z_n^0}{\bar{z}_n^0 z_n - 1}, \qquad g(z_1, \ldots, z_n) := (z_1, \ldots, z_{n-1}, T(z_n)).$$

$g : P \to P$ is a biholomorphic mapping with $g(0, \ldots, 0, z_n^0) = 0$. If $U = U(z_n^0) \subset \{z_n \in \mathbb{C} : r_n^0 < |z_n| < 1\}$ is an open neighborhood, then $E_{(1)} \times \cdots \times E_{(n-1)} \times U \subset G$, and therefore $E_{(1)} \times \cdots \times E_{(n-1)} \times T(U) \subset g(G)$. Choose

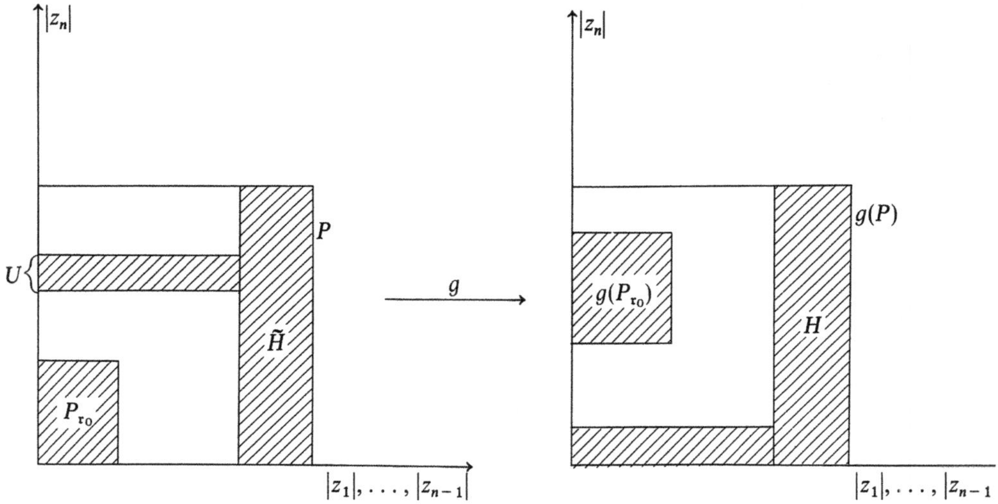

Figure 10. The proof of Theorem 1.3.

real numbers $q_1, \ldots, q_n$ with $r_\nu^0 < q_\nu < 1$ for $\nu = 1, \ldots, n - 1$ and $\{w_n : |w_n| < q_n\} \subset T(U)$. Then

$$H := \{\mathfrak{w} \in P : q_1 < |w_1|\} \cup \{\mathfrak{w} \in P : |w_\mu| < q_\mu \text{ for } \mu = 2, \ldots, n\}$$

is contained in $g(G)$ and (P, H) is a Euclidean Hartogs figure. $(\tilde{P}, \tilde{H})$ with $\tilde{P} := g^{-1}(P) = P$ and $\tilde{H} := g^{-1}(H)$ is a general Hartogs figure with $\tilde{H} \subset G$.

Moreover, $\tilde{P} \cap G = G$ is connected. The proposition now follows from the continuity theorem. □

The preceding theorem is a special case of the so-called *Kugelsatz*:

Let $n \geqslant 2$, $G \subset \mathbb{C}^n$ a domain, $K \subset G$ a compact subset, $G - K$ connected. Then every function holomorphic in $G - K$ can be uniquely extended to a function holomorphic on G.

The proof of the *Kugelsatz* is substantially more difficult than that of the preceding theorem. An important tool in its proof is the Bochner–Martinelli integral formula, which is a generalization of the Cauchy integral formula to a domain with piecewise smooth boundary.

Theorem 1.4. *Let $n \geqslant 2$, $B \subset \mathbb{C}^n$ be a region, and $\mathfrak{z}_0 \in B$. Let f be holomorphic in $B' := B - \{\mathfrak{z}_0\}$. Then f has a unique holomorphic extension on B. (For $n \geqslant 2$ there are no isolated singularities.)*

Proof. Without loss of generality we assume that $\mathfrak{z}_0 = 0$. Let P be a polycylinder about $\mathfrak{z}_0$ with $P \subset B$, $P' := P - \{\mathfrak{z}_0\}$. This is the situation of Theorem 1.3; so there is a holomorphic function F' in P with $F'|P' = f|P'$.

Let
$$F(\mathfrak{z}) := \begin{cases} F'(\mathfrak{z}) & \mathfrak{z} \in P \\ f(\mathfrak{z}) & \mathfrak{z} \in B'. \end{cases}$$

F is the holomorphic continuation of f to B. □

Def. 1.2. Let $G \subset \mathbb{C}^{n-1}$ be a domain, $g: G \to \mathbb{C}$ a continuous function. Then $\mathscr{F} := \{\mathfrak{z} \in \mathbb{C} \times G : z_1 = g(z_2, \ldots, z_n)\}$ is called a *real* $(2n-2)$-*dimensional surface*. If g is holomorphic, then $\mathscr{F}$ is called an *analytic surface*.

Theorem 1.5. *Let $G \subset \mathbb{C}^{n-1}$, $G_1 \subset \mathbb{C}$ be domains, $g: G \to \mathbb{C}$ be a continuous function with $g(G) \subset G_1$ and $\mathfrak{z}_0 \in \mathscr{F} = \text{graph}(g)$. If $U = U(\mathfrak{z}_0) \subset \hat{G} := G_1 \times G$ is an open neighborhood and f is a holomorphic function on $S := (\hat{G} - \mathscr{F}) \cup U$, then f has a unique holomorphic extension to $\hat{G}$.*

Proof. The uniqueness of the extension follows from the identity theorem because $\hat{G}$ is a domain. For the proof of existence we treat only the case $G = \{\mathfrak{z}^* \in \mathbb{C}^{n-1} : |\mathfrak{z}^*| < 1\}$, $G_1 = E_{(1)}$ (then $\hat{G} = P$, the unit polycylinder in $\mathbb{C}^n$), and in addition assume that $|g(\mathfrak{z}^*)| < q < 1$ for fixed $q \in \mathbb{R}$ and all $\mathfrak{z}^* \in G$. The proof is in two steps:

1. $S = (\hat{G} - \mathscr{F}) \cup U$ is connected.

a. Let $\mathfrak{z}_1, \mathfrak{z}_2$ be points in $\hat{G} - \mathscr{F}$. Then define

$$\mathfrak{z}_1^* := \left(\frac{1+q}{2}, z_2^{(1)}, \ldots, z_n^{(1)}\right), \qquad \mathfrak{z}_2^* := \left(\frac{1+q}{2}, z_2^{(2)}, \ldots, z_n^{(2)}\right).$$

$\mathfrak{z}_\lambda$ and $\mathfrak{z}_\lambda^*$ lie in the punctured disk

$$(E_{(1)} - \{g(z_2^{(\lambda)}, \ldots, z_n^{(\lambda)})\}) \times \{(z_2^{(\lambda)}, \ldots, z_n^{(\lambda)})\}$$

and can therefore be connected by a path which does not cross $\mathscr{F}$. The line segment connecting $\mathfrak{z}_1^*$ and $\mathfrak{z}_2^*$ also lies in $\hat{G} - \mathscr{F}$, so we can join $\mathfrak{z}_1$ and $\mathfrak{z}_2$ by a path in $\hat{G} - \mathscr{F}$.

b. If $\mathfrak{z}_1 \in U$, $\mathfrak{z}_2 \in \hat{G} - \mathscr{F}$, let U_1 be the connected component of $\mathfrak{z}_1$ in U. Since $U_1 - \mathscr{F}$ is non-empty, we can join $\mathfrak{z}_1$ in U_1 with a point $\mathfrak{z}_1^* \in U_1 - \mathscr{F}$. In particular, $\mathfrak{z}_1^*$ then lies in $\hat{G} - \mathscr{F}$ and by case (a) we can join it with $\mathfrak{z}_2$. If $\mathfrak{z}_1, \mathfrak{z}_2 \in U$ then both points can be connected with a point $\mathfrak{z}_0 \in \hat{G} - \mathscr{F}$ and therefore with one another.

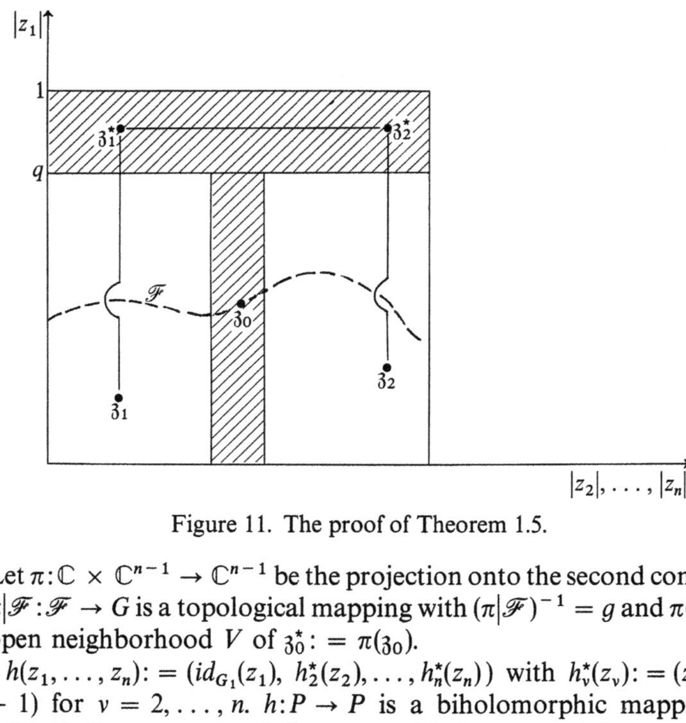

Figure 11. The proof of Theorem 1.5.

2. Let $\pi: \mathbb{C} \times \mathbb{C}^{n-1} \to \mathbb{C}^{n-1}$ be the projection onto the second component. Then $\pi|\mathscr{F}: \mathscr{F} \to G$ is a topological mapping with $(\pi|\mathscr{F})^{-1} = g$ and $\pi(\mathscr{F} \cap U)$ is an open neighborhood V of $\mathfrak{z}_0^* := \pi(\mathfrak{z}_0)$.

Let $h(z_1, \ldots, z_n) := (id_{G_1}(z_1), h_2^*(z_2), \ldots, h_n^*(z_n))$ with $h_\nu^*(z_\nu) := (z_\nu - z_\nu^0)/(\bar{z}_\nu^0 z_\nu - 1)$ for $\nu = 2, \ldots, n$. $h: P \to P$ is a biholomorphic mapping with $h(0) = (0, \mathfrak{z}_0^*)$. Set $q_1 := q$ and choose q_ν with $0 < q_\nu < 1$ for $\nu = 2, \ldots, n$ so that $h(\{(w_1, \ldots, w_n) \in P: |w_\nu| < q_\nu \text{ for } \nu = 2, \ldots, n\})$ is contained in $E_{(1)} \times V$.

Let $H := \{\mathfrak{w} \in P: |w_\nu| < q_\nu \text{ for } \nu = 2, \ldots, n\} \cup \{\mathfrak{w} \in P: q_1 < |w_1|\}$. Then (P, H) is a Euclidean Hartogs figure and $(P, \tilde{H})$ with $\tilde{H} := h(H)$ is a general Hartogs figure. Clearly $\tilde{H} \subset (E_{(1)} \times V) \cup \{\mathfrak{z} \in P: q_1 < |z_1|\} \subset S$ and by (1) $P \cap S = S$ is connected. The proposition follows from the continuity theorem. □

Remark. If g is holomorphic, therefore $\mathscr{F}$ an analytic surface, then there is a holomorphic function f on $\hat{G} - \mathscr{F}$ which does not permit a holomorphic extension beyond $\mathscr{F}$. For example set

$$f(z_1, \mathfrak{z}^*) := \frac{1}{z_1 - g(\mathfrak{z}^*)}.$$

PROOF. Assume there existed a point $\mathfrak{z}_0 \in \mathscr{F}$ and an open neighborhood $U = U(\mathfrak{z}_0) \subset \hat{G}$ such that f had a holomorphic extension F defined on $(\hat{G} - \mathscr{F}) \cup U$. Then there would be a sequence $(\mathfrak{z}_j)$ of points of $(\hat{G} - \mathscr{F})$ which converged to $\mathfrak{z}_0$ and clearly as $\mathfrak{z}_j \to \mathfrak{z}_0$, $|f(\mathfrak{z}_j)|$ would tend to infinity. But since F is continuous at $\mathfrak{z}_0$ we would have $\lim_{j\to\infty} f(\mathfrak{z}_j) = \lim_{j\to\infty} F(\mathfrak{z}_j) = F(\mathfrak{z}_0)$ and that would be a contradiction. □

With much more effort, one can prove the converse:

If $\mathscr{F} \subset \hat{G}$ is a real $(2n - 2)$-dimensional surface and there is a holomorphic function f in $\hat{G} - \mathscr{F}$ which is not holomorphically continuable to $\hat{G}$, then $\mathscr{F}$ is an analytic surface.

2. Pseudoconvexity

Def. 2.1. Let $B \subset \mathbb{C}^n$ be a region. B is called *pseudoconvex* if for all general Hartogs figures (P, H) with $H \subset B$, all of P lies in B.

Def. 2.2. Let $B \subset \mathbb{C}^n$ be a region. f holomorphic in B, $\mathfrak{z}_0 \in \partial B$ a point. f is called *completely singular at* $\mathfrak{z}_0$ if there exists a neighborhood $V = V(\mathfrak{z}_0)$ such that for any connected neighborhood $U = U(\mathfrak{z}_0)$ with $U \subset V$. There is no holomorphic function F which in a non-empty open subset of $U \cap B$ coincides with f.

Def. 2.3. Let $B \subset \mathbb{C}^n$ be a non-empty open set. B is called a *region of holomorphy* if there is a function f holomorphic in B which is completely singular at every point $\mathfrak{z}_0 \in \partial B$. If in addition B is connected, then B is called a domain of holomorphy.

EXAMPLES

1. Since $\mathbb{C}^n$ has no boundary it trivially satisfies the requirements of Def. 2.3. Therefore $\mathbb{C}^n$ is a domain of holomorphy.

2. The unit disk $E_{(1)} \subset \mathbb{C}$ is a domain of holomorphy, as is shown in 1-dimensional theory.

3. The dicylinder $E_{(1)} \times E_{(1)}$ is a domain of holomorphy: If $f: E_{(1)} \to \mathbb{C}$ is a holomorphic function which is completely singular on $\partial E_{(1)}$, then $g: E_{(1)} \times E_{(1)} \to \mathbb{C}$ with $g(z_1, z_2) := f(z_1) + f(z_2)$ is a holomorphic function which is completely singular on $\partial(E_{(1)} \times E_{(1)})$.

4. Let (P, H) be a Euclidean Hartogs figure, $\mathfrak{z}_0 \in \partial H \cap P$. For every function f holomorphic in H there exists a function F holomorphic in P with $F|H = f$. If V is an arbitrary open neighborhood of $\mathfrak{z}_0$ which is entirely

contained in P and U is the connected component of $\mathfrak{z}_0$ in V, then $F|V$ is holomorphic, $U \cap H \neq \emptyset$, and $F|U \cap H = f|U \cap H$. Therefore H is not a domain of holomorphy.

Theorem 2.1. *Let $B \subset \mathbb{C}^n$ be a region, $G \subset \mathbb{C}^n$ a domain with $B \cap G \neq \emptyset$ and $(\mathbb{C}^n - B) \cap G \neq \emptyset$. Then for each connected component Q of $B \cap G$*

$$G \cap \partial Q \cap \partial B \neq \emptyset.$$

PROOF. We have $G = Q \cup (G - Q)$. Q is open and not empty, and because $(\mathbb{C}^n - B) \cap G \neq \emptyset$, $G - Q$ is also non-empty. Since G is a domain it does not split into two non-empty open subsets. Hence $G - Q$ is not open. Let $\mathfrak{z}_1 \in G - Q$ not be an interior point. Then for every arbitrary neighborhood $U(\mathfrak{z}_1) \subset G$ it is true that $U \cap Q \neq \emptyset$. Therefore $\mathfrak{z}_1$ lies in ∂Q. If $\mathfrak{z}_1 \in B$ then there is a connected neighborhood $V(\mathfrak{z}_1) \subset B \cap G$ (with $V \cap Q \neq \emptyset$ also). But then $Q \cup V$ is an open connected set in $B \cap G$ which properly contains Q. Since Q is a connected component this is a contradiction. Therefore $\mathfrak{z}_1$ does not lie in B. Hence it follows that $\mathfrak{z}_1 \in \partial Q \cap \partial B \cap G$. □

Theorem 2.2. *Let G be a domain of holomorphy. Then G is pseudoconvex.*

PROOF. Assume that G is not pseudoconvex. Then there is a Hartogs figure (P, H) with $H \subset G$ but $P \cap G \neq P$. We choose an arbitrary $\mathfrak{z}_0$ in H and set $Q := C_{P \cap G}(\mathfrak{z}_0)$. Since H lies in $P \cap G$ and is connected it follows that $H \subset Q$. Furthermore, $Q \subsetneq P$.

Since $P \cap G \neq \emptyset$ and $(\mathbb{C}^n - G) \cap P \neq \emptyset$ there is by Theorem 2.1 a point $z_1 \in \partial Q \cap \partial G \cap P$.

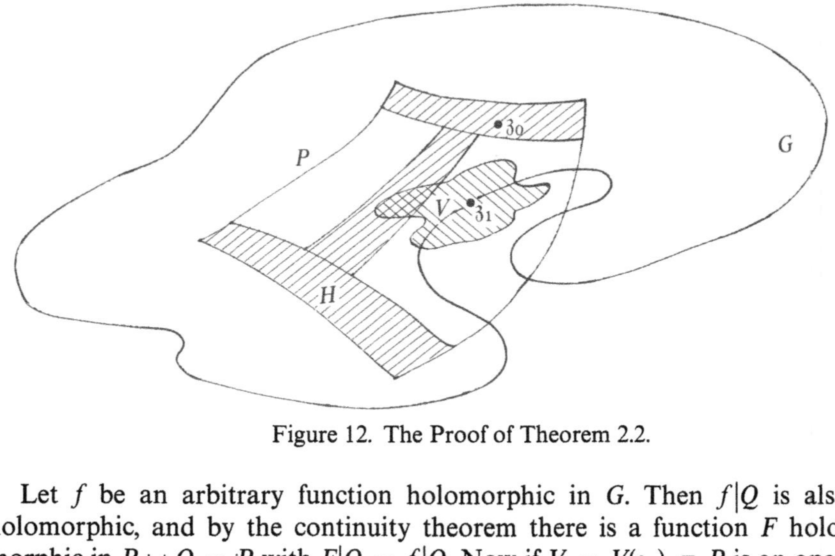

Figure 12. The Proof of Theorem 2.2.

Let f be an arbitrary function holomorphic in G. Then $f|Q$ is also holomorphic, and by the continuity theorem there is a function F holomorphic in $P \cup Q = P$ with $F|Q = f|Q$. Now if $V = V(\mathfrak{z}_1) \subset P$ is an open connected subset, then $F|V$ is holomorphic, $Q \cap V$ is open and non-empty, and $F|Q \cap V = f|Q \cap V$. Therefore G is not a domain of holomorphy. This completes the proof by contradiction. □

In 1910 the converse of the above theorem was proven in special cases by E. E. Levi. The so-called *Levi Conjecture*, that this converse holds without additional assumptions was first proved in 1942 by Oka for $n = 2$ and in 1954 for $n > 2$ simultaneously by Oka and by Norguet and Bremermann. The proof is very deep and will not be presented here (see, for example, [7]).

To conclude this section, we will sketch the connection between the pseudoconvexity of a domain G and the curvature of its boundary.

Let $B \subset \mathbb{C}^n$ be a region, $\mathfrak{z}_0 \in B$ and $\varphi: B \to \mathbb{R}$ a twice continuously differentiable function. One can regard B as a subset of $\mathbb{R}^{2n}$ and consider the tangent space $T_{\mathfrak{z}_0}$ and the space $T^*_{\mathfrak{z}_0}$ of the Pfaffian forms (see [21], [22]). The total differential of φ at the point $\mathfrak{z}_0$ is the linear form

$$(d\varphi)_{\mathfrak{z}_0} = \sum_{\nu=1}^{n} \varphi_{x_\nu}(\mathfrak{z}_0)\, dx_\nu + \sum_{\nu=1}^{n} \varphi_{y_\nu}(\mathfrak{z}_0)\, dy_\nu \in T^*_{\mathfrak{z}_0}.$$

If $f = g + ih$ is a complex-valued differentiable function, set $df := dg + idh$. Then $dz_\nu = dx_\nu + idy_\nu$, $d\bar{z}_\nu = dx_\nu - idy_\nu$ and we can write the differential in the form

$$(d\varphi)_{\mathfrak{z}_0} = \sum_{\nu=1}^{n} \varphi_{z_\nu}(\mathfrak{z}_0)\, dz_\nu + \sum_{\nu=1}^{n} \varphi_{\bar{z}_\nu}(\mathfrak{z}_0)\, d\bar{z}_\nu.$$

Def. 2.4. A *domain with C^2 boundary* is a domain $G \subset \mathbb{C}^n$ with the following properties:

1. G is bounded.
2. For each point $\mathfrak{z}_0 \in \partial G$ there is an open neighborhood $U = U(\mathfrak{z}_0) \subset \mathbb{C}^n$ and a twice continuously differentiable function $\varphi: U \to \mathbb{R}$ for which

   a. $U \cap G = \{\mathfrak{z} \in U : \varphi(\mathfrak{z}) < 0\}$
   b. $(d\varphi)_{\mathfrak{z}} \neq 0$ for all $\mathfrak{z} \in U$.

Remark. Under the conditions of Def. 2.4 the implicit function theorem implies

1. $\partial G \cap U = \{\mathfrak{z} \in U : \varphi(\mathfrak{z}) = 0\}$;
2. there is (after a reduction of U if necessary) a C^2-diffeomorphism $\Phi: U \to B$, where $B \subset \mathbb{C}^n$ is a region such that $\Phi(U \cap G) = \{\mathfrak{z} \in B : x_1 < 0\}$ and $\Phi(U \cap \partial G) = \{\mathfrak{z} \in B : x_1 = 0\}$.

We say that $(G, \partial G)$ is a *differentiable manifold with boundary.*

Theorem 2.3. *Let $G \subset \mathbb{C}^n$ be a domain with C^2 boundary, U an open set with $U \cap \partial G \neq \emptyset$. Let φ, ψ be two functions on U which satisfy the conditions of Def. 2.4. Then there is a uniquely determined positive differentiable function h on U such that $\varphi \doteq h \cdot \psi$.*

PROOF. We only need to show that for each $\mathfrak{z}_0 \in U \cap \partial G$ there is a neighborhood $V(\mathfrak{z}_0) \subset U$ and in V exactly one differentiable function h with $\varphi|V = h \cdot (\psi|V)$. Therefore let $\mathfrak{z}_0 \in U \cap \partial G$ and $W(\mathfrak{z}_0) \subset U$ be chosen so that there is a C^2-diffeomorphism $\Phi: W \to B \subset \mathbb{C}^n$ with $\Phi(W \cap G) = \{\mathfrak{z} \in B: x_1 < 0\}$, $\Phi(W \cap \partial G) = \{\mathfrak{z} \in B: x_1 = 0\}$. Then the functions $\tilde{\varphi}: = \varphi \circ \Phi^{-1}$, $\tilde{\psi}: = \psi \circ \Phi^{-1}$ are twice differentiable in B. Without loss of generality we assume that $\Phi(\mathfrak{z}_0) = 0$ and B is convex (in the sense that for any two points in B the connecting line segment lies in B). Define

$$h_1(x_1, \ldots, x_n, y_1, \ldots, y_n): = \int_0^1 \frac{\partial \tilde{\varphi}}{\partial x_1}(tx_1, x_2, \ldots, y_n)\, dt,$$

$$h_2(x_1, \ldots, x_n, y_1, \ldots, y_n): = \int_0^1 \frac{\partial \tilde{\psi}}{\partial x_1}(tx_1, x_2, \ldots, y_n)\, dt.$$

Then $\tilde{\varphi} = h_1 \cdot x_1$ and $\tilde{\psi} = h_2 \cdot x_1$. Since $(d\varphi)_{\mathfrak{z}_0} \neq 0$ and $(d\psi)_{\mathfrak{z}_0} \neq 0$, near $\partial\tilde{\varphi}/\partial x_1$ and $\partial\tilde{\psi}/\partial x_1$ have no zeroes, $0 \in B$ and the same holds for h_1, h_2. Set $h: = (h_1/h_2) \circ \Phi$ in a neighborhood of $\mathfrak{z}_0$. Then

$$\begin{aligned} h(\mathfrak{z}) \cdot \psi(\mathfrak{z}) &= (h \circ \Phi^{-1})(\Phi(\mathfrak{z})) \cdot (\psi \circ \Phi^{-1})(\Phi(\mathfrak{z})) = \left(\frac{h_1}{h_2}\right)(\Phi(\mathfrak{z})) \cdot \tilde{\psi}(\Phi(\mathfrak{z})) \\ &= (h_1 \cdot x_1)(\Phi(\mathfrak{z})) = \tilde{\varphi}(\Phi(\mathfrak{z})) = \varphi(\mathfrak{z}). \end{aligned}$$

Here h is continuously differentiable and, near $\mathfrak{z}_0$, has no zeroes. h is uniquely determined, for outside ∂G we have $h = \varphi/\psi$. □

Def. 2.5. Let $B \subset \mathbb{C}^n$ be a region, $\varphi: B \to \mathbb{R}$ be twice continuously differentiable, $\mathfrak{z}_0 \in B$. Then the quadratic form $L_{\varphi,\mathfrak{z}_0}$ with $L_{\varphi,\mathfrak{z}_0}(\mathfrak{w}): = \sum_{i,j=1}^{n} \varphi_{z_i\bar{z}_j}(\mathfrak{z}_0) w_i \bar{w}_j$ is called the *Levi form of* φ *at* $\mathfrak{z}_0$. φ satisfies the Levi condition if the following holds: If $\mathfrak{w} \in \mathbb{C}^n$ and $\sum_{i=1}^{n} \varphi_{z_i}(\mathfrak{z}_0) w_i = 0$, then $L_{\varphi,\mathfrak{z}_0}(\mathfrak{m}) \geqslant 0$.

Theorem 2.4. *Let $G \subset \mathbb{C}^n$ be a domain with C^2 boundary, $\mathfrak{z}_0 \in \partial G$ and $U = U(\mathfrak{z}_0)$ an open neighborhood. Let φ, ψ be two functions on U which satisfy the conditions of Def. 2.4. If φ satisfies the Levi condition at $\mathfrak{z}_0$, so does ψ.*

PROOF. We can find a twice continuously differentiable positive function h on U with $\psi = h \cdot \varphi$. Now let $\mathfrak{w} \in \mathbb{C}^n$ and $\sum_{i=1}^{n} \psi_{z_i}(\mathfrak{z}_0) w_i = 0$. Then at $\mathfrak{z}_0$

$$\begin{aligned} 0 &= \sum_{i=1}^{n} (h \cdot \varphi_{z_i} + \varphi \cdot h_{z_i}) w_i \\ &= h \cdot \sum_{i=1}^{n} \varphi_{z_i} w_i \text{ (because of } \varphi|\partial G = 0), \qquad \text{so} \qquad \sum_{i=1}^{n} \varphi_{z_i} w_i = 0. \end{aligned}$$

It follows that:

$$L_\psi(\mathfrak{w}) = \sum_{i,j=1}^{n} \psi_{z_i\bar{z}_j} w_i \bar{w}_j = \sum_{i,j=1}^{n} (h_{z_i}\varphi_{\bar{z}_j} + \varphi_{z_i\bar{z}_j} h + \varphi_{z_i} h_{\bar{z}_j}) w_i \bar{w}_j$$

$$= h \cdot L_\varphi(\mathfrak{w}) + \sum_{i=1}^{n} \left(\sum_{j=1}^{n} \overline{\varphi_{z_j} w_j} \right) h_{z_i} w_i + \sum_{j=1}^{n} \left(\sum_{i=1}^{n} \varphi_{z_i} w_i \right) h_{\bar{z}_j} \bar{w}_j,$$

where the last two terms vanish, as was shown above. Since h is positive, the proposition follows. □

Def. 2.6. For a domain $G \subset \mathbb{C}^n$ with C^2 boundary the Levi condition is satisfied at a point $\mathfrak{z}_0 \in \partial G$ if there is an open neighborhood $U = U(\mathfrak{z}_0)$ and a function φ on U which satisfies the conditions required by Def. 2.4 so that at $\mathfrak{z}_0$ φ satisfies the Levi condition.

Theorem 2.5. *Let $G \subset \mathbb{C}^n$ be a domain with C^2 boundary. Then G is pseudoconvex if and only if the Levi condition is satisfied for every boundary point of G.*

This theorem will not be proved here.

3. Holomorphic Convexity

We will investigate whether there is a relationship between pseudoconvexity and the usual convexity of sets. We start with some observations about convex domains in $\mathbb{R}^2$.

Let L be the set of linear mappings $\ell: \mathbb{R}^2 \to \mathbb{R}$ with

$$\ell(\mathfrak{x}) = ax_1 + bx_2 + c, \qquad a, b, c \in \mathbb{R}.$$

A line g in $\mathbb{R}^2$ is a set of points $\mathfrak{x} = \mathfrak{x}_0 + t\mathfrak{v}$ with $t \in \mathbb{R}$ and appropriate fixed vectors $\mathfrak{x}_0, \mathfrak{v} \in \mathbb{R}^2$, $\mathfrak{v} \neq 0$,

$$g = \{\mathfrak{x} \in \mathbb{R}^2 : \mathfrak{x} = \mathfrak{x}_0 + t\mathfrak{v}, \qquad t \in \mathbb{R}\}.$$

Now let $\ell \in L$ with $\ell(\mathfrak{x}) = ax_1 + bx_2 + c$ and $(a, b) \neq (0, 0)$. For $b \neq 0$ let $\mathfrak{x}_0 := (0, -c/b)$, $\mathfrak{v} := (1, -a/b)$; for $b = 0$ and $a \neq 0$ let $\mathfrak{x}_0 := (-c/a, 0)$, $\mathfrak{v} := (0, 1)$. Then

$$\{\mathfrak{x} \in \mathbb{R}^2 : \ell(\mathfrak{x}) = 0\} = \{\mathfrak{x} \in \mathbb{R}^2 : \mathfrak{x} = \mathfrak{x}_0 + t\mathfrak{v}, \qquad t \in \mathbb{R}\} = g.$$

We therefore have two distinct ways of describing a straight line. We shall use whichever description is most suitable.

Let $g = \{\mathfrak{x} \in \mathbb{R}^2 : \mathfrak{x} = \mathfrak{x}_0 + t\mathfrak{v}, t \in \mathbb{R}\}$ be a line. We denote the positive ray $\{\mathfrak{x} \in \mathbb{R}^2 : \mathfrak{x} = \mathfrak{x}_0 + t\mathfrak{v}, t \geqslant 0\}$ by g^+ and the negative ray $\{\mathfrak{x} \in \mathbb{R}^2 : \mathfrak{x} = \mathfrak{x}_0 + t\mathfrak{v}, t \leqslant 0\}$ by g^-. If g is represented by the mapping ℓ, then we define

$$H_g^+ := \{\mathfrak{x} \in \mathbb{R}^2 : \ell(\mathfrak{x}) > 0\}, \qquad H_g^- := \{\mathfrak{x} \in \mathbb{R}^2 : (\mathfrak{x}) < 0\}.$$

These are the two half-planes determined by g.

We shall use the following terminology: A set K lies *relatively compact* in a set B ($K \subset\subset B$) if $\bar{K}$ is compact and contained in B.

Def. 3.1. Let $M \subset \mathbb{R}^2$ be a subset. M is called geometrically *convex* if for each point $\mathfrak{x} \in \mathbb{R}^2 - M$ there is a line g with $\mathfrak{x} \in g$ and $M \subset H_g^-$.

Remark. The intersection of convex sets is again convex.

Def. 3.2. Let $M \subset \mathbb{R}^2$ be an arbitrary subset. Then $\hat{M}_e := \{\mathfrak{x} \in \mathbb{R}^2 : \ell(\mathfrak{x}) \leqslant \sup \ell(M)$ for all $\ell \in L\}$ is called the *geometrically convex hull of* M.

Theorem 3.1 (The properties of the geometrically convex hull). *Let $M \subset \mathbb{R}^2$ be an arbitrary subset. Then:*

1. $M \subset \hat{M}_e$.
2. $\hat{M}_e$ *is closed and geometrically convex.*
3. $\hat{\hat{M}}_e = \hat{M}_e$.
4. *Let $M_1 \subset M_2 \subset \mathbb{R}^2$. Then $(\hat{M}_1)_e \subset (\hat{M}_2)_e$.*
5. *If M is closed and geometrically convex, then $M = \hat{M}_e$.*
6. *If M is bounded, then $\hat{M}_e$ is also bounded.*

PROOF

1. Let $\mathfrak{x} \in M$. Then for each $\ell \in L$, $\ell(\mathfrak{x}) \leqslant \sup \ell(M)$. Therefore $\mathfrak{x}$ lies in $\hat{M}_e$.

2. Let $\mathfrak{x}_0 \notin \hat{M}_e$. Then there exists an $\ell \in L$ with $\ell(\mathfrak{x}_0) > \sup \ell(M)$. Since ℓ is continuous, it is also true that in an entire neighborhood of $\mathfrak{x}_0$ we have $\ell(\mathfrak{x}) > \sup \ell(M)$. Therefore $\hat{M}_e$ is closed. ℓ^* with $\ell^*(\mathfrak{x}) := \ell(\mathfrak{x}) - \ell(\mathfrak{x}_0)$ is in L and $\ell^*(\mathfrak{x}_0) = 0$, $\sup \ell^*(\hat{M}_e) = \sup \ell^*(M) = \sup \ell(M) - \sup \ell(M) = 0$. Therefore $g = \{\mathfrak{r} \in \mathbb{R}^2 : \ell^*(\mathfrak{x}) = 0\}$ is a line with $\mathfrak{x}_0 \in g$ and $\hat{M}_e \subset H_g^-$.

3. By (1) we have $\hat{M}_e \subset \hat{\hat{M}}_e$. But for $\mathfrak{x} \in \hat{\hat{M}}_e$, $\ell(\mathfrak{x}) \leqslant \sup \ell(M_e) \leqslant \sup \ell(M)$ for $\ell \in L$. Hence it is also true that $\hat{\hat{M}}_e \subset \hat{M}_e$.

4. $\sup \ell(M_1) \leqslant \sup \ell(M_2)$, for all $\ell \in L$, so $(\hat{M}_1)_e \subset (\hat{M}_2)_e$.

5. Let $\mathfrak{x}_0 \notin M$. Since M is closed, there is an $\mathfrak{x}_1 \in M$ with minimal distance from $\mathfrak{x}_0$. If $\mathfrak{x}_2$ is the midpoint of the line segment between $\mathfrak{x}_0$ and $\mathfrak{x}_1$, then $\mathfrak{x}_2 \notin M$, and there is an $\ell \in L$ with $\ell(\mathfrak{x}_2) = 0$, $\ell|M < 0$. Thus $\sup \ell(M) \leqslant 0$, but $\ell(\mathfrak{x}_0) > 0$. Therefore $\mathfrak{x}_0 \notin \hat{M}_e$ and it follows that $\hat{M}_e \subset M$.

6. If M is bounded, then there is a closed rectangle Q with $M \subset Q$. For each $\mathfrak{x} \in \mathbb{R}^2 - Q$ there is a line g through $\mathfrak{x}$ with $Q \subset H_g^-$, and therefore an $\ell \in L$ with $\ell(\mathfrak{x}) = 0$ and $\sup \ell(M) \leqslant \sup \ell(Q) < 0$. That is, $\mathbb{R}^2 - Q \subset \mathbb{R}^2 - \hat{M}_e$, there $\hat{M}_e \subset Q$. □

Remark. $\hat{M}_e$ is the smallest closed geometrically convex set which contains M. (If $M \subset K$, K closed and geometrically convex, then $\hat{M}_e \subset \hat{K}_e = K$.)

Theorem 3.2. *Let $B \subset \mathbb{R}^2$ be an open subset. B is geometrically convex if and only if $K \subset\subset B$ implies $\hat{K}_e \subset\subset B$.*

PROOF

1. Let B be convex. $K \subset\subset B$ means that $\bar{K}$ is compact and lies in B. Therefore K and also $\hat{K}_e$ is bounded. Since $\hat{K}_e$ is closed it follows that $\hat{K}_e$ is compact. It remains to show that $\hat{K}_e$ lies in B.

We assume that there exists an $\mathfrak{x}_0 \in \hat{K}_e - B$. Since B is convex there is and $\ell \in L$ with $\ell(\mathfrak{x}_0) = 0$ and $\ell(\mathfrak{x}) < 0$ for $\mathfrak{x} \in B$. ℓ attains its supremum on $\bar{K}$ so it is even true that $\ell(\mathfrak{x}_0) > \sup \ell(\bar{K}) \geqslant \sup \ell(K)$. However, that contradicts the fact that $\mathfrak{x}_0$ lies in $\hat{K}_e$. Hence $\hat{K}_e - B = \varnothing$.

2. Now we assume that $\mathfrak{x}_0$ does not lie in B. First we show that for every line g which contains $\mathfrak{x}_0$ either $g^+ \cap B = \varnothing$ or $g^- \cap B = \varnothing$. From that we shall deduce finally that there is a line g_0 through $\mathfrak{x}_0$ which does not intersect B at all. We obtain g_0 by rotating the above line g about $\mathfrak{x}_0$ until the desired effect occurs.

a. Assume that there exists a line $g = \{\mathfrak{x} \in \mathbb{R}^2 : \mathfrak{x} = \mathfrak{x}_0 + t\mathfrak{v}, t \in \mathbb{R}\}$ with $g^+ \cap B \neq \varnothing$ and $g^- \cap B \neq \varnothing$. Then let $\mathfrak{x}_1 = \mathfrak{x}_0 + t_1\mathfrak{v} \in g^+ \cap B$ and $\mathfrak{x}_2 = \mathfrak{x}_0 + t_2\mathfrak{v} \in g^- \cap B$. The connecting line segment S between $\mathfrak{x}_1$ and $\mathfrak{x}_2$ is given by

$$\begin{aligned} S &= \{\mathfrak{x} = \mathfrak{x}_1 + t(\mathfrak{x}_2 - \mathfrak{x}_1) : t \in [0, 1]\} \\ &= \{\mathfrak{x} = t^*\mathfrak{x}_1 + t^{**}\mathfrak{x}_2 \text{ with } t^*, t^{**} \geqslant 0, \qquad t^* + t^{**} = 1\}. \end{aligned}$$

Now let $t_0^* := -t_2/(t_1 - t_2)$ and $t_0^{**} := 1 - t_0^* = t_1/(t_1 - t_2)$. Then $\mathfrak{x}_0^* := t_0^*\mathfrak{x}_1 + t_0^{**}\mathfrak{x}_2 \in S$ and $\mathfrak{x}_0 = \mathfrak{x}_0^*$. Let $\ell \in L$ be arbitrary. We shall show that $\ell(\mathfrak{x}_0) \leqslant m = \max(\ell(\mathfrak{x}_1), \ell(\mathfrak{x}_2))$. Clearly, we can restrict ourselves to homogeneous functions $\ell : \ell(\mathfrak{x}) = ax_1 + bx_2$. Then $\ell(\mathfrak{x}_0) = \ell(t_0^*\mathfrak{x}_1 + t_0^{**}\mathfrak{x}_2) = t_0^*\ell(\mathfrak{x}_1) + t_0^{**}\ell(\mathfrak{x}_2) \leqslant (t_0^* + t_0^{**})m = m$.

Now let $K := \{\mathfrak{x}_1, \mathfrak{x}_2\}$. Then $K \subset\subset B$ and therefore, $\hat{K}_e \subset\subset B$. Because $\ell(\mathfrak{x}_0) \leqslant \max(\ell(\mathfrak{x}_1), \ell(\mathfrak{x}_2)) = \sup \ell(K)$ for each $\ell \in L$ it follows that $\mathfrak{x}_0 \in \hat{K}_e$. That means $\mathfrak{x}_0 \in B$, which is a contradiction.

b. Now let such a g be given. If $g^+ \cap B = \varnothing$ and $g^- \cap B = \varnothing$ we are done. We assume that $g^+ \cap B \neq \varnothing$. Let θ_0 be the angle between g and the x_1-axis, $\theta_1 := \sup\{\theta : \theta_0 \leqslant \theta \leqslant \theta_0 + \pi, g_\theta^+ \cap B \neq \varnothing\}$, where g_θ denotes the line which makes the angle θ with the x_1-axis.

Case 1. $g_{\theta_1}^+ \cap B \neq \varnothing$. Then $\theta_1 < \theta_0 + \pi$. If $\mathfrak{x}_1 \in g_{\theta_1}^+ \cap B$, then there exists an $\varepsilon > 0$ such that $U_\varepsilon(\mathfrak{x}_1)$ lies in B. We can now find a θ_2 with $\theta_1 < \theta_2 < \theta_0 + \pi$ such that $g_{\theta_2}^+$ still intersects $U_\varepsilon(\mathfrak{x}_1)$ and of course B as well. That contradicts the definition of θ_1, so Case 1 can be discarded.

Case 2. $g_{\theta_1} \cap B \neq \varnothing$. We proceed in exactly the same manner as above to obtain a contradiction.

c. Let H^+ and H^- be the two half-planes belonging to g_{θ_1}. From (b) $B \subset H^+ \cup H^-$. But from (a) B must lie on exactly *one* side of g_{θ_1}. Suitable choice of the orientation of g_{θ_1} yields that B lies in H^- □

One could use the conditions of Theorem 3.1 as the definition of convexity.

We now come to the notion of *holomorphic convexity* by replacing linear functions by holomorphic functions.

Def. 3.3. Let $B \subset \mathbb{C}^n$ be a region, $K \subset B$ a subset. Then $\hat{K}_B := \{\mathfrak{z} \in B : |f(\mathfrak{z})| \leqslant \sup|f(K)|$ for every holomorphic function f in $B\}$ is called the *holomorphically convex hull of* K *in* B. When no misunderstanding can arise, we write $\hat{K}$ instead of $\hat{K}_B$.

Theorem 3.3 (The properties of the holomorphically convex hull). *Let* $B \subset \mathbb{C}^n$ *be a region,* $K \subset B$ *a subset. Then*:

1. $K \subset \hat{K}$
2. $\hat{K}$ *is closed in* B.
3. $\hat{\hat{K}} = \hat{K}$
4. *Let* $K_1 \subset K_2 \subset B$. *Then* $\hat{K}_1 \subset \hat{K}_2$.
5. *If* K *is bounded, then* $\hat{K}$ *is also bounded.*

PROOF

1. For $\mathfrak{z} \in K$, $|f(\mathfrak{z})| \leqslant \sup|f(K)|$.

2. Let $\mathfrak{z} \in B - \hat{K}$. Then there exists a holomorphic function f on B with $|f(\mathfrak{z})| > \sup|f(K)|$. Since $|f|$ is continuous, these inequalities hold on an entire neighborhood $U(\mathfrak{z}) \subset B$ which is contained in $B - \hat{K}$. Therefore $B - \hat{K}$ is open.

3. $\sup|f(K)| = \sup|f(\hat{K})|$.

4. The statement is trivial.

5. If K is bounded then there exists an $R > 0$ such that K is contained in the set $\{\mathfrak{z} = (z_1, \ldots, z_n) : |z_\nu| \leqslant R\}$. The coordinate functions $f_\nu(\mathfrak{z}) \equiv z_\nu$ are holomorphic in B, and therefore for $\mathfrak{z} \in \hat{K}$, $|z_\nu| = |f_\nu(\mathfrak{z})| \leqslant \sup|f_\nu(K)| \leqslant R$. Hence $\hat{K}$ is also bounded. □

Def. 3.4. Let $B \subset \mathbb{C}^n$ be a region. B is called *holomorphically convex* if $K \subset\subset B$ implies $\hat{K} \subset\subset B$.

Remark. In $\mathbb{C}$ every domain is holomorphically convex.

PROOF. Let $K \subset\subset G$. Then K is bounded, and therefore $\hat{K}$ also. Hence $\overline{\hat{K}}$ is compact and it only remains to show that $\overline{\hat{K}} \subset G$. If there is a point $z_0 \in \overline{\hat{K}} - G$, then z_0 lies in $\partial\hat{K} \cap \partial G$. But then $f(z) = 1/(z - z_0)$ is holomorphic in G.

Now let (z_ν) be a sequence in $\hat{K}$ with $\lim_{\nu \to \infty} z_\nu = z_0$. From the definition of $\hat{K}$, $|f(z_\nu)| \leqslant \sup|f(K)| \leqslant \sup|f(\bar{K})|$, contradicting the fact that $\{|f(z_\nu)| : \nu \in \mathbb{N}\}$ is unbounded. □

By no means is every domain in $\mathbb{C}^n$ holomorphically convex. However, we have

Theorem 3.4. *Let* $B \subset \mathbb{C}^n$ *be a region. If* B *is geometrically convex, then* B *is also holomorphically convex.*

PROOF. We must first clarify when a region in $\mathbb{C}^n$ is geometrically convex.
Let $\ell:\mathbb{C}^n \to \mathbb{R}$ be a homogeneous linear mapping of the form

$$\ell(\mathfrak{z}) = \sum_{\nu=1}^{n} a_\nu x_\nu + \sum_{\nu=1}^{n} b_\nu y_\nu = \sum_{\nu=1}^{n} \alpha_\nu z_\nu + \sum_{\nu=1}^{n} \beta_\nu \bar{z}_\nu.$$

Since we are supposed to have $\overline{\ell(\mathfrak{z})} = \ell(\mathfrak{z})$ it follows that $\beta_\nu = \bar{\alpha}_\nu$, and therefore

$$\ell(\mathfrak{z}) = \sum_{\nu=1}^{n} \alpha_\nu z_\nu + \sum_{\nu=1}^{n} \bar{\alpha}_\nu \bar{z}_\nu = 2 \cdot \operatorname{Re}\left(\sum_{\nu=1}^{n} \alpha_\nu z_\nu\right).$$

B is geometrically convex if $K \subset\subset B$ implies $\hat{K}_e \subset\subset B$, where we define $\hat{K}_e := \{\mathfrak{z} \in \mathbb{C}^n : \ell(\mathfrak{z}) \leqslant \sup \ell(K)$ for all homogeneous linear mappings $\ell\}$. $\hat{K}_e$ has the properties required by Theorem 3.1.

Now let $K \subset\subset B$. Then $\hat{K}_e \subset\subset B$. Let $\mathfrak{z}_0 \in B - \hat{K}_e$. Then there exists a linear homogeneous mapping ℓ with $\ell(\mathfrak{z}) = 2 \cdot \operatorname{Re} \sum_{\nu=1}^{n} \alpha_\nu z_\nu$ and $\ell(\mathfrak{z}_0) >$ $\sup \ell(K)$.

Now we define a function f holomorphic on B by

$$f(\mathfrak{z}) := \exp\left(2 \cdot \sum_{\nu=1}^{n} \alpha_\nu z_\nu\right).$$

Then

$$|f(\mathfrak{z})| = \exp\left(2 \cdot \operatorname{Re}\left(\sum_{\nu=1}^{n} \alpha_\nu z_\nu\right)\right) = \exp \circ\, \ell(\mathfrak{z}),$$

therefore

$$|f(\mathfrak{z}_0)| = \exp \circ\, \ell(\mathfrak{z}_0) > \sup((\exp \circ\, \ell)(K)) = \sup|f(K)|.$$

Thus $\mathfrak{z}_0 \in B - \hat{K}_B$, and we have shown $\hat{K}_B \subset \hat{K}_e \subset\subset B$. This proves $\hat{K}_B \subset\subset B$.

In general holomorphic convexity is a much weaker property than geometric convexity.

4. The Thullen Theorem

Let $M \subset \mathbb{C}^n$ be an arbitrary non-empty subset. If $\mathfrak{z}_0 \in \mathbb{C}^n - M$ is a point, then $\operatorname{dist}'(\mathfrak{z}_0, M) := \inf_{\mathfrak{z} \in M} |\mathfrak{z} - \mathfrak{z}_0|$ is a non-negative real number. If $K \subset \mathbb{C}^n - M$ is a compact set and M closed, then

$$\operatorname{dist}'(K, M) := \inf_{\mathfrak{z} \in K} \operatorname{dist}'(\mathfrak{z}, M)$$

is a positive number.

Def. 4.1. Let $B \subset \mathbb{C}^n$ be a region, $\varepsilon > 0$. We define

$$B_\varepsilon := \{\mathfrak{z} \in B : \operatorname{dist}'(\mathfrak{z}, \mathbb{C}^n - B) \geqslant \varepsilon\}.$$

Remarks

1. $\{\mathfrak{z}\}$ is compact, $\mathbb{C}^n - B$ is closed, so for $\mathfrak{z} \in B$ $\operatorname{dist}'(\mathfrak{z}, \mathbb{C}^n - B) > 0$.
2. If $\mathfrak{z} \in B$, then $\mathfrak{z} \in B_\varepsilon$ for $\varepsilon := \operatorname{dist}'(\mathfrak{z}, \mathbb{C}^n - B)$. Therefore $B = \bigcup_{\varepsilon > 0} B_\varepsilon$.
3. $\varepsilon_1 \leqslant \varepsilon_2 \Rightarrow B_{\varepsilon_1} \supset B_{\varepsilon_2}$.

Theorem 4.1. *B_ε is closed.*

PROOF. Let $\mathfrak{z}_0 \in \mathbb{C}^n - B_\varepsilon$. We define $\delta := \operatorname{dist}'(\mathfrak{z}_0, \mathbb{C}^n - B)$. $\varepsilon > \delta \geqslant 0$, so $\varepsilon - \delta > 0$. Let $U := U'_{\varepsilon-\delta}(\mathfrak{z}_0) = \{\mathfrak{z} : |\mathfrak{z} - \mathfrak{z}_0| < \varepsilon - \delta\}$. For $\mathfrak{z} \in U$ we have $\operatorname{dist}'(\mathfrak{z}, \mathbb{C}^n - B) \leqslant \operatorname{dist}'(\mathfrak{z}, \mathfrak{z}_0) + \operatorname{dist}'(\mathfrak{z}_0, \mathbb{C}^n - B) \leqslant \varepsilon - \delta + \delta = \varepsilon$. Therefore U lies in $\mathbb{C}^n - B_\varepsilon$, that is, $\mathbb{C}^n - B_\varepsilon$ is open. □

We need the following terminology. Let $M \subset \mathbb{C}^n$ be an arbitrary non-empty set. A function f is called *holomorphic in M* if f is defined and holomorphic in an open set $U = U(M)$ with $U \supset M$.

Theorem 4.2. *Let B be a region, f holomorphic in $\bar{B}$, $|f(\bar{B})| \leqslant M$, $\varepsilon > 0$, and $\mathfrak{z}_0 \in B_\varepsilon$ a point. In a neighborhood $U = U(\mathfrak{z}_0) \subset B$, let f have the power series expansion $f(\mathfrak{z}) = \sum_{\nu=0}^{\infty} a_\nu(\mathfrak{z} - \mathfrak{z}_0)^\nu$. Then for all ν,*

$$|a_\nu| \leqslant \frac{M}{\varepsilon^{|\nu|}}.$$

PROOF. Let $P := \{\mathfrak{z} \in \mathbb{C}^n : \operatorname{dist}'(\mathfrak{z}, \mathfrak{z}_0) < \varepsilon\}$. Then for $\mathfrak{z} \in P$, $\operatorname{dist}'(\mathfrak{z}, \mathbb{C}^n - B) \geqslant \operatorname{dist}'(\mathfrak{z}_0, \mathbb{C}^n - B) - \operatorname{dist}'(\mathfrak{z}, \mathfrak{z}_0) > \varepsilon - \varepsilon = 0$. Therefore P lies in B, that is $\bar{P} \subset \bar{B} \subset V(\bar{B})$, where V is an open neighborhood of $\bar{B}$ and f is defined and holomorphic on V. Then

$$|a_{\nu_1, \ldots, \nu_n}| = \left| \frac{1}{(2\pi i)^n} \int_T \frac{f(\xi_1, \ldots, \xi_n)\, d\xi_1 \ldots d\xi_n}{(\xi_1 - z_1^{(0)})^{\nu_1+1} \cdots (\xi_n - z_n^{(0)})^{\nu_n+1}} \right|$$

where T is the n-dimensional torus $T := \{(\xi_1, \ldots, \xi_n) : \xi_\nu = z_\nu^{(0)} + \varepsilon e^{i\theta_\nu}, 0 \leqslant \theta_\nu \leqslant 2\pi\}$. Because $d\xi_\nu = \varepsilon \cdot e^{i\theta_\nu} \cdot i\, d\theta_\nu = i(\xi_\nu - z_\nu^{(0)})\, d\theta_\nu$,

$$\begin{aligned}
|a_\nu| &= \frac{1}{(2\pi)^n} \left| \int_0^{2\pi} \frac{d\theta_1}{(\xi_1 - z_1^{(0)})^{\nu_1}} \int_0^{2\pi} \cdots \int_0^{2\pi} \frac{d\theta_n}{(\xi_n - z_n^{(0)})^{\nu_n}} \cdot f(\xi_1, \ldots, \xi_n) \right| \\
&\leqslant \frac{1}{(2\pi)^n} \int_0^{2\pi} \frac{d\theta_1}{|\xi_1 - z_1^{(0)}|^{\nu_1}} \int_0^{2\pi} \cdots \int_0^{2\pi} \frac{d\theta_n}{|\xi_n - z_n^{(0)}|^{\nu_n}} |f(\xi_1, \ldots, \xi_n)| \\
&\leqslant \frac{1}{(2\pi)^n} \cdot (2\pi)^n \frac{M}{\varepsilon^{\nu_1 + \cdots + \nu_n}} = \frac{M}{\varepsilon^{|\nu|}}
\end{aligned}$$

holds. □

Theorem 4.3 *Let $B \subset \mathbb{C}^n$ be a region, f holomorphic in B, $\varepsilon > 0$, and $K \subset B_\varepsilon$ compact. Then for every δ with $0 < \delta < \varepsilon$ there exists an $M > 0$ such that*

$$\sup_{\mathfrak{z} \in K} |a_\nu(\mathfrak{z})| \leqslant \frac{M}{\delta^{|\nu|}}.$$

(We denote by $a_\nu(\mathfrak{z}_0)$ the coefficients a_ν of the power series expansion $f(\mathfrak{z}) = \sum_{\nu=0}^{\infty} a_\nu(\mathfrak{z} - \mathfrak{z}_0)^\nu$.)

PROOF

1. Set $B^*: = \widehat{(\mathring{B}_{\varepsilon-\delta})}$. We claim that K lies in $(B^*)_\delta$, that is, that for $\mathfrak{z}_0 \in K$ we have $\text{dist}'(\mathfrak{z}_0, \mathbb{C}^N - B^*) \geqslant \delta$. Assume there is a $\mathfrak{z}_1 \in \mathbb{C}^N - B^*$ and a δ with $0 < \delta' < \delta$ such that $\text{dist}'(\mathfrak{z}_0, \mathfrak{z}_1) < \delta'$. Since $\mathfrak{z}_1$ does not lie in B^*, $\mathfrak{z}_1$ is not an interior point of $B_{\varepsilon-\delta}$. Therefore there still are points of $\mathbb{C}^n - B_{\varepsilon-\delta}$ arbitrarily close to $\mathfrak{z}_1$. Now let $\varepsilon' > 0$ be given. Then there is a $\mathfrak{z}_2 \in \mathbb{C}^N - B_{\varepsilon-\delta}$ such that $\text{dist}'(\mathfrak{z}_1, \mathfrak{z}_2) < \varepsilon'$. Since $\text{dist}'(\mathfrak{z}_2, \mathbb{C}^n - B) < \varepsilon - \delta$ it follows that there exists a $\mathfrak{z}_3 \in \mathbb{C}^n - B$ such that $\text{dist}'(\mathfrak{z}_2, \mathfrak{z}_3) < \varepsilon - \delta$. Therefore $\text{dist}'(\mathfrak{z}_0, \mathfrak{z}_3) \leqslant \text{dist}'(\mathfrak{z}_0, \mathfrak{z}_1) + \text{dist}'(\mathfrak{z}_1, \mathfrak{z}_2) + \text{dist}'(\mathfrak{z}_2, \mathfrak{z}_3) < \delta' + \varepsilon' + \varepsilon - \delta$. This holds for every $\varepsilon' > 0$. Therefore

$$\text{dist}'(\mathfrak{z}_0, \mathbb{C}^N - B) \leqslant (\delta' - \delta) + \varepsilon < \varepsilon.$$

So $\mathfrak{z}_0$ does not lie in B_ε, contrary to our assumption. So K must lie in $(B^*)_\delta$.

2. K is bounded, so there exists a polycylinder $P = P(0)$ with $K \subset P$. We can choose P in such a way that $\text{dist}'(K, \mathbb{C}^n - P) > \delta$. Then let $B': = P \cap B^*$. B' is open and non-empty.

We shall apply Theorem 4.2 to the region B'. Clearly $\bar{B}'$ is compact. Moreover, $\bar{B}' \subset \bar{P} \cap \bar{B}^* \subset \bar{P} \cap B_{\varepsilon-\delta} \subset B$. Therefore f is holomorphic in $\bar{B}'$ and can be bounded there by a constant M.

Because $\text{dist}'(K, \mathbb{C}^n - P) > \delta$ and $K \subset (B^*)_\delta$, $K \subset B'_\delta$. Therefore

$$|a_\nu(\mathfrak{z}_0)| \leqslant \frac{\sup|(f|B')|}{\delta^{|\nu|}} \leqslant \frac{M}{\delta^{|\nu|}}$$

for every point $\mathfrak{z}_0 \in K$; in particular

$$\sup_{\mathfrak{z} \in K} |a_\nu(\mathfrak{z})| \leqslant \frac{M}{\delta^{|\nu|}}.$$

□

Theorem 4.4 (Cartan–Thullen). *If $B \subset \mathbb{C}^n$ is a region of holomorphy, then B is holomorphically convex.*

PROOF. Let $K \subset\subset B$. We want to show that $\hat{K} \subset\subset B$. Let $\varepsilon: = \text{dist}'(K, \mathbb{C}^n - B) \geqslant \text{dist}'(\bar{K}, \mathbb{C}^n - B) > 0$. Clearly K lies in B_ε.

1. We assert that even the holomorphic convex hull $\hat{K}$ lies in B_ε. Suppose this is not so. Then there is a $\mathfrak{z}_0 \in \hat{K} - B_\varepsilon$. Since B is a region of holomorphy, there is a function f holomorphic in B which is completely singular at each

point $\mathfrak{z} \in \partial B$. In a neighborhood $U = U(\mathfrak{z}_0) \subset B$ f has the expansion

$$f(\mathfrak{z}) = \sum_{\nu=0}^{\infty} a_\nu (\mathfrak{z} - \mathfrak{z}_0)^\nu.$$

$$a_\nu(\mathfrak{z}) = \frac{1}{\nu_1! \cdots \nu_n!} \cdot \frac{\partial^{|\nu|} f(\mathfrak{z})}{\partial z_1^{\nu_1} \cdots \partial z_n^{\nu_n}}$$

is holomorphic in B by the theorem on partial derivatives, and $a_\nu(\mathfrak{z}_0) = a_\nu$. Because $\mathfrak{z}_0 \in \hat{K}$, $|a_\nu(\mathfrak{z}_0)| \leqslant \sup_{\mathfrak{z} \in K} |a_\nu(\mathfrak{z})|$. And by Theorem 4.3 for every δ with $0 < \delta < \varepsilon$ there exists an $M > 0$ with

$$\sup_{\mathfrak{z} \in K} |a_\nu(\mathfrak{z})| \leqslant \frac{M}{\delta^{|\nu|}}.$$

Therefore

$$\sum := M \cdot \sum_{\nu=0}^{\infty} \left(\frac{|z_1 - z_1^{(0)}|}{\delta}\right)^{\nu_1} \cdots \left(\frac{|z_n - z_n^{(0)}|}{\delta}\right)^{\nu_n}$$

dominates $\sum_{\nu=0}^{\infty} |a_\nu(\mathfrak{z} - \mathfrak{z}_0)^\nu|$. Now let $P_\delta(\mathfrak{z}_0)$ be the polycylinder about $\mathfrak{z}_0$ with radius δ. For $\mathfrak{z} \in P_\delta(\mathfrak{z}_0)$, $\sum$ is a geometric series, and therefore convergent. Hence $\sum_{\nu=0}^{\infty} a_\nu(\mathfrak{z} - \mathfrak{z}_0)^\nu$ converges on the interior of $P_\delta(\mathfrak{z}_0)$.

Let $P_\varepsilon := \{\mathfrak{z} : |z_\nu - z_\nu^{(0)}| < \varepsilon\}$. The sets P_δ with $0 < \delta < \varepsilon$ exhaust P_ε, hence $\sum_{\nu=0}^{\infty} a_\nu(\mathfrak{z} - \mathfrak{z}_0)^\nu$ is convergent in P_ε, say to the holomorphic function $\hat{f}$. Near $\mathfrak{z}_0$, $f = \hat{f}$. If $Q := C_{P_\varepsilon \cap B}(\mathfrak{z}_0)$ is the component of $\mathfrak{z}_0$ in $P_\varepsilon \cap B$, then $f = \hat{f}$ in Q. There exists a point $\mathfrak{z}_1 \in P_\varepsilon \cap \partial Q \cap \partial B$. If $U = U(\mathfrak{z}_1) \subset P_\varepsilon$ is an open connected neighborhood, then $\hat{f}$ is holomorphic in U, $U \cap Q$ is open in $U \cap B$ and $f | U \cap Q = \hat{f} | U \cap Q$. That is a contradiction, for f is supposed to be completely singular at $\mathfrak{z}_1$. Therefore $\hat{K} \subset B_\varepsilon$.

2. Since $\mathbb{C}^n - \hat{K} = (B - \hat{K}) \cup (\mathbb{C}^n - B_\varepsilon)$ is open, it follows that $\hat{K}$ is closed. Since $\bar{K}$ is compact, K is bounded, and by Theorem 3.3 $\hat{K}$ is also bounded. Hence $\hat{K}$ is compact. This completes the proof. □

In the next section we shall show that the converse of this theorem also holds.

5. Holomorphically Convex Domains

Theorem 5.1. *Let $B \subset \mathbb{C}^n$ be a region. Then there exists a sequence of subsets $K_\nu \subset B$ with the following properties.*

1. *K_ν is compact for all $\nu \in \mathbb{N}$.*
2. $\bigcup_{\nu=1}^{\infty} K_\nu = B$.
3. *$K_\nu \subset \mathring{K}_{\nu+1}$ for all $\nu \in \mathbb{N}$.*

PROOF. It is clear how the K_ν should be chosen: If $\bar{P}_\nu := \{\mathfrak{z} : |z_\lambda| \leqslant \nu$ for all $\lambda\}$, we define $K_\nu := \bar{P}_\nu \cap B_{1/\nu}$. Obviously K_ν is compact and lies in B. Let $\mathfrak{z} \in B$. Then $\varepsilon := \operatorname{dist}'(\mathfrak{z}, \mathbb{C}^n - B) > 0$ and there exists a $\nu_0 \in \mathbb{N}$ with $\mathfrak{z} \in \bar{P}_{\nu_0}$. Let $\nu \geqslant \max(\nu_0, 1/\varepsilon)$. Then $\mathfrak{z} \in \bar{P}_\nu \cap B_{1/\nu} = K_\nu$. Therefore $B = \bigcup_{\nu=1}^{\infty} K_\nu$. Let $\mathfrak{z}_0 \in B_{1/\nu}$. Then

$$\operatorname{dist}'(\mathfrak{z}_0, \mathbb{C}^n - B) \geqslant \frac{1}{\nu} > \frac{1}{\nu+1},$$

and

$$U = U(\mathfrak{z}_0) := \left\{\mathfrak{z} \in \mathbb{C}^n : \operatorname{dist}'(\mathfrak{z}, \mathfrak{z}_0) < \frac{1}{\nu} - \frac{1}{\nu+1}\right\}$$

is an open neighborhood of $\mathfrak{z}_0$. For $\mathfrak{z} \in U$, however,

$$\begin{aligned}\operatorname{dist}'(\mathfrak{z}, \mathbb{C}^n - B) &\geqslant \operatorname{dist}'(\mathfrak{z}_0, \mathbb{C}^n - B) \\ &\quad - \operatorname{dist}'(\mathfrak{z}, \mathfrak{z}_0) > \frac{1}{\nu} - \left(\frac{1}{\nu} - \frac{1}{\nu+1}\right) = \frac{1}{\nu+1}.\end{aligned}$$

Therefore U lies in $B_{1/(\nu+1)}$. Hence $\mathfrak{z}_0$ is an interior point of $B_{1/(\nu+1)}$ and $B_{1/\nu} \subset \mathring{B}_{1/(\nu+1)}$. Because $\bar{P}_\nu \subset \mathring{\bar{P}}_{\nu+1}$ it follows that $K_\nu \subset \mathring{K}_{\nu+1}$. □

Remark. It is actually true that

$$B = \bigcup_{\nu=1}^{\infty} \mathring{K}_\nu, \qquad \text{since} \qquad B = \bigcup_{\nu=1}^{\infty} K_\nu = \bigcup_{\nu=2}^{\infty} K_{\nu-1} \subset \bigcup_{\nu=2}^{\infty} \mathring{K}_\nu \subset B.$$

In the remainder of this section we shall call any sequence of compact subsets of a region B which satisfy the conditions of Theorem 5.1 a *normal exhaustion of* B. We define $M_1 := K_1$ and $M_\nu := K_\nu - K_{\nu-1}$ for $\nu \geqslant 2$. Then:

1. $M_\nu \cap M_\mu = \emptyset$ for $\nu \neq \mu$,
2. $\bigcup_{\nu=1}^{\infty} M_\nu = B$,
3. $\bigcup_{\nu=1}^{\mu} M_\nu = K_\mu$

Theorem 5.2. *Let $B \subset \mathbb{C}^n$ be a region and (K_ν) a normal exhaustion of B. Then there exists a strictly monotonic increasing subsequence (λ_μ) of the natural numbers and a sequence $(\mathfrak{z}_\mu)$ of points in B such that*

1. $\mathfrak{z}_\mu \in M_{\lambda_\mu}$
2. *If $G \subset \mathbb{C}^n$ is a domain, $G \cap B \neq \emptyset$, $G \cap (\mathbb{C}^n - B) \neq \emptyset$ and G_1 a connected component of $G \cap B$, then G_1 contains infinitely many points of the sequence $(\mathfrak{z}_\mu)_{\mu \in \mathbb{N}}$.*

Proof

1. A point $\mathfrak{z} = (z_1, \ldots, z_n) \in \mathbb{C}^n$ is called rational if

$$z_\nu = x_\nu + iy_\nu \qquad \text{with } x_\nu, y_\nu \in \mathbb{Q} \text{ for all } \nu.$$

The set of $U_\varepsilon(\mathfrak{z})$ with rational $\mathfrak{z} \in \mathbb{C}^n$ and $\varepsilon \in \mathbb{Q}$ forms a countable basis for the topology of $\mathbb{C}^n$; we denote this basis by $\mathfrak{W} = \{W_\kappa : \kappa \in \mathbb{N}\}$.

Now let $\mathfrak{V} := \{W_\kappa \in \mathfrak{W} : W_\kappa \cap B \neq \varnothing$ and $W_\kappa \cap (\mathbb{C}^n - B) \neq \varnothing\}$. If $W_\kappa \in \mathfrak{V}$, then $W_\kappa \cap B$ has countably many components, as each contains at least one rational point.

Let $\mathfrak{B} := \{B_\mu$: There is a $\kappa \in \mathbb{N}$ such that $W_\kappa \in \mathfrak{V}$ and B_μ is a component of $W_\kappa \cap B\}$.

$\mathfrak{B}$ is now a countable system $\{B_\mu : \mu \in \mathbb{N}\}$ of connected sets, and for each $\mu \in \mathbb{N}$ there is a $\kappa = \kappa(\mu)$ such that $B_\mu \subset W_\kappa \cap B$.

2. The sequences (λ_μ) and $(\mathfrak{z}_\mu)$ are now constructed inductively: Let $\mathfrak{z}_1$ be arbitrary in B_1. Then $B_1 \subset W_{\kappa(1)} \cap B \subset B$ and $B = \bigcup_{\nu=1}^{\infty} K_\nu$. Therefore there exists a $\nu(1) \in \mathbb{N}$ such that $\mathfrak{z}_1$ lies in $K_{\nu(1)}$. Since the system of M_ν is a decomposition of B, there is a $\lambda(1) \leqslant \nu(1)$ such that $\mathfrak{z}_1 \in M_{\lambda(1)}$.

Now suppose $\mathfrak{z}_1, \ldots, \mathfrak{z}_{\mu-1}$ have been constructed so that $\mathfrak{z}_\iota \in K_{\nu(\iota)} \cap B_\iota$ and $\lambda(\iota)$ is chosen so that $\mathfrak{z}_\iota \in M_{\lambda(\iota)}$, $\iota = 1, \ldots, \mu - 1$. Choose $\mathfrak{z}_\mu \in B_\mu - K_{\nu(\mu-1)}$ arbitrarily. That is possible since there is a point $\mathfrak{z}_\mu^* \in W_{\kappa(\mu)} \cap \partial B \cap \partial B_\mu$. $\mathbb{C}^n - K_{\nu(\mu-1)}$ is an open neighborhood of $\mathfrak{z}_\mu^*$ and contains points of B_μ. These points lie in $B_\mu - K_{\nu(\mu-1)}$. Now there is a $\nu(\mu) \in \mathbb{N}$ with $\mathfrak{z}_\mu \in K_{\nu(\mu)}$. Therefore $\mathfrak{z}_\mu \in K_{\nu(\mu)} \cap B_\mu$, and there exists exactly one $\lambda(\mu) \leqslant \nu(\mu)$ with $\mathfrak{z}_\mu \in M_{\lambda(\mu)}$.

3. If $\lambda(\mu) \leqslant \nu(\mu - 1)$, we would have $\mathfrak{z}_\mu \in M_{\lambda(\mu)} \subset K_{\nu(\mu-1)}$ contrary to construction. Therefore $\nu(\mu - 1) < \lambda(\mu) < \nu(\mu)$; the sequences $\nu(\mu)$ and $\lambda(\mu)$ are strictly monotone increasing.

4. Now let $G \subset \mathbb{C}^n$ be a domain, $G \cap B \neq \varnothing$, $G \cap (\mathbb{C}^n - B) \neq \varnothing$, and G_1 a component of $G \cap B$. We assume that only finitely many $\mathfrak{z}_\mu$ lie in G_1, say $\mathfrak{z}_1, \ldots, \mathfrak{z}_m$. Then let

$$G^* := G - \{\mathfrak{z}_1, \ldots, \mathfrak{z}_m\},$$
$$G_1^* := G_1 - \{\mathfrak{z}_1, \ldots, \mathfrak{z}_m\}.$$

G^* and G_1^* are again domains and $G_1^* \subset G^* \cap B$. Let $\mathfrak{z}_1 \in G_1^*$, $\mathfrak{z} \in G^* \cap B$, $\mathfrak{z}_1$ and $\mathfrak{z}$ be connected by a path in $G^* \cap B$. Then they can be connected by a path in $G \cap B$, and $\mathfrak{z}$ belongs to $G_1 \cap G^* = G_1^*$. It follows that G_1^* is a component of $G^* \cap B$.

Now let $\mathfrak{z}_0 \in G^* \cap \partial G_1^* \cap \partial B$. There is a $\kappa \in \mathbb{N}$ such that $W_\kappa \in \mathfrak{V}$ and $\mathfrak{z}_0 \in W_\kappa$ and such that $W_\kappa \cap B \subset G^* \cap B$. Moreover $W_\kappa \cap B$ must contain points of G_1^*.

Now let $\mathfrak{z}_1 \in W_\kappa \cap B \cap G_1^*$ (and $B^* := C_{W_\kappa \cap B}(\mathfrak{z}_1)$. Because $G_1^* = C_{G^* \cap B}(\mathfrak{z}_1)$ it follows that $B^* \subset G_1^*$. B^* is an element of $\mathfrak{B}$ and therefore contains a $\mathfrak{z}_\mu$. That is a contradiction. The assumption was false and we have proved the theorem. □

Theorem 5.3 *Let $B \subset \mathbb{C}^n$ be a region and (K_ν) be a normal exhaustion of B. In addition, suppose that, for each $\nu \in \mathbb{N}$, $K_\nu = \hat{K}_\nu$.*

Let (λ_μ) be a strictly monotone increasing sequence of natural numbers and $(\mathfrak{z}_\mu)$ a sequence of points with $\mathfrak{z}_\mu \in M_{\lambda_\mu}$.

Then there exists a holomorphic function f in B such that $|f(\mathfrak{z}_\mu)|$ is unbounded.

Proof. We represent f as the limit function of an infinite series $f = \sum_{\mu=1}^{\infty} f_\mu$; we define the summands f_μ by induction.

1. Let $f_1 := 1$. Now suppose $f_1, \ldots, f_{\mu-1}$ are constructed. Since $\mathfrak{z}_\mu$ does not lie in $K_{\lambda(\mu)-1} = \hat{K}_{\lambda(\mu)-1}$, there exists a function g holomorphic in B such that $|g(\mathfrak{z}_\mu)| > q$, where $q := \sup|(g|K_{\lambda(\mu)-1})|$. By normalization one can make $g(\mathfrak{z}_\mu) = 1$. Hence $q < 1$. Now let $a_\mu := \sum_{\nu=1}^{\mu-1} f_\nu(\mathfrak{z}_\mu)$ and m be chosen so that $q^m < 1/(\mu + |a_\mu|) \cdot 2^{-\mu}$. This is possible since q^m tends to zero. We set $f_\mu := (\mu + |a_\mu|) \cdot g^m$. Then f_μ is holomorphic in B, $f_\mu(\mathfrak{z}_\mu) = \mu + |a_\mu|$ and $\sup|(f_\mu|K_{\lambda(\mu)-1})| < 2^{-\mu}$.

2. We assert that $\sum_{\mu=1}^{\infty} f_\mu$ converges uniformly in the interior of B. Let $K \subset B$ be compact. There exists a $\nu_0 \in \mathbb{N}$ such that $K \subset \mathring{K}_{\nu_0-1}$. Now let $\mu_0 \in \mathbb{N}$ be chosen so that $\lambda(\mu_0) \geqslant \nu_0$. Then $K_{\lambda(\mu)} \supset K_{\nu_0}$ for $\mu \geqslant \mu_0$, that is, $K_{\lambda(\mu)-1} \supset K_{\nu_0-1}$. By construction $\sup|(f_\mu|K_{\lambda(\mu)-1})| < 2^{-\mu}$, therefore in particular $\sup|(f_\mu|K)| < 2^{-\mu}$. As $\sum_{\mu=1}^{\infty} 2^{-\mu}$ dominates $\sum_{\mu=1}^{\infty} f_\mu$ in K, the series converges uniformly in K. Therefore $f = \sum_{\mu=1}^{\infty} f_\mu$ is holomorphic in B.

3. $|f(\mathfrak{z}_\mu)| = \left|\sum_{\nu=1}^{\infty} f_\nu(\mathfrak{z}_\mu)\right| \geqslant |f_\mu(\mathfrak{z}_\mu)| - \left|\sum_{\nu=1}^{\mu-1} f_\nu(\mathfrak{z}_\mu)\right| - \sum_{\nu=\mu+1}^{\infty} |f_\nu(\mathfrak{z}_\mu)| \geqslant \mu + |a_\mu| - |a_\mu| - \sum_{\nu=\mu+1}^{\infty} |f_\nu(\mathfrak{z}_\mu)|$. Because $\mathfrak{z}_\mu \in K_{\lambda(\mu)} \subset K_{\lambda(\nu)-1}$ for $\nu \geqslant \mu + 1$, we have

$$|f(\mathfrak{z}_\mu)| \geqslant \mu - \sum_{\nu=\mu+1}^{\infty} 2^{-\nu} \geqslant \mu - 1.$$

$|f(\mathfrak{z}_\mu)| \to \infty$ for $\mu \to \infty$ follows. □

Theorem 5.4. *Let $B \subset \mathbb{C}^n$ be a region. If B is holomorphically convex then there exists a normal exhausting (K_ν) of B with the property that $K_\nu = \hat{K}_\nu$ for every $\nu \in \mathbb{N}$.*

Proof. Let (K_ν) be any normal exhaustion of B. Then for all ν, $K_\nu \subset\subset B$ and as B is holomorphically convex, it follows that $\hat{K}_\nu \subset\subset B$. $\hat{K}_\nu$ is therefore a compact subset of B. We now construct a subsequence of the $\hat{K}_\nu$.

Let $K_1^* := \hat{K}_1$.

Suppose $K_1^*, \ldots, K_{\nu-1}^*$ have been constructed ($K_{\nu-1}^*$ compact and $\hat{K}_{\nu-1}^* = K_{\nu-1}^*$). Then there exists a $\lambda(\nu) \in \mathbb{N}$ such that $K_{\nu-1}^* \subset \mathring{K}_{\lambda(\nu)}$. Let $K_\nu^* := \hat{K}_{\lambda(\nu)}$ Clearly the K_ν^* are compact subsets of B with $\hat{K}_\nu^* = K_\nu^*$. Moreover $\bigcup_{\nu=1}^{\infty} K_\nu^* = \bigcup_{\nu=1}^{\infty} \hat{K}_{\lambda(\nu)} \supset \bigcup_{\nu=1}^{\infty} K_{\lambda(\nu)} = B$ and $K_\nu^* \subset \mathring{K}_{\lambda(\nu+1)} \subset \mathring{K}_{\nu+1}^*$.

We now come to the main theorem of this section.

Theorem 5.5. *Let $B \subset \mathbb{C}^n$ be a region. If B is holomorphically convex, then B is a region of holomorphy.*

PROOF. By the preceding theorem there is a normal exhausting (K_ν) of B with $K_\nu = \hat{K}_\nu$ for all ν and hence sequences (λ_μ) and $(\mathfrak{z}_\mu)$ in the sense of Theorem 5.2 and a function f holomorphic in B with $|f(\mathfrak{z}_\mu)| \to \infty$ for $\mu \to \infty$.

We now show that f is completely singular at every boundary point of B. Assume that there exists a point $\mathfrak{z}_0 \in \partial B$ at which f has no essential singularity; that is, that there is an open connected neighborhood $U = U(\mathfrak{z}_0)$ and a function $\hat{f}$ holomorphic in U such that $f = \hat{f}$ in near some point $\mathfrak{z}_1 \in U \cap B$.

Let $U_1 := C_{U \cap B}(\mathfrak{z}_1)$ be the connected component of $\mathfrak{z}_1$ in $U \cap B$. There is a point $\mathfrak{z}_2 \in U \cap \partial U_1 \cap \partial B$. Let $V = V(\mathfrak{z}_2)$ be an open connected neighborhood of $\mathfrak{z}_2$ with $V \subset \subset U$.

$V \cap U_1$ contains a point $\mathfrak{z}_3$. Let $V_1 := C_{V \cap B}(\mathfrak{z}_3)$. If $\mathfrak{z}$ lies in V_1, then $\mathfrak{z}$ can be joined to $\mathfrak{z}_3$ in $V \cap B \subset U \cap B$, and $\mathfrak{z}_3$ lies in U_1 so that it, too, can be joined to $\mathfrak{z}_1$ in $U \cap B$. Hence $V_1 \subset U_1$.

Because "$f = \hat{f}$ in the region of $\mathfrak{z}_1$", it follows that $f = \hat{f}$ in U_1 and from this that $f = \hat{f}$ in V_1 also. On the other hand, infinitely many points of the sequence $(\mathfrak{z}_\mu)$ lie in V_1. That is, $\hat{f}$ is unbounded in V_1. That leads to a contradiction, since $\hat{f}$ is holomorphic in U, $\bar{V}$ is compact and therefore $\sup|(\hat{f}|V_1)| \leqslant \sup|(\hat{f}|V)| \leqslant \sup|(\hat{f}|\bar{V})| < \infty$. Therefore f is completely singular in ∂B. □

Def. 5.1. Let $M \subset \mathbb{C}^n$ be an arbitrary subset. $D \subset M$ is called *discrete* if D has no cluster points in M.

Theorem 5.6. *Let $B \subset \mathbb{C}^n$ be a region. B is holomorphically convex if and only if for every infinite set M which is discrete in B there exists a function f holomorphic in B such that $|f|$ is unbounded on D.*

(This theorem permits a simpler definition of holomorphically convex. It holds in complete generality, both on complex manifolds and complex spaces.)

PROOF

1. Let B be holomorphically convex, $D \subset B$ infinite and discrete. Moreover, let (K_ν) be a normal exhaustion of B with $K_\nu = \hat{K}_\nu$. Then $K_\nu \cap D$ is finite for every $\nu \in \mathbb{N}$.

We construct a sequence $(\mathfrak{z}_\mu)$ of points in D by induction:

Let $\mathfrak{z}_1 \in D$ be arbitrary, $\nu(1) \in \mathbb{N}$ minimal with the property that $\mathfrak{z}_1$ lies in $K_{\nu(1)}$.

Now suppose the points $\mathfrak{z}_1, \ldots, \mathfrak{z}_{\mu-1}$ have been constructed. Then we choose $\mathfrak{z}_\mu \in D - K_{\nu(\mu-1)}$ where $\nu(\mu - 1)$ is to be the smallest number with the property that $\mathfrak{z}_{\mu-1}$ lies in $K_{\nu(\mu-1)}$. That is possible, for $K_\nu \cap D$ is finite, so $D - K_\nu$ contains infinitely many points.

Thus, $\nu(\mu)$ is a strictly monotone increasing sequence of natural numbers such that $\mathfrak{z}_\mu$ lies in $M_{\nu(\mu)}$.

By Theorem 5.3 there is a function f holomorphic in B such that $|f(\mathfrak{z}_\mu)|$ is unbounded. Therefore $|f|$ is unbounded on D.

2. Now let the criterion be satisfied. We assume B not to be holomorphically convex: that is, we assume there is a $K \subset\subset B$ such that $\hat{K}$ is not relatively compact in B. We construct an appropriate set D.

Let (K_ν) be a normal exhaustion of B. Clearly $\hat{K} - K_\nu \neq \varnothing$ for all ν, otherwise we would have $\hat{K} \subset\subset B$. We define D by induction as a point sequence. Let $\mathfrak{z}_1 \in \hat{K}$ be arbitrary and $\nu(1)$ minimal such that $\mathfrak{z}_1 \in K_{\nu(1)}$. Suppose $\mathfrak{z}_1, \ldots, \mathfrak{z}_{\mu-1}$ have been constructed, and for $1 \leqslant \lambda \leqslant \mu - 1$ let $\nu(\lambda)$ always be the smallest number such that $\mathfrak{z}_\lambda \in K_{\nu(\lambda)}$. Then we choose $\mathfrak{z}_\mu$ arbitrarily in $\hat{K} - K_{\nu(\mu-1)}$. Then $\nu(\lambda)$ is strictly monotone increasing and $\mathfrak{z}_\mu \in K_{\nu(\mu)}$.

Let D be the set of points $\mathfrak{z}_\mu$, $\mu \in \mathbb{N}$. If $\mathfrak{z}_0 \in B$, then there exists a $\mu \in \mathbb{N}$ such that $\mathfrak{z}_0$ lies in $K_\mu \subset \mathring{K}_{\mu+1} \subset \mathring{K}_{\nu(\mu+1)}$. $\mathring{K}_{\nu(\mu+1)}$ is an open neighborhood of $\mathfrak{z}_0$, which contains only the points $\mathfrak{z}_1, \ldots, \mathfrak{z}_{\mu+1}$. Therefore $\mathfrak{z}_0$ is not a cluster point of D. The set D is discrete in B. By assumption there is a function f holomorphic in B which is unbounded on D. But then there exists a $\mu \in \mathbb{N}$ such that $|f(\mathfrak{z}_\mu)| > \sup|f(K)|$. That means that $\mathfrak{z}_\mu$ does not lie in $\hat{K}$, contrary to construction. Therefore B must be holomorphically convex. □

6. Examples

Theorem 6.1. *Let $B \subset \mathbb{C}$ be a region. Then B is a region of holomorphy. (Hence for every open set B in $\mathbb{C}$ there exists a holomorphic function which cannot be extended to any proper open superset of B.)*

Proof. It was shown in Section 3, that every region in $\mathbb{C}$ is holomorphically convex. From Theorem 5.4 it follows that B is a region of holomorphy. □

In $\mathbb{C}^n$ we have the following theorem:

Theorem 6.2. *Let $B \subset \mathbb{C}^n$ be a region. Then the following statements are are equivalent:*

1. *B is pseudo-convex.*
2. *B is a region of holomorphy.*
3. *B is holomorphically convex.*

4. *For every infinite set D discrete in B there exists a function f holomorphic in B such that f is unbounded on D.*

PROOF. The statements have all been proved in the preceding paragraphs (apart from the solution of the *Levi problem*: if B is pseudoconvex, then B is a region of holomorphy). □

Theorem 6.3. *If $G \subset \mathbb{C}^n$ is a geometrically convex domain then G is a domain of holomorphy.*

The n-fold cartesian product of open sets is a further example.

Theorem 6.4. *Let $V_1, \ldots, V_n \subset \mathbb{C}$ be regions. Then $V := V_1 \times \cdots \times V_n \subset \mathbb{C}^n$ is a region of holomorphy.*

PROOF. Let $D \subset V$ be a discrete infinite set and $(\mathfrak{z}_\mu)$ a sequence of distinct points of D with $\mathfrak{z}_\mu = (z_1^{(\mu)}, \ldots, z_n^{(\mu)})$. If the sequence $(z_1^{(\mu)})$ in V_1 has a cluster point $z_1^{(0)}$, then there exists a subsequence $(z_1^{(\mu_1(\nu))})$ which converges to $z_1^{(0)}$. If the sequence $(z_2^{(\mu_1(\nu))})$ in V_2 has a cluster point $z_2^{(0)}$, then there exists a subsequence $(z_2^{(\mu_2(\nu))})$ which converges to $z_2^{(0)}$. Continuing this way until n-th component (thus obtaining a subsequence $(z_n^{(\mu_n(\nu))})$ which converges to a point $z_n^{(0)} \in V_n$), then the sequence $(\mathfrak{z}_{\mu_n(\nu)})$ converges to $\mathfrak{z}_0 := (z_1^{(0)}, \ldots, z_n^{(0)}) \in V$. This is a contradiction, because D is discrete in V.

Therefore there is a $q \in \{1, \ldots, n\}$ and a subsequence $z_{\mu_\nu} = (z_1^{(\mu_\nu)}, \ldots, z_n^{(\mu_\nu)})$ of the sequence $(\mathfrak{z}_\mu)$ such that the sequence $(z_q^{(\mu_\nu)})$ has no cluster point in V_q.

By Theorems 6.1 and 6.2 there exists a function f holomorphic in V_q for which $f(z_q^{(\mu_\nu)})$ is unbounded. Now $g(z_1, \ldots, t_n) := f(z_q)$ is a holomorphic function on V which is unbounded on D. Therefore V is holomorphically convex. □

Def. 6.1. Let $B \subset \mathbb{C}^n$ and $V_1, \ldots, V_k \subset \mathbb{C}$ be regions, $f_1, \ldots, f_k$ holomorphic functions in B, and $U \subset B$ an open subset. The set $P := \{\mathfrak{z} \in U : f_j(\mathfrak{z}) \in V_j$

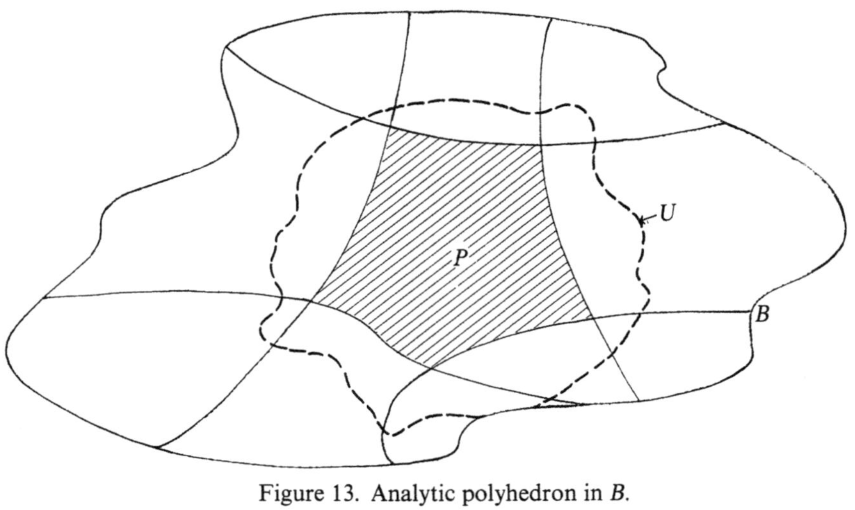

Figure 13. Analytic polyhedron in B.

for $j = 1, \ldots, k\}$ is called an *analytic polyhedron in B* if $P \subset\subset U$. If, in addition, $V_1 = \cdots = V_k = \{z \in \mathbb{C} : |z| < 1\}$, then one speaks of a *special analytic polyhedron in B*.

Theorem 6.5. *Let $B \subset \mathbb{C}^n$ be a region. Then every analytic polyhedron in B is a region of holomorphy.*

PROOF. Let $U, V_1, \ldots, V_k, f_1, \ldots, f_k$ and P be given as in Def. 6.1. Then $F: = (f_1|U, \ldots, f_k|U): U \to \mathbb{C}^k$ is a holomorphic mapping and $P = F^{-1}(V_1 \times \cdots \times V_k)$. By Theorem 6.4 $V: = V_1 \times \cdots \times V_k$ is a region of holomorphy. Let $D \subset P$ be an infinite discrete set. It suffices to show that $F(D) \subset V$ is infinite and discrete. For then there exists a function holomorphic in V which is unbounded on $F(D)$ and the function $g: = f \circ F$ satisfies the corresponding conditions in P.

Let $(\mathfrak{z}_j)$ be a subsequence of pairwise distinct points of D. $F(\mathfrak{z}_j)$ has a cluster point $\mathfrak{w}_0$ in V. Then there exists a subsequence $F(\mathfrak{z}_{j_\nu})$ which converges to $\mathfrak{w}_0$. The points $\mathfrak{z}_{j_\nu}$ lie in P, and by assumption $\bar{P}$ is compact. Hence $(\mathfrak{z}_{j_\nu})$ has a cluster point $\mathfrak{z}_1$ in $\bar{P}$, and there is therefore a subsequence $(\mathfrak{z}_{j_{\nu(\mu)}})$, which converges to $\mathfrak{z}_1 \in \bar{P} \subset U$. $F(\mathfrak{z}_{j_{\nu(\mu)}})$ then converges to $F(\mathfrak{z}_1)$ and to $\mathfrak{w}_0$ simultaneously; that is, $F(\mathfrak{z}_1) = \mathfrak{w}_0$ lies in V. This means that $\mathfrak{z}_1$ lies in $F^{-1}(V) = P$, which is a contradiction to the assumption that D is discrete in P. Hence it follows that $F(\mathfrak{z}_j)$ has no cluster point in V, and we are done. □

EXAMPLE. Let $q < 1$ be a positive real number and

$$P: = \{\mathfrak{z} \in \mathbb{C}^2 : |z_1| < 1, |z_2| < 1, |z_1 \cdot z_2| < q\}.$$

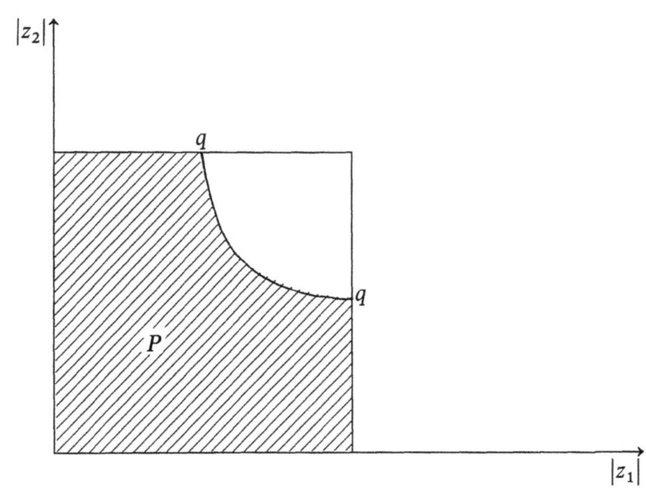

Figure 14. Example of a nontrivial analytic polyhedron.

P is clearly an analytic polyhedron, but neither a geometrically convex region nor a cartesian product of regions. The analytic polyhedrons therefore enrich our stock of examples of regions of holomorphy.

We shall show that every region of holomorphy is "almost" an analytic polyhedron:

Theorem 6.6. *Every region of holomorphy $B \subset \mathbb{C}^n$ can be exhausted by analytic polyhedra in the sense that there exists a sequence (P_j) of special analytic polyhedra in B with $P_j \subset\subset P_{j+1}$ and* $\bigcup_{j+1}^{\infty} P_j = B$.

PROOF

1. Let (K_j) be a normal exhaustion of B with $K_j = \hat{K}_j$. If $\mathfrak{z} \in \partial K_{j+1}$ is an arbitrary point, then $\mathfrak{z}$ does not lie in $K_j \subset \mathring{K}_{j+1}$, and therefore not in $\hat{K}_j$. Hence there exists a function f holomorphic in B for which $q := \sup|f(K_j)| < |f(\mathfrak{z})|$. By multiplication with a suitable constant we obtain $q < 1 < |f(\mathfrak{z})|$, and then there is an entire neighborhood $U = U(\mathfrak{z})$ such that $|(f|U)| > 1$.

Since the boundary ∂K_{j+1} is compact, we can find finitely many open sets $U(\mathfrak{z}_1^{(j)}), \ldots, U(\mathfrak{z}_{k_j}^{(j)})$ and corresponding functions $f_1^{(j)}, \ldots, f_{k_j}^{(j)}$ holomorphic in B such that $\partial K_{j+1} \subset \bigcup_{p=1}^{k_j} U(\mathfrak{z}_p^{(j)})$ and $|(f_p^{(j)}|U(\mathfrak{z}_p^{(j)}))| > 1$. Let $P_j :=$ $\{\mathfrak{z} \in \mathring{K}_{j+1} : |f_p^{(j)}(\mathfrak{z})| < 1 \text{ for } p = 1, \ldots, k_j\}$.

2. Clearly $K_j \subset P_j \subset \mathring{K}_{j+1}$. But beyond that $M := K_{j+1} - \bigcup_{p=1}^{k_j} U(\mathfrak{z}_p^{(j)}) =$ $K_{j+1} \cap \left(\mathbb{C}^n - \bigcup_{p=1}^{k_j} U(\mathfrak{z}_p^{(j)})\right)$ is a compact set with $P_j \subset M \subset \mathring{K}_{j+1}$. Consequently $\bar{P}_j \subset \bar{M} = M \subset \mathring{K}_{j+1}$; that is, $P_j \subset\subset K_{j+1}$. Thus P_j is a special analytic polyhedron in B. It follows trivially from the relation "$K_j \subset P_j \subset\subset \mathring{K}_{j+1}$" that the sequence (P_j) exhausts the region B. □

In the theory of Stein manifolds one can prove the converse of this theorem.

7. Riemann Domains over $\mathbb{C}^n$

If G is a domain in $\mathbb{C}^n$, we can ask if there exists a largest set M with $G \subset M$ over which every function holomorphic in G can be (holomorphically) extended. It turns out that we cannot restrict ourselves to subsets of $\mathbb{C}^n$. We must consider domains covering $\mathbb{C}^n$:

Def. 7.1. A (*Riemann*) *domain over* $\mathbb{C}^n$ is a pair (G, π) with the following properties:

1. G is a connected topological space.
2. For every two points $x_1, x_2 \in G$ with $x_1 \neq x_2$ there are open neighborhoods $U_1 = U_1(x_1) \subset G$, $U_2 = U_2(x_2) \subset G$ with $U_1 \cap U_2 = \varnothing$ (that is, G is a Hausdorff space).
3. $\pi: G \to \mathbb{C}^n$ is a locally topological mapping (that is: If $x \in G$ and $\mathfrak{z} := \pi(x)$ is the "base point of x", then there exist open neighborhoods $U = U(x) \subset G$ and $V = V(\mathfrak{z}) \subset \mathbb{C}^n$ such that $\pi|U: U \to V$ is topological).

Remarks

a. The mapping π is in particular continuous.

b. G is path-connected.

Take $x_0 \in G$ and let $Z := \{x \in G: x \text{ can be joined with } x_0 \text{ in } G\}$.

1. $x_0 \in Z$, therefore $Z \neq \emptyset$.
2. Z is open, since G is locally homeomorphic to $\mathbb{C}^n$ and therefore locally pathwise connected.
3. Z is closed: If $x_1 \in \partial Z$, then there exists a neighborhood $U(x_1) \subset G$ homeomorphic to $\mathbb{C}^n$ with $U \cap Z \neq \emptyset$. We can join x_1 in U with a point $x_2 \in U \cap Z$, and from the definition of Z we can join x_2 with x_0. Therefore x_1 also belongs to Z.

It follows from statements 1, 2, and 3 that $Z = G$.

c. If (G_ν, π_ν) are domains over $\mathbb{C}^n$ for $\nu = 1, \ldots, \ell$ and $x_\nu \in G_\nu$ are points with $\pi_\nu(x_\nu) = \mathfrak{z}_0$ for $\nu = 1, \ldots, \ell$, then there are open neighborhoods $U_\nu(x_\nu) \subset G_\nu$ and a connected open neighborhood $V(\mathfrak{z}_0) \subset \mathbb{C}^n$ such that for $\nu = 1, \ldots, \ell\, \pi_\nu | U_\nu : U_\nu \to V$ is topological.

PROOF. Choose neighborhoods $\tilde{U}_\nu(x_\nu) \subset G_\nu$, $\tilde{V}_\nu(\mathfrak{z}_0) \subset \mathbb{C}^n$ such that $\pi_\nu | \tilde{U}_\nu : \tilde{U}_\nu \to \tilde{V}_\nu$ is topological. Then let V be the component of $\mathfrak{z}_0$ in $\tilde{V} := \bigcap_{\nu=1}^{\ell} \tilde{V}_\nu$ and $U_\nu := (\pi_\nu | \tilde{U}_\nu)^{-1}(V)$ for $\nu = 1, \ldots, \ell$. □

EXAMPLES

1. Domains in $\mathbb{C}^n$. Let $G \subset \mathbb{C}^n$ be a domain. $\pi := \mathrm{id}_G$ the natural inclusion. Clearly (G, π) is a domain over $\mathbb{C}^n$ in the sense of Def. 7.1.

2. The Riemann surface of $\sqrt{z}$. Let $G := \{(w, z) \in \mathbb{C}^2 : w^2 = z, z \neq 0\}$ be provided with the relative topology induced from $\mathbb{C}^2$. G is a Hausdorff space. The mapping $\varphi : \mathbb{C} - \{0\} \to G$ defined by $t \mapsto (t, t^2)$ is bijective and continuous. G is therefore connected.

Now let $\hat{\pi} : \mathbb{C}^2 \to \mathbb{C}$ be defined by $\hat{\pi}(w, z) := z$. Then $\pi := \hat{\pi} | G : G \to \mathbb{C}$ is continuous. If $(w_0, z_0) \in G$ is an arbitrary point, then $z_0 \neq 0$, and we can find a simply connected neighborhood $V(z_0) \subset \mathbb{C} - \{0\}$. From the theory of a single variable we know that there exists a holomorphic function f in V with $f^2(z) \equiv z$ and $f(z_0) = w_0$. We denote f by $\sqrt{z}$. Then $\pi^{-1}(V)$ can be written as the union of the disjoint open sets $U_+ := U = \{(f(z), z) : z \in V\}$ and $U_- := \{(-f(z), z) : z \in V\}$. Let $\hat{f}(z) := (f(z), z)$. Then $(\pi | U)^{-1} = \hat{f}$, that is $\pi | U$ is topological. Hence (G, π) is a domain over $\mathbb{C}$, the so-called "Riemann surface of $\sqrt{z}$".

G can be visualized in the following manner: We cover $\mathbb{C}$ with two additional copies of $\mathbb{C}$, cut both these "sheets" along the positive real axis and paste them crosswise to one another (this is not possible in $\mathbb{R}^3$ without self intersection, but in higher dimensions, it is).

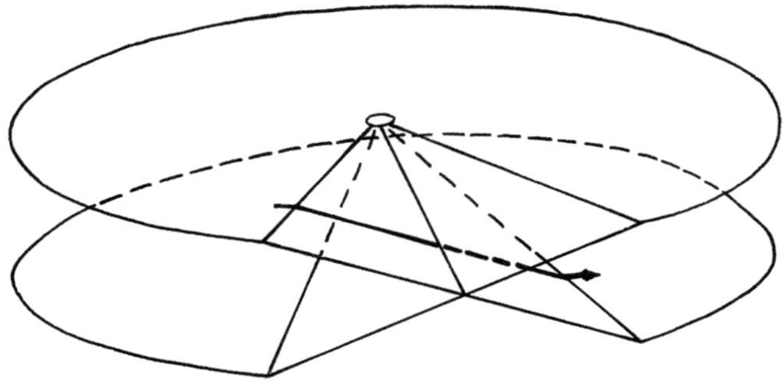

Figure 15. The Riemann surface of $\sqrt{z}$.

Next we consider Riemann domains with a distinguished point.

Def. 7.2. Let $\mathfrak{z}_0 \in \mathbb{C}^n$ be fixed. Then a (*Riemann*) *domain over* $\mathbb{C}^n$ *with base point* is a triple $\mathfrak{G} = (G, \pi, x_0)$ for which:

1. (G, π) is a domain over $\mathbb{C}^n$.
2. $\pi(x_0) = \mathfrak{z}_0$.

The point x_0 is called the *base point*.

Def. 7.3. Let $\mathfrak{G}_j = (G_j, \pi_j, x_j)$ be domains with base point over $\mathbb{C}^n$. We say $\mathfrak{G}_1 < \mathfrak{G}_2$ ("$\mathfrak{G}_1$ *is contained in* $\mathfrak{G}_2$") if there is a continuous mapping $\varphi: G_1 \to G_2$ with the following properties:

1. $\pi_1 = \pi_2 \circ \varphi$ ("φ preserves fibers")
2. $\varphi(x_1) = x_2$.

Theorem 7.1 (Uniqueness of lifting). *Let* $\mathfrak{G} = (G, \pi, x_0)$ *be a domain over* $\mathbb{C}^n$ *with base point, Y a connected topological space and* $y_0 \in Y$ *a point. If* $\psi_1, \psi_2: Y \to G$ *are continuous mappings with* $\psi_1(y_0) = \psi_2(y_0) = x_0$ *and* $\pi \circ \psi_1 = \pi \circ \psi_2$, *then* $\psi_1 = \psi_2$.

Proof. Let $M := \{y \in Y: \psi_1(y) = \psi_2(y)\}$. By assumption $y_0 \in M$, so $M \neq \varnothing$. Since G is a Hausdorff space it follows immediately that M is closed. Now let $y_1 \in M$ be chosen arbitrarily, $x_1 := \psi_1(y_1) = \psi_2(y_1)$ and $\mathfrak{z}_1 := \pi(x_1)$. There are open neighborhoods $U(x_1)$, $V(\mathfrak{z}_1)$ such that $\pi|U: U \to V$ is topological and there are open neighborhoods $Q_1(y_1)$, $Q_2(y_1)$ with $\psi_\lambda(Q_\lambda) \subset U$ for $\lambda = 1, 2$. Let $Q := Q_1 \cap Q_2$. Then

$$\psi_1|Q = (\pi|U)^{-1} \circ \pi \circ \psi_1|Q = (\pi|U)^{-1} \circ \pi \circ \psi_2|Q = \psi_2|Q,$$

therefore $Q \subset M$. Hence M is also open, and since Y is connected, it follows that $M = Y$. □

Theorem 7.2. *Let* $\mathfrak{G}_j = (G_j, \pi_j, x_j)$ *be domains with base point over* $\mathbb{C}^n$ *for* $j = 1, 2$. *Then there exists at most one continuous fiber-preserving mapping* $\varphi: G_1 \to G_2$ *with* $\varphi(x_1) = x_2$.

PROOF. If there are two continuous mappings $\varphi, \psi: G_1 \to G_2$ with $\pi_2 \circ \varphi = \pi_1 = \pi_2 \circ \psi$ and $\varphi(x_1) = \psi(x_1) = x_2$, then it follows from Theorem 7.1 that $\varphi = \psi$. □

Theorem 7.3. *The relation "$<$" is a weak ordering; that is:*

1. $\mathfrak{G} < \mathfrak{G}$;
2. $\mathfrak{G}_1 < \mathfrak{G}_2$ *and* $\mathfrak{G}_2 < \mathfrak{G}_3 \Rightarrow \mathfrak{G}_1 < \mathfrak{G}_3$.

The proof is trivial.

Def. 7.4. Two domains $\mathfrak{G}_1, \mathfrak{G}_2$ with base point over $\mathbb{C}^n$ are called *isomorphic* (symbolically $\mathfrak{G}_1 \simeq \mathfrak{G}_2$) if $\mathfrak{G}_1 < \mathfrak{G}_2$ and $\mathfrak{G}_2 < \mathfrak{G}_1$.

Theorem 7.4. *Two domains $\mathfrak{G}_j = (G_j, \pi_j, x_j)$, $j = 1, 2$, are isomorphic if and and if there exists a topological fiber preserving mapping $\varphi: G_1 \to G_2$* with $\varphi(x_1) = x_2$.

PROOF. $\mathfrak{G}_1 \simeq \mathfrak{G}_2$ means that there exist continuous fiber-preserving mappings $\varphi_1: G_1 \to G_2$ with $\varphi_1(x_1) = x_2$ and $\varphi_2: G_2 \to G_1$ with $\varphi_2(x_2) = x_1$. Then $\varphi_2 \circ \varphi_1: G_1 \to G_1$ is continuous and both $\pi_1 \circ (\varphi_2 \circ \varphi_1) = (\pi_1 \circ \varphi_2) \circ \varphi_1 = \pi_2 \circ \varphi_1 = \pi_1$ and $\varphi_2 \circ \varphi_1(x_1) = \varphi_2(x_2) = x_1$. From the uniqueness theorem (Theorem 7.2) it follows that $\varphi_2 \circ \varphi_1 = \mathrm{id}_{G_1}$. Analogously one shows $\varphi_1 \circ \varphi_2 = \mathrm{id}_{G_2}$. Hence φ_1 is bijective and $(\varphi_1)^{-1} = \varphi_2$. We set $\varphi: = \varphi_1$. To prove the converse we set $\varphi_1: = \varphi$ and $\varphi_2: = \varphi^{-1}$. □

Def. 7.5. A domain $\mathfrak{G} = (G, \pi, x_0)$ over $\mathbb{C}^n$ with base point is called *schlicht* if:

1. $G \subset \mathbb{C}^n$;
2. $\pi = \mathrm{id}_G$ is the natural inclusion. (In particular then $x_0 = \mathfrak{z}_0$.)

Theorem 7.5. $\mathfrak{G}_1 < \mathfrak{G}_2 \Leftrightarrow G_1 \subset G_2$ if $\mathfrak{G}_1, \mathfrak{G}_2$ are *schlicht domains.* The proof is trivial.

EXAMPLE. Let $\mathfrak{G}_1: = (G, \pi, x_0)$ be the Riemann surface of $\sqrt{z}$ with the base point $x_0: = (1, 1)$, $\mathfrak{G}_2: = (\mathbb{C}, \mathrm{id}_{\mathbb{C}}, 1)$. Then $\varphi: = \pi: G \to \mathbb{C}$ is a continuous mapping with $\mathrm{id}_{\mathbb{C}} \circ \varphi = \pi$ and $\varphi(x_0) = 1$. Therefore $\mathfrak{G}_1 < \mathfrak{G}_2$.

Next we consider systems of domains over $\mathbb{C}^n$. Let I be a set, $(\mathfrak{G}_\iota)_{\iota \in I}$ a family of domains over $\mathbb{C}^n$ with base point.

If $\iota \in I$, $\mathfrak{G}_\iota = (G_\iota, \pi_\iota, x_\iota)$, then $\pi_\iota(x_\iota) = \mathfrak{z}_0$.

Let $X: = \dot{\bigcup\limits_{\iota \in I}} G_\iota = \bigcup\limits_{\iota \in I} (G_\iota \times \{\iota\})$ be the disjoint union of the spaces G_ι. Let K be another set, $(N_\kappa)_{\kappa \in K}$ a family of sets. For each $\kappa \in K$, let

$$\mathfrak{X}_\kappa: = \{X_{\nu_\kappa}^{(\kappa)}: \nu_\kappa \in N_\kappa\}$$

be a partition of X (that is:

1. $X_{\nu_\kappa}^{(\kappa)} \subset X$
2. $\bigcup_{\nu_\kappa} X_{\nu_\kappa}^{(\kappa)} = X$
3. For $\nu_\kappa \neq \mu_\kappa$, $X_{\nu_\kappa}^{(\kappa)} \cap X_{\mu_\kappa}^{(\kappa)} = \varnothing$).

Let $\mathfrak{X} := (\mathfrak{X}_\kappa)_{\kappa \in K}$ be the family of these partitions. We show that there exists a partition $\mathfrak{X}^*$ of X which is finer than any $\mathfrak{X}_\kappa$ with $\kappa \in K$. (That is, there exists a partition $\mathfrak{X}^* = \{X_\nu : \nu \in N\}$ for which, for any $\nu \in N$ and $\kappa \in K$, there exists a $\nu_\kappa \in N_\kappa$ with $X_\nu \subset X_{\nu_\kappa}^{(\kappa)}$.)

Let $N := \prod_{\kappa \in K} N_\kappa$ and $X_\nu := \bigcap_{\kappa \in K} X_{\nu_\kappa}^{(\kappa)}$ for $\nu = (\nu_\kappa)_{\kappa \in K} \in N$, $\mathfrak{X}^* := \{X_\nu : \nu \in N\}$. Then for each $\nu \in N$, $X_\nu \subset X$, and

$$\bigcup_{\nu \in N} X_{\nu_\kappa} = \bigcup_{\nu \in \Pi N_\kappa} \left(\bigcap_{\kappa \in K} X_{\nu_\kappa}^{(\kappa)} \right) = \bigcap_{\kappa \in K} \left(\bigcup_{\nu_\kappa \in N} X_{\nu_\kappa}^{(\kappa)} \right) = \bigcap_{\kappa \in K} X = X$$

and

$$X_\nu \cap X_\mu = \left(\bigcap_{\kappa \in K} X_{\nu_\kappa}^{(\kappa)} \right) \cap \left(\bigcap_{\kappa \in K} X_{\mu_\kappa}^{(\kappa)} \right) = \bigcap_{\kappa \in K} (X_{\nu_\kappa}^{(\kappa)} \cap X_{\mu_\kappa}^{(\kappa)}) = \varnothing,$$

if $\nu_\kappa \neq \mu_\kappa$ for some $\kappa \in K$. Therefore $\mathfrak{X}^*$ is a partition, and clearly for fixed ν, $X_\nu = \bigcap_{\kappa \in K} X_{\nu_\kappa}^{(\kappa)} \subset X_{\nu_\kappa}^{(\kappa)}$. That is, $\mathfrak{X}^*$ is finer than any partition $\mathfrak{X}_\kappa$, $\kappa \in K$.

Definition. We say that the *equivalence relation* $\sim$ *on* X *has property* (P) if for all $\iota_1, \iota_2 \in I$ it is true that

1. $(x_{\iota_1}, \iota_1) \sim (x_{\iota_2}, \iota_2)$
2. If $\psi:[0, 1] \to G_{\iota_1}$, $\varphi:[0, 1] \to G_{\iota_2}$ are paths (=continuous mappings) with $(\psi(0), \iota_1) \sim (\varphi(0), \iota_2)$ and $\pi_{\iota_1} \circ \psi = \pi_{\iota_2} \circ \varphi$ then $(\psi(1), \iota_1) \sim (\varphi(1), \iota_2)$.

Example. Let $(y, \iota_1) \sim (y', \iota_2)$ if $\pi_{\iota_1}(y) = \pi_{\iota_2}(y')$. Clearly $\sim$ is an equivalence relation on X and $\sim$ has property (P).

Now let K be the set of all equivalence relations on X which have property (P). For $\kappa \in K$ let $\mathfrak{X}_\kappa$ be the partition of X corresponding to the equivalence relation κ, that is the set of equivalence classes.

For the partition system $\mathfrak{X} = \{\mathfrak{X}_\kappa : \kappa \in K\}$ one can construct a refinement $\mathfrak{X}^* = \{X_\nu : \nu \in N\}$ as above.

Lemma 1. *The equivalence relation* $\sim$ *defined on* X *by* $\mathfrak{X}^*$ *has property* (P). *Furthermore, the equivalence classes* X_ν *in each case contain only points over the same fundamental point* $\mathfrak{z}_\nu \in \mathbb{C}^n$.

Proof. The equivalence relation $\kappa \in K$ will also be denoted by "$\tilde{\kappa}$". Then $(x_{\iota_1}, \iota_1) \tilde{\kappa} (x_{\iota_2}, \iota_2)$ holds for each $\kappa \in K$, $\iota_1, \iota_2 \in I$. Therefore, for each $\kappa \in K$

there exists a $\nu_\kappa \in N_\kappa$, such that (x_{ι_1}, ι_1), (x_{ι_2}, ι_2) simultaneously lie in the set $X^{(\kappa)}_{\nu_\kappa}$. But then the points also lie in the set $\bigcap_{\kappa \in K} X^{(\kappa)}_{\nu_\kappa} = X_\nu$ for $\nu = (\nu_\kappa)_{\kappa \in K} \in \prod_{\kappa \in K} N_\kappa$, that is $(x_{\iota_1}, \iota_1) \sim (x_{\iota_2}, \iota_2)$. One shows similarly that the second requirement of (P) is satisfied. $\mathfrak{X}^*$ is therefore the finest partition of X which defines an equivalence relation with property (P). If two points (x, ι_1), (x', ι_2) lie in $X_\nu = \bigcap_{\kappa \in K} X^{(\kappa)}_{\nu_\kappa}$, then for every $\kappa \in K$ $(x, \iota_1)\tilde{\kappa}(x', \iota_2)$, in particular for the equivalence relation given in the example. But then $\pi_{\iota_1}(x) = \pi_{\iota_2}(x')$. The fundamental point uniquely determined by X_ν will be denoted by $\mathfrak{z}_\nu$. □

Definition. Let $\mathfrak{X}^* = (X_\nu)_{\nu \in N}$ be the finest partition of X which defines an equivalence relation on X with property (P). Then let $\tilde{G} := \{X_\nu : \nu \in N\}$, and let the mapping $\tilde{\pi}: \tilde{G} \to \mathbb{C}^n$ be defined by $\tilde{\pi}(X_\nu) := \mathfrak{z}_\nu$. Further, let $\tilde{x}_0 = X_{\nu_0}$ be the equivalence class which contains all the base points (x_ι, ι), $\iota \in I$.

Subsequently it will be shown that $\tilde{\mathfrak{G}} = (\tilde{G}, \tilde{\pi}, \tilde{x}_0)$ can be given such a topology that $\tilde{\mathfrak{G}}$ is a Riemann domain and $\mathfrak{G}_\iota < \tilde{\mathfrak{G}}$ for all $\iota \in I$.

Definition. For $\iota \in I$ let a mapping $\varphi_\iota: G_\iota \to \tilde{G}$ be defined as follows: If $y \in G_\iota$, then let $\varphi_\iota(y) = X_{\nu(\iota)} \in \tilde{G}$ be that equivalence class which contains y. Clearly $\tilde{\pi} \circ \varphi_\iota(y) = \pi_\iota(y)$ and $\varphi_\iota(x_\iota) = \tilde{x}_0$.

It suffices, therefore, to give $(\tilde{G}, \tilde{\pi}, \tilde{x}_0)$ a Hausdorff topology so that all mappings φ_ι are continuous.

Lemma 2. *Let $(y_1, \iota_1), (y_2, \iota_2) \in X$ be equivalent, $\mathfrak{z}_1 \in \mathbb{C}^n$ the common fundamental point, $V = V(\mathfrak{z}_1) \subset \mathbb{C}^n$ a connected open neighborhood and $U_i = U_i(y_i) \subset G_{\iota_i}$ open neighborhoods such that $\pi_{\iota_i}|U_i : U_i \to V$ is topological (for $i = 1, 2$). Then $((\pi_{\iota_1}|U_1)^{-1}(\mathfrak{z}), \iota_1) \sim ((\pi_{\iota_2}|U_2)^{-1}(\mathfrak{z}), \iota_2)$ for every $\mathfrak{z} \in V$.*

Proof. Let φ be a path in V which joins $\mathfrak{z}_1$ with $\mathfrak{z}$. Then $\psi_1 := (\pi_{\iota_1}|U_{\iota_1})^{-1} \circ \varphi$ and $\psi_2 := (\pi_{\iota_2}|U_{\iota_2})^{-1} \circ \varphi$ are paths in U_1, resp. U_2, which connect y_1 with $(\pi_{\iota_1}|U_{\iota_1})^{-1}(\mathfrak{z})$, resp. y_2 with $(\pi_{\iota_2}|U_{\iota_2})^{-1}(\mathfrak{z})$. The initial points are equivalent, and therefore so are the endpoints. □

Lemma 3. *For all $\iota_1, \iota_2 \in I$ it is true that: If $M \subset G_{\iota_1}$ is open, then $\varphi_{\iota_2}^{-1}(\varphi_{\iota_1}(M)) \subset G_{\iota_2}$ is open.*

Proof. $\varphi_{\iota_2}^{-1}(\varphi_{\iota_1}(M)) = \{x \in G$: There is a $y \in M$ with $\varphi_{\iota_1}(y) = \varphi_{\iota_2}(x)\} = \{x \in G_{\iota_2}$: There is a $y \in M$ with $(y, \iota_1) \sim (x, \iota_2)\}$.

Let $x \in \varphi_{\iota_2}^{-1}(\varphi_{\iota_1}(M))$ be given, $y \in M$ with $(y, \iota_1) \sim (x, \iota_2)$ and $\mathfrak{z} := \pi_{\iota_1}(y) = \pi_{\iota_2}(x)$. There exist open neighborhoods $U_1 = U_1(y)$, $U_2 = U_2(x)$

and a connected open neighborhood $V = V(\mathfrak{z})$ such that $\pi_{\iota_1}|U_1: U_1 \to V$, $\pi_{\iota_2}|U_2: U_2 \to V$ are topological mappings. Let $\varphi := (\pi_{\iota_2}|U_2)^{-1} \circ (\pi_{\iota_1}|U_1): U_1 \to U_2$. From Lemma 2 it follows that for $y' \in U_1$

$$(\varphi(y'), \iota_2) \sim (y', \iota_1).$$

M is open, and so are $U_1' := M \cap U_1$ and $U_2' := \varphi(U_1') \subset U_2$.
But $x \in U_2' \subset \varphi_{\iota_2}^{-1}(\varphi_{\iota_1}(M))$. Hence x is an interior point, which was to be shown. □

Lemma 4. *Let $M_{\iota_1} \subset G_{\iota_1}$, $M_{\iota_2} \subset G_{\iota_2}$ be arbitrary subsets. Then $\varphi_{\iota_1}(M_{\iota_1}) \cap \varphi_{\iota_2}(M_{\iota_2}) = \varphi_{\iota_2}(M_{\iota_2} \cap \varphi_{\iota_2}^{-1}(\varphi_{\iota_1}(M_{\iota_1})))$.*

PROOF

1. Let $y \in \varphi_{\iota_1}(M_{\iota_1}) \cap \varphi_{\iota_2}(M_{\iota_2})$. Then there are points $y_1 \in M_{\iota_1}$, $y_2 \in M_{\iota_2}$ with $\varphi_{\iota_1}(y_1) = \varphi_{\iota_2}(y_2) = y$. Clearly $y_2 \in \varphi_{\iota_2}^{-1}(\varphi_{\iota_1}(M_{\iota_1})) \cap M_{\iota_2}$.

2. Let $y \in \varphi_{\iota_2}(M_{\iota_2} \cap \varphi_{\iota_2}^{-1}(\varphi_{\iota_1}(M_{\iota_1})))$. Then there is a point $y_2 \in M_{\iota_2} \cap \varphi_{\iota_2}^{-1}(\varphi_{\iota_1}(M_{\iota_1}))$ with $\varphi_{\iota_2}(y_2) = y$ and furthermore there is also a point $y_1 \in M_{\iota_1}$ with $\varphi_{\iota_2}(y_2) = \varphi_{\iota_1}(y_1)$. Therefore $y \in \varphi_{\iota_1}(M_{\iota_1}) \cap \varphi_{\iota_2}(M_{\iota_2})$. □

Now we can introduce a topology on $\tilde{G}$:

Let $\mathfrak{T}' := \{A \subset \tilde{G}$: There exists an $\iota \in I$, $M_\iota \subset G_\iota$ open, such that $\varphi_\iota(M_\iota) = A\} \cup \{\tilde{G}\}$. Then:

1. $\emptyset = \varphi_\iota(\emptyset)$ for every $\iota \in I$, so $\emptyset \in \mathfrak{T}'$
2. $\tilde{G} \in \mathfrak{T}'$ by definition
3. $A_1, A_2 \in \mathfrak{T}' \Rightarrow A_1 \cap A_2 \in \mathfrak{T}'$, from Lemmas 4 and 3.

$\mathfrak{T}'$ satisfies the axioms for the basis of a topology. Let $\mathfrak{T}$ be the corresponding topology on $\tilde{G}$, that is, the set of arbitrary unions of elements of $\mathfrak{T}'$.

Theorem 7.6. *Let $\{\mathfrak{G}_\iota = (G_\iota, \pi_\iota, x_\iota): \iota \in I\}$ be a family of domains over $\mathbb{C}^n$ with base point, $X = \dot{\bigcup}_{\iota \in I} G_\iota$ the disjoint union of the spaces G_ι, and $\mathfrak{X}^* = (X_\nu)_{\nu \in N}$ the finest partition of X which defines an equivalence relation with (P). Let $\tilde{G} := \{X_\nu: \nu \in N\}$ be the set of classes of $\mathfrak{X}^*$. Let the point $\tilde{x}_0 \in \tilde{G}$ and the mappings $\tilde{\pi}: \tilde{G} \to \mathbb{C}^n$, $\varphi_\iota: G_\iota \to \tilde{G}$ be defined as above, and $\tilde{G}$ be provided with the topology given above. Then:*

1. *$\tilde{\mathfrak{G}} = (\tilde{G}, \tilde{\pi}, \tilde{x}_0)$ is a domain over $\mathbb{C}^n$ with base point.*
2. *For every $\iota \in I$, $\mathfrak{G}_\iota < \tilde{\mathfrak{G}}$.*
3. *If $\mathfrak{G}^* = (G^*, \pi^*, x_0^*)$ is a domain over $\mathbb{C}^n$ and $\mathfrak{G}_\iota < \mathfrak{G}^*$ for all $\iota \in I$, then also $\tilde{\mathfrak{G}} < \mathfrak{G}^*$.*

($\tilde{\mathfrak{G}}$ is the smallest Riemann domain over $\mathbb{C}^n$, which contains all domains $\mathfrak{G}_\iota$.)

PROOF

1a. $\tilde{G}$ is a topological space and $\tilde{\pi}(\tilde{x}_0) = \mathfrak{z}_0 = \pi_\iota(x_\iota)$.

b. $\tilde{G}$ is connected: If $y \in \tilde{G}$, then there is an $\iota \in I$ and a $y_\iota \in G_\iota$ such that $y = \varphi_\iota(y_\iota)$. Let ψ be a path in G_ι which connects y_ι with x_ι. Then $\varphi_\iota \circ \psi: [0, 1] \to \tilde{G}$ is a mapping with $\varphi_\iota \circ \psi(0) = y, \varphi_\iota \circ \psi(1) = \tilde{x}_0$. φ_ι (and hence

$\varphi_\iota \circ \psi$ also) is continuous: if $M \subset \tilde{G}$ is open, then $M = \bigcup_{\iota \in I} \varphi_\iota(M_\iota)$, where $M_\iota \subset G_\iota$ is open (possibly empty) for every ι.

It follows that, for $\iota_0 \in I$, $\varphi_{\iota_0}^{-1}(M) = \bigcup_{\iota \in I} \varphi_{\iota_0}^{-1}(\varphi_\iota(M))$ is open in G_{ι_0}. We can therefore connect every point to the base point by a path in $\tilde{G}$.

c. $\tilde{G}$ is a Hausdorff space: Let $y_1, y_2 \in \tilde{G}$ with $y_1 \neq y_2$.

Case 1. $\tilde{\pi}(y_1) := \mathfrak{z}_1 \neq \mathfrak{z}_2 =: \tilde{\pi}(y_2)$. Then there are open neighborhoods $V(\mathfrak{z}_1)$, $V'(\mathfrak{z}_2)$ with $V \cap V' = \varnothing$, and $\tilde{\pi}^{-1}(V) \cap \tilde{\pi}^{-1}(V') = \varnothing$. Therefore it suffices to show that $\tilde{\pi}$ is continuous. Let $V \subset \mathbb{C}^n$ be open, $M := \tilde{\pi}^{-1}(V)$, $\iota \in I$. Then $\varphi_\iota^{-1}(M) = (\tilde{\pi} \circ \varphi_\iota)^{-1}(V) = \pi_\iota^{-1}(V)$ is open in G_ι, therefore $M = \bigcup_{\iota \in I} \varphi_\iota \varphi_\iota^{-1}(M)$ is open in $\tilde{G}$.

Case 2. Let $\mathfrak{z} := \tilde{\pi}(y_1) = \tilde{\pi}(y_2)$. There are elements $\hat{y}_1 \in G_{\iota_1}$, $\hat{y}_2 \in G_{\iota_2}$ with $\varphi_{\iota_1}(\hat{y}_1) = y_1$ and $\varphi_{\iota_2}(\hat{y}_2) = y_2$. Furthermore we can find open neighborhoods $U_1(\hat{y}_1) \subset G_{\iota_1}$, $U_2(\hat{y}_2) \subset G_{\iota_2}$ and a connected open neighborhood $V(\mathfrak{z}) \subset \mathbb{C}^n$ such that $\pi_{\iota_1}|U_1 : U_1 \to V$ and $\pi_{\iota_2}|U_2 : U_2 \to V$ are topological mappings. The points $(\hat{y}_1, \iota_1)$, $(\hat{y}_2, \iota_2)$ are not equivalent, so by Lemma 2 it must be that $\varphi_{\iota_1}(U_1) \cap \varphi_{\iota_2}(U_2) = \varnothing$, and we have found disjoint neighborhoods.

d. $\tilde{\pi}$ is locally topological. Let $y \in \tilde{G}$, $\iota \in I$, $y_\iota \in G_\iota$ be such that $\varphi_\iota(y_\iota) = y$. Let $\mathfrak{z} := \tilde{\pi}(y) = \pi_\iota(y_\iota)$. Then there exist open neighborhoods $U_\iota(y_\iota)$ and $V(\mathfrak{z})$ such that $\pi_\iota|U_\iota : U_\iota \to V$ is topological. $U := \varphi_\iota(U_\iota)$ is an open neighborhood of y, $\tilde{\pi}|U : U \to V$ is continuous and surjective. From the equality $(\tilde{\pi}|U) \circ (\varphi_\iota|U_\iota) = \pi_\iota|U_\iota$ it follows that $\tilde{\pi}|U$ is also injective and $(\tilde{\pi}|U)^{-1}$ is continuous.

2. The mappings $\varphi_\iota : G_\iota \to \tilde{G}$ are fiber-preserving and by (1b) are also continuous. Therefore $\mathfrak{G}_\iota < \tilde{\mathfrak{G}}$.

3. If $\mathfrak{G}^*$ is given, then there exists a fiber-preserving mapping $\varphi_\iota^* : G_\iota \to G^*$. With the help of the statement "$(y, \iota_1) \simeq (y', \iota_2)$ if and only if $\varphi_{\iota_1}^*(y) = \varphi_{\iota_2}^*(y')$" we can introduce an equivalence relation on X, which because of the uniqueness lifting also has property (P): Namely, if $\psi : [0, 1] \to G_{\iota_1}$, $\varphi : [0, 1] \to G_{\iota_2}$ are two paths with $(\psi(0), \iota_1) \simeq (\varphi(0), \iota_2)$ and $\pi_{\iota_1} \circ \psi = \pi_{\iota_2} \circ \varphi$, then $\varphi_{\iota_1}^* \circ \psi(0) = \varphi_{\iota_2}^* \circ \varphi(0)$ and (because $\pi^* \circ \varphi_\iota^* = \pi_\iota$) also $\pi^* \circ (\varphi_{\iota_1}^* \circ \psi) = \pi^* \circ (\varphi_{\iota_2}^* \circ \varphi)$. Hence $\varphi_{\iota_1}^* \circ \psi = \varphi_{\iota_2}^* \circ \varphi$, by Theorem 7.1.

In particular it follows that $(\psi(1), \iota_1) \simeq (\varphi(1), \iota_2)$. But that means that a mapping $\varphi : \tilde{G} \to G^*$ is defined by $\varphi \circ \varphi_\iota = \varphi_\iota^*$. φ is continuous and fiber-preserving. □

Def. 7.6. The domain $\tilde{\mathfrak{G}}$ described in Theorem 7.6 is called the *union of the domains* $\mathfrak{G}_\iota$, $\iota \in I$, and we write

$$\tilde{\mathfrak{G}} = \bigcup_{\iota \in I} \mathfrak{G}_\iota.$$

Special Cases

1. From $\mathfrak{G}_1 < \mathfrak{G}$ and $\mathfrak{G}_2 < \mathfrak{G}$ it follows that $\mathfrak{G}_1 \cup \mathfrak{G}_2 < \mathfrak{G}$
2. From $\mathfrak{G}_1 < \mathfrak{G}_2$ it follows that $\mathfrak{G}_1 \cup \mathfrak{G}_2 \simeq \mathfrak{G}_2$

3. $\mathfrak{G} \cup \mathfrak{G} \simeq \mathfrak{G}$
4. $\mathfrak{G}_1 \cup \mathfrak{G}_2 \simeq \mathfrak{G}_2 \cup \mathfrak{G}_1$
5. $\mathfrak{G}_1 \cup (\mathfrak{G}_2 \cup \mathfrak{G}_3) \simeq (\mathfrak{G}_1 \cup \mathfrak{G}_2) \cup \mathfrak{G}_3$

EXAMPLE. Let $\mathfrak{G}_1 = (G_1, \pi_1, x_1)$ be the Riemann surface of $w = \sqrt{z}$, with $x_1 = (1, 1)$ as base point and with the canonical projection $\pi_1:(w, z) \mapsto z$. Let $\mathfrak{G}_2 = (G_2, \pi_2, x_2)$ be given by $G_2 := \{z \in \mathbb{C}: \frac{1}{2} < |z| < 2\}, x_2 := \pi_1(x_1) = 1$ and $\pi_2 := \mathrm{id}_{G_2}$.

Then $\mathfrak{G}_1 \cup \mathfrak{G}_2 = (\tilde{G}, \tilde{\pi}, \tilde{x}_0)$, where $\tilde{G} = (G_1 \dot{\cup} G_2)/\sim$ is the set of all equivalence classes $[(x, \imath)]$, $\imath \in \{1, 2\}$ with respect to the "finest" equivalence relation with propery (P).

1. Let $y \in \pi_1^{-1}(G_2) \subset G_1$. Then we can connect y with the point x_1 by a path ψ in $\pi_1^{-1}(G_2)$. The path $\pi_1 \circ \psi$ then connects $\pi_1(y)$ in G_2 with x_2. But now $(x_1, 1) \sim (x_2, 2)$, so $(y, 1) \sim (\pi_1(y), 2)$ as well. On the other hand, the equivalence classes contain only points over the same fundamental point, so it follows that over each point of G_2 there is exactly one equivalence class.

2. Let $z \in \mathbb{C} - \{0\}$ be arbitrary. The line through z and 0 contains a segment $\varphi:[0, 1] \to \mathbb{C} - \{0\}$ which connects a point $z^* \in G_2$ with z. Then there exist two paths ψ_1, ψ_2 in G_1 with $\pi_1 \circ \psi_1 = \pi_1 \circ \psi_2 = \varphi$ and $(\psi_1(0), 1) \sim (\psi_2(0), 1) \sim (z^*, 2)$. Hence it follows that the points $(\psi_1(1), 2), (\psi_2(1), 2)$ over z are equivalent. From (1) and (2) we have:

$$\mathfrak{G}_1 \cup \mathfrak{G}_2 = (\mathbb{C} - \{0\}, \mathrm{id}_{\mathbb{C}-\{0\}}, 1).$$

8. Holomorphic Hulls

Def. 8.1. Let (G, π) be a domain over $\mathbb{C}^n$, $f:G \to \mathbb{C}$ a function. f is called *holomorphic at a point* $x_0 \in G$ if there exist open neighborhoods $U = U(x_0)$ and $V = V(\pi(x_0))$ such that $\pi|U:U \to V$ is topological and $f \circ (\pi|U)^{-1}: V \to \mathbb{C}$ is holomorphic. f is called *holomorphic on* G if f is holomorphic at every point $x_0 \in G$.

Remarks

1. Holomorphy at a point does not depend on the neighborhood.
2. For schlicht domains the new notion of holomorphy agrees with the previous one.
3. If f is holomorphic on G, then f is continuous.

Lemma 1. *Let* $(G_1, \pi_1, y_1), \ldots, (G_\ell, \pi_\ell, y_\ell)$, (G, π, y) *be domains with base point over* $\mathbb{C}^n$ *and let* $\mathfrak{z} = \pi(y)$. *If* $\varphi_i:G \to G_i$ *are fiber-preserving mappings with* $\varphi_i(y) = y_i$ *for* $i = 1, \ldots, \ell$, *then there exist open neighborhoods* $U = U(y)$, $v = V(\mathfrak{z})$ *and* $U_i = U_i(y_i)$ *such that for every i all the mappings in the following commutative diagram are topological*

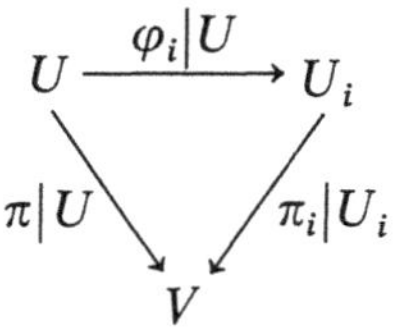

PROOF. We can find open neighborhoods $\hat{U}(y)$, $\hat{V}(\mathfrak{z})$, and $\hat{U}_i(y_i)$ such that the mappings $\pi|\hat{U}:\hat{U} \to \hat{V}$ and $\pi_i|\hat{U}_i:\hat{U}_i \to V$ are topological. Since φ_i is continuous, there is an open neighborhood $U(y) \subset \hat{U}(y)$ with $\varphi_i(U) \subset \hat{U}_i$. If we set $V(\mathfrak{z}) := \pi(U)$ and $U_i := \varphi_i(U)$, we obtain the desired result. □

Def. 8.2. Let $\mathfrak{G}_i = (G_i, \pi_i, x_i)$, $i = 1, 2$ be domains with base points and $\mathfrak{G}_1 < \mathfrak{G}_2$ by virtue of a continuous mapping $\varphi: G_1 \to G_2$. If f is a complex valued function on G_2, then we define $f|G_1 := f \circ \varphi$.

Theorem 8.1. *Under the conditions of Def. 8.2, $f|G_1$ is holomorphic on G_1 whenever f is holomorphic on G_2.*

PROOF. Let $y_1 \in G_1$ be arbitrary, $y_2 := \varphi(y_1) \in G_2$ and $\mathfrak{z}_1 := \pi_1(y_1) = \pi_2(y_2)$. By Lemma 1 we obtain a commutative diagram of topological mappings:

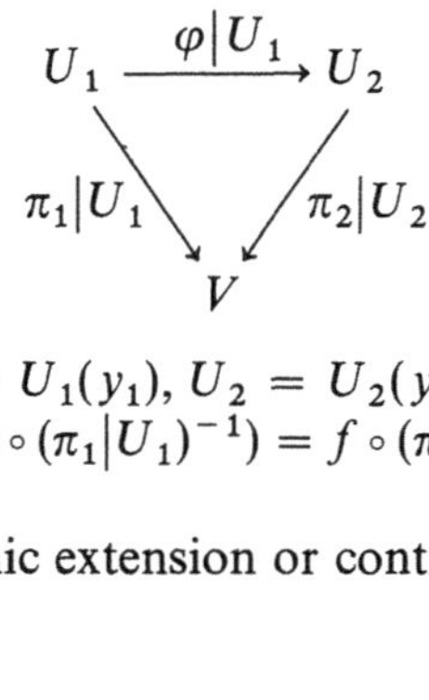

(with neighborhoods $U_1 = U_1(y_1)$, $U_2 = U_2(y_2)$ and $V = V(\mathfrak{z}_1)$). But then $(f|G_1) \circ (\pi_1|U_1)^{-1} = f \circ (\varphi \circ (\pi_1|U_1)^{-1}) = f \circ (\pi_2|U_2)^{-1}$. □

f is called a holomorphic extension or continuation of $f|G_1$ to G_2.

Def. 8.3

1. Let (G, π) be a domain over $\mathbb{C}^n$. If $x \in G$ is a point and f a holomorphic function defined near x, then the pair (f, x) is called a *locally holomorphic function* at x.

2. Let (G_1, π_1), (G_2, π_2) be domains over $\mathbb{C}^n$, $y_1 \in G_1$ and $y_2 \in G_2$ points with $\pi_1(y_1) = \pi_2(y_2) = :\mathfrak{z}$. Two local functions (f_1, y_1), (f_2, y_2) are called *equivalent* (symbolically $(f_1)_{y_1} = (f_2)_{y_2}$) if there exist open neighborhoods $U_1(y_1)$, $U_2(y_2)$, $V(\mathfrak{z})$ and topological mappings $\pi_1|U_1: U_1 \to V$, $\pi_2|U_2: U_2 \to V$ with $f_1 \circ (\pi_1|U_1)^{-1} = f_2 \circ (\pi_2|U_2)^{-1}$.

Remark. If $(f_1)_{y_1} = (f_2)_{y_2}$, then clearly $f_1(y_1) = f_2(y_2)$. In particular if $G_1 = G_2$, $\pi_1 = \pi_2$ and $y_1 = y_2$ then it follows that f_1 and f_2 coincide in an open neighborhood of $y_1 = y_2$.

Theorem 8.2. *Let (G_1, π_1), (G_2, π_2) be domains over $\mathbb{C}^n$, $\psi_\lambda:[0, 1] \to G_\lambda$ be paths with $\pi_1 \circ \psi_1 = \pi_2 \circ \psi_2$. Additionally, let f_λ be holomorphic on G_λ, $\lambda = 1, 2$. If $(f_1)_{\psi_1(0)} = (f_2)_{\psi_2(0)}$, then also $(f_1)_{\psi_1(1)} = (f_2)_{\psi_2(1)}$.*

PROOF

1. Let $x_1 \in G_1$, $x_2 \in G_2$ be points with $\pi_1(x_1) = \pi_2(x_2) = \mathfrak{z}_0$. Then there are open neighborhoods $U_1(x_1)$, $U_2(x_2)$ and an open connected neighborhood $V(\mathfrak{z}_0)$ such that the mappings $\pi_\lambda|U_\lambda: U_\lambda \to V$ are topological.

If there exist points $x_1' \in U_1$, $x_2' \in U_2$ with $\pi_1(x_1') = \pi_2(x_2') = \mathfrak{z}$ and $(f_1)_{x_1'} = (f_2)_{x_2'}$ then $f_1 \circ (\pi_1|U_1)^{-1} = f_2 \circ (\pi_2|U_2)^{-1}$ near $\mathfrak{z} \in V$ and therefore, by the identity theorem, in all of V. It follows that if $(f_1)_{x_1} = (f_2)_{x_2}$, then $(f_1)_{x_1'} = (f_2)_{x_2'}$ for all $x_1' \in U_1, x_2' \in U_2$ with $\pi_1(x_1') = \pi_2(x_2')$. If $(f_1)_{x_1} \neq (f_2)_{x_2}$, then $(f_1)_{x_1'} \neq (f_2)_{x_2'}$ for all $x_1' \in U_1$, $x_2' \in U_2$ with $\pi_1(x_1') = \pi_2(x_2')$.

2. Let $W: = \{t \in [0, 1]:(f_1)_{\psi_1(t)} = (f_2)_{\psi_2(t)}\}$

a. By assumption $W \neq \varnothing$, as 0 lies in W.

b. If $t_1 \in W$, then one sets $x_\lambda: = \psi_\lambda(t_1)$. By (1) there exist open neighborhoods $U_1(x_1)$, $U_2(x_2)$ such that $(f_1)_{x_1'} = (f_2)_{x_2'}$ for all $x_1' \in U_1$, $x_2' \in U_2$ with $\pi_1(x_1') = \pi_2(x_2')$. Since the mapping ψ_λ are continuous, there exists a neighborhood $Q(t_1) \subset [0,1]$ with $\psi_\lambda(Q) \subset U_\lambda$, $\lambda = 1,2$. Therefore $(f_1)_{\psi(t)} = (f_2)_{\psi(t)}$ for $t \in Q$. This means that W is open.

c. One shows that $[0, 1] - W$ is open in exactly the same way. Since $[0, 1]$ is connected, it follows that $W = [0, 1]$. □

Theorem 8.3. *Let $\mathfrak{G}_\lambda = (G_\lambda, \pi_\lambda, x_\lambda)$ be domains over $\mathbb{C}^n$ with $\pi_\lambda(x_\lambda) = \mathfrak{z}_0$, $\lambda = 1, 2$, and with $\mathfrak{G}_1 < \mathfrak{G}_2$.*

Let f be a holomorphic function on G_1, F a holomorphic extension of f to G_2. Then F is uniquely determined by f.

PROOF. Let F_1, F_2 be holomorphic extensions of f to G_2. By Lemma 1 there exist neighborhoods $U_\lambda(x_\lambda)$ such that the restriction of the canonical mapping $\varphi: G_1 \to G_2$ to U_1 maps the set U_1 topologically onto U_2. For $\nu = 1, 2$ it is true that $F_\nu \circ \varphi|U_1 = f|U_1$, consequently $F_1|U_2 = F_2|U_2$, and therefore $(F_1)_{x_2} = (F_2)_{x_2}$. Since each point $x \in G_2$ can be joined to x_2, the equality $F_1 = F_2$ follows from Theorem 8.2. □

For $j = 1, \ldots, n$ let $\mathrm{pr}_j:\mathbb{C}^n \to \mathbb{C}$ be the projection onto the j-th component. If (G, π) is a domain over $\mathbb{C}^n$, then $z_j: = \mathrm{pr}_j \circ \pi$ is a holomorphic function on G, so the set $A(G)$ of all holomorphic functions on G contains more than the constant functions.

Def. 8.4. Let $\mathfrak{G} = (G, \pi, x_0)$ be a domain over $\mathbb{C}^n$ with base point $\mathscr{F}$ a nonempty set of holomorphic functions on G. Let $\{\mathfrak{G}_\iota, \iota \in I\}$ be the set of domains over $\mathbb{C}^n$ with the following properties:

1. $\mathfrak{G} < \mathfrak{G}_\iota$ for all $\iota \in I$.
2. If $f \in \mathscr{F}$, then for every $\iota \in I$ there is an $F_\iota \in A(G_\iota)$ with $F_\iota|G = f$.

Then $H_{\mathscr{F}}(\mathfrak{G}) := \bigcup_{\iota \in I} \mathfrak{G}_\iota$ is called the *holomorphic hull* of $\mathfrak{G}$ relative to $\mathscr{F}$. If $\mathscr{F} = A(G)$, then $H(\mathfrak{G}) := H_{A(G)}(\mathfrak{G})$ is called the (*absolute*) holomorphic hull of $\mathfrak{G}$. If $\mathscr{F} = \{f\}$, then $H_f(\mathfrak{G}) := H_{\{f\}}(\mathfrak{G})$ is called the *domain of holomorphy* of f.

Theorem 8.4. *Let $\mathfrak{G} = (G, \pi, x_0)$ be a domain over $\mathbb{C}^n$, $\mathscr{F}$ a non-empty set of functions holomorphic on G and $H_{\mathscr{F}}(\mathfrak{G}) = (\hat{G}, \hat{\pi}, \hat{x})$ the holomorphic hull of G relative of $\mathscr{F}$. Then $\mathfrak{G} < H_{\mathscr{F}}(\mathfrak{G})$, and for each function $f \in \mathscr{F}$ there exists exactly one function $F \in A(\hat{G})$ with $F|G = f$. If $\mathfrak{G}_1 = (G_1, \pi_1, x_1)$ is a domain over $\mathbb{C}^n$ with $\mathfrak{G} < \mathfrak{G}_1$ and the property that every function $f \in \mathscr{F}$ can be holomorphically extended to G_1, then $\mathfrak{G}_1 < H_{\mathscr{F}}(\mathfrak{G})$.*

PROOF

1. Let "$\sim$" be the finest equivalence relation on $X := \bigcup_{\iota \in I} G_\iota$ with property (P). Then $\hat{G}$ is the set of equivalence classes of X relative to $\sim$. We now define a new equivalence relation on X: $(y, \iota_1) \simeq (y', \iota_2)$, if and only if:

a. $\pi_{\iota_1}(y) = \pi_{\iota_2}(y')$.

b. If $f \in \mathscr{F}$ and $f_1 \in A(G_{\iota_1})$, $f_2 \in A(G_{\iota_2})$ are holomorphic extensions of f, then $(f_1)_y = (f_2)_{y'}$. "$\simeq$" is an equivalence relation and has property (P).

a. For each $\iota \in I$ there exists a continuous fiber-preserving mapping φ_ι: $G \to G_\iota$ with $\varphi_\iota(x_0) = x_\iota$. We can find open neighborhoods $U(x_0)$, $U_1(x_{\iota_1})$, $U_2(x_{\iota_2})$ and $V(\pi(x_0))$ such that all mappings are topological in the two commutative diagrams below.

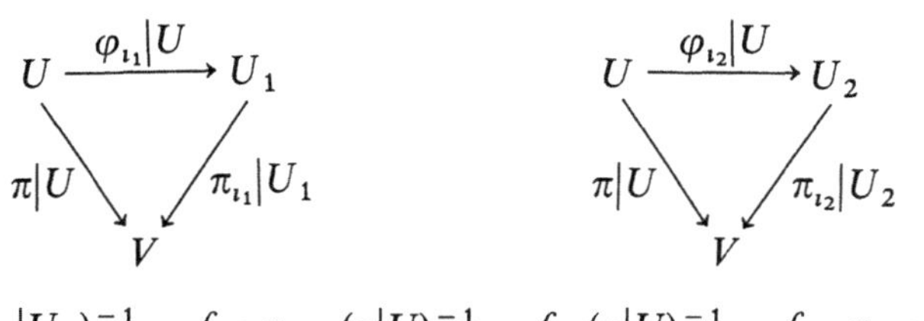

Then $f_2 \circ (\pi_{\iota_2}|U_2)^{-1} = f_2 \circ \varphi_{\iota_2} \circ (\pi|U)^{-1} = f \circ (\pi|U)^{-1} = f_1 \circ \varphi_{\iota_1} \circ (\pi|U)^{-1} = f_1 \circ (\pi_{\iota_1}|U_1)^{-1}$; that is the base points are equivalent.

b. If $\psi_\lambda : [0, 1] \to G_{\iota_\lambda}$ are paths with $(\psi_1(0), \iota_1) \simeq (\psi_2(0), \iota_2)$ and $\pi_{\iota_1} \circ \psi_1 = \pi_{\iota_2} \circ \psi_2$, then $(f_1)_{\psi_1(0)} = (f_2)_{\psi_2(0)}$. It follows from Theorem 8.2 that: $(f_1)_{\psi_1(1)} = (f_2)_{\psi_2(1)}$, and, therefore $(\psi_1(1), \iota_1) \simeq (\psi_2(1), \iota_2)$. Since "$\sim$" is the finest partition with property (P), $(y, \iota_1) \sim (y, \iota_2)$ implies $(y, \iota_1) \simeq (y, \iota_2)$.

2. For all $\iota \in I$, $\mathfrak{G} < \mathfrak{G}_\iota < \bigcup_{\iota \in I} \mathfrak{G}_\iota = H_{\mathfrak{F}}(\mathfrak{G})$. Let $\hat{\varphi}_\iota : G_\iota \to \hat{G}$ and $\hat{\varphi} : G \to \hat{G}$ with $\hat{\varphi} = \hat{\varphi}_\iota \circ \varphi_\iota$ be the canonical mappings. Let $f \in \mathscr{F}$ be given. For $\hat{y} \in \hat{G}$ there exists an $\iota \in I$ and a $y_\iota \in G_\iota$ with $\hat{\varphi}_\iota(y_\iota) = \hat{y}$. Let $F_\iota \in A(G_\iota)$ be a holomorphic extension of f. Then we set $F(\hat{y}) := F_\iota(y_\iota)$. If $\kappa \in I$, $y_\kappa \subset G_\kappa$, $\hat{\varphi}_\kappa(y_\kappa) = \hat{y}$, and if $F_\kappa \in A(G_\kappa)$ is a holomorphic extension of f, then $(y_\iota, \iota) \sim (y_\kappa, \kappa)$. Hence $(y_\iota, \iota) \simeq (y_\kappa, \kappa)$ as well, so that $(F_\iota)_{y_\iota} = (F_\kappa)_{y_\kappa}$. It follows that $F_\iota(y_\iota) = F_\kappa(y_\kappa)$. So F is well defined. Also, $F \circ \hat{\varphi} = F \circ \hat{\varphi}_\iota \circ \varphi_\iota = F_\iota \circ \varphi_\iota = f$, so F is an extension of f. It remains to show that F is holomorphic:

Let $\hat{y} \in \hat{G}$. $\hat{y} = \hat{\varphi}_\iota(y_\iota)$ and $\mathfrak{z} = \hat{\pi}(\hat{y})$. Then there exist open neighborhoods $U_1(y_\iota)$, $U_2(\hat{y})$, $V(\mathfrak{z})$ and a commutative diagram of topological mappings:

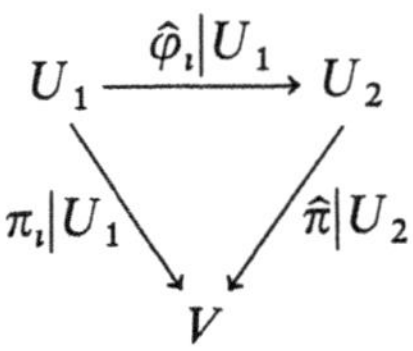

It follows that $F \circ (\hat{\pi}|U_2)^{-1} = F \circ \hat{\varphi}_\iota \circ (\pi_\iota|U_1)^{-1} = F_\iota \circ (\pi_\iota|U_1)^{-1}$; the last is a holomorphic function.

3. The "maximality" of $H_{\mathcal{F}}(\mathfrak{G})$ follows immediately from the construction. The holomorphic hull $H_{\mathcal{F}}(\mathfrak{G})$ is therefore the largest domain into which all functions $f \in \mathcal{F}$ can be holomorphically extended. □

Theorem 8.5. *Let $\mathfrak{G}_\lambda = (G_\lambda, \pi_\lambda, x)$, $\lambda = 1, 2$ be domains over $\mathbb{C}^n$ with $\mathfrak{G}_1 \cup \mathfrak{G}_2 = (\tilde{G}, \tilde{\pi}, \tilde{x})$, and $f_1: G_1 \to \mathbb{C}$, $f_2: G_2 \to \mathbb{C}$ be holomorphic functions. If there is a domain $\mathfrak{G} = (G, \pi, x_0)$ with $\mathfrak{G} < \mathfrak{G}_\lambda$ for $\lambda = 1, 2$ and $f_1|G = f_2|G$, then there is a function $\tilde{f}$ holomorphic on $\tilde{G}$ with $\tilde{f}|G_\lambda = f_\lambda$ for $\lambda = 1, 2$.*

Proof. Let $f := f_1|G = f_2|G$, $\mathcal{F} := \{f\}$. Then f_1 is a holomorphic extension of f to G_1 and f_2 is a holomorphic extension of f to G_2. Therefore by Theorem 8.4: $\mathfrak{G}_1 < H_{\mathcal{F}}(\mathfrak{G})$ and $\mathfrak{G}_2 < H_{\mathcal{F}}(\mathfrak{G})$. But then by Theorem 7.6 $\mathfrak{G}_1 \cup \mathfrak{G}_2 < H_{\mathcal{F}}(\mathfrak{G})$. Let $\hat{f}$ be the holomorphic extension of f to $H_{\mathcal{F}}(\mathfrak{G})$ and $\tilde{f} := \hat{f}|\tilde{G}$. For $\lambda = 1, 2$, $\tilde{f}|G = (\hat{f}|\tilde{G})|G = \hat{f}|G = f = f_\lambda|G$, therefore $\tilde{f}|G_\lambda = f_\lambda$. □

Now let $P \subset \mathbb{C}^n$ be the unit polycylinder, (P, H) a Euclidean Hartogs figure, $\Phi: P \to B \subset \mathbb{C}^n$ a biholomorphic mapping. $(B, \Phi(H))$ is then a generalized Hartogs figure. Since P and H are connected Hausdorff spaces and Φ is, in particular, locally topological, it follows that $\mathfrak{P} = (P, \Phi, 0)$ and $\mathfrak{H} = (H, \Phi, 0)$ are domains over $\mathbb{C}^n$ with base point and we have $\mathfrak{H} < \mathfrak{P}$. We regard the pair $(\mathfrak{P}, \mathfrak{H})$ as a generalized Hartogs figure.

Theorem 8.6. *Let (G, π) be a domain over $\mathbb{C}^n$, $(\mathfrak{P}, \mathfrak{H})$ a generalized Hartogs figure, and $x_0 \in G$ a point for which $\mathfrak{H} < \mathfrak{G} := (G, \pi, x_0)$.*

Then every function $f \in A(G)$ can be extended holomorphically to $\mathfrak{G} \cup \mathfrak{P}$.

Proof. $f|H$ has a holomorphic extension $F \in A(G)$. Let $\mathfrak{G}_1 := \mathfrak{G}$, $\mathfrak{G}_2 := \mathfrak{P}$, $f_1 := f$, $f_2 := F$. Because $\mathfrak{H} < \mathfrak{G}_1$, $\mathfrak{H} < \mathfrak{G}_2$ and $f_1|H = f_2|H$, the proposition follows from Theorem 8.5. □

Def. 8.5. A domain (G, π) over $\mathbb{C}^n$ is called *pseudoconvex* if the fact that $(\mathfrak{P}, \mathfrak{H})$ is a generalized Hartogs figure and $x_0 \in G$ a point with $\mathfrak{H} < \mathfrak{G} := (G, \pi, x_0)$ implies $\mathfrak{G} \cup \mathfrak{P} \cong \mathfrak{G}$.

Def. 8.6. A domain $\mathfrak{G} = (G, \pi, x_0)$ is called a *domain of holomorphy* if there exists an $f \in A(G)$ with $H_f(\mathfrak{G}) = \mathfrak{G}$. In the schlicht case this definition agrees with the old one.

Theorem 8.7.

1. *If* $\mathfrak{G} = (G, \pi, x_0)$ *is a domain over* $\mathbb{C}^n$ *and* F *a non-empty set of functions holomorphic on* G, *then* $H_{\mathfrak{F}}(\mathfrak{G})$ *is a pseudoconvex domain.*
2. *Every domain of holomorphy is pseudoconvex.*

The proof is trivial.

The definition of holomorphic convexity can be extended from the schlicht case; then we have

Theorem 8.8. (Oka, 1953). *If* $\mathfrak{G}$ *is pseudoconvex then* $\mathfrak{G}$ *is holomorphically convex and is a domain of holomorphy.*

The proof is tedious.

At present the concept of a holomorphic hull is only of theoretical interest, although it is possible to construct the holomorphic hull by adjoining Hartogs figures and it is conceivable that such a construction is realized with the help of a computer. Quicker methods have been found in only a few special cases, as for example, in connection with the Edge-of-the-Wedge theorem which in quantum field theory serves as a proof of the *PCT* theorem ("Under certain assumptions the product *PCT* of space reflection *P*, time reversal *T*, and charge conjunction *C* is a symmetry in the sense of field theory").

CHAPTER III

The Weierstrass Preparation Theorem

1. The Algebra of Power Series

In this chapter we shall deal more extensively than before with power series in $\mathbb{C}^n$. Our objective is to find a division algorithm for power series which will facilitate our investigation of the zero sets of holomorphic functions.

Let $\mathbb{N}_0 := \mathbb{N} \cup \{0\}$, $\mathbb{N}_0^n := \{(v_1, \ldots, v_n) : v_i \in \mathbb{N}_0\}$. We denote by $\mathbb{C}\{\mathfrak{z}\}$ the integral domain of formal power series about 0 with variables $z_1, \ldots, z_n$ and coefficients in $\mathbb{C}$. Let $\mathbb{R}_+^n$ be the set of n-tuples of positive real numbers. An element $f \in \mathbb{C}\{\mathfrak{z}\}$ can be written as $f = \sum_{v=0}^{\infty} a_v \mathfrak{z}^v$.

Def. 1.1. Let $\mathfrak{t} = (t_1, \ldots, t_n) \in \mathbb{R}_+^n$ and $f = \sum_{v=0}^{\infty} a_v \mathfrak{z}^v \in \mathbb{C}\{\mathfrak{z}\}$. The *norm of f with respect to* $\mathfrak{t}$ is the "number"

$$\|f\|_{\mathfrak{t}} := \sum_{v=0}^{\infty} |a_v| \mathfrak{t}^v \in \mathbb{R}_+ \cup \{0\} \cup \{\infty\}.$$

One can introduce a weak ordering on $\mathbb{R}_+^n$ if one defines $(t_1, \ldots, t_n) \leqslant (t_1^*, \ldots, t_n^*)$ if and only if $t_i \leqslant t_i^*$ for $i = 1, \ldots, n$. The norm of f relative to $\mathfrak{t}$ is then monotone in $\mathfrak{t}$: If $\mathfrak{t} \leqslant \mathfrak{t}^*$, then $\|f\|_{\mathfrak{t}} \leqslant \|f\|_{\mathfrak{t}^*}$.

Def. 1.2. A formal power series $f \in \mathbb{C}\{\mathfrak{z}\}$ is called *convergent* if $f(\mathfrak{z}) = \sum_{v=0}^{\infty} a_v \mathfrak{z}^v$ is convergent in a polycylinder about 0. (The definition of this convergence was given in Chapter I.) We denote the ring of convergent power series by H_n.

Theorem 1.1. *$f \in \mathbb{C}\{\mathfrak{z}\}$ is convergent if and only if there is a $\mathfrak{t} \in \mathbb{R}_+^n$ with $\|f\|_{\mathfrak{t}} < \infty$.*

PROOF

1. Let $f(\mathfrak{z}) = \sum_{\nu=0}^{\infty} a_\nu \mathfrak{z}^\nu$ be convergent in the polycylinder P. Then there exists a $\mathfrak{t} \in \mathbb{R}_+^n$ with $P_{\mathfrak{t}} \subset P$, and therefore $\|f\|_{\mathfrak{t}} < \infty$.

2. If $\|f\|_{\mathfrak{t}} = \sum_{\nu=0}^{\infty} |a_\nu| \mathfrak{t}^\nu$ is convergent, then $f(\mathfrak{z})$ is convergent at the point $\mathfrak{t}$, so f converges on all of $P_{\mathfrak{t}}$.

Def. 1.3. For $\mathfrak{t} \in \mathbb{R}_+^n$, let $B_{\mathfrak{t}} := \{f \in \mathbb{C}\{\mathfrak{z}\} : \|f\|_{\mathfrak{t}} < \infty\}$.

Def. 1.4. A set B is called a (*complex*) *Banach algebra* if

1. There are operations $+ : B \times B \to B$, $\cdot : \mathbb{C} \times B \to B$ and $\circ : B \times B \to B$ such that

    a. $(B, +, \cdot)$ is a $\mathbb{C}$-vector space
    b. $(B, +, \circ)$ is a commutative ring with 1
    c. For all $f, g \in B$ and all $c \in \mathbb{C}$, $c \cdot (f \circ g) = (c \cdot f) \circ g = f \circ (c \cdot g)$.

2. To every $f \in B$ a number $\|f\| \in \mathbb{R}_+ \cup \{0\}$ is assigned with the properties of a norm:

    a. $\|c \cdot f\| = |c| \cdot \|f\|$ for $c \in \mathbb{C}$, $f \in B$.
    b. $\|f + g\| \leqslant \|f\| + \|g\|$ for $f, g \in B$.
    c. $\|f\| = 0 \Leftrightarrow f = 0$.

3. $\|f \circ g\| \leqslant \|f\| \cdot \|g\|$ for $f, g \in B$.
4. As a normed $\mathbb{C}$-vector space, B is complete; that is, every Cauchy sequence (f_ν) of elements of B converges to an element f of B.

Theorem 1.2. *$B_{\mathfrak{t}}$ is a Banach algebra for every $\mathfrak{t} \in \mathbb{R}_+^n$.*

PROOF. Clearly $\mathbb{C}\{\mathfrak{z}\}$ is a $\mathbb{C}$-algebra. In order to show that $B_{\mathfrak{t}}$ is a $\mathbb{C}$-algebra, it suffices to show that $B_{\mathfrak{t}}$ is closed under the operations:

$$c \cdot \sum_{\nu=0}^{\infty} a_\nu \mathfrak{z}^\nu = \sum_{\nu=0}^{\infty} (c \cdot a_\nu) \mathfrak{z}^\nu,$$

$$\sum_{\nu=0}^{\infty} a_\nu \mathfrak{z}^\nu + \sum_{\nu=0}^{\infty} b_\nu \mathfrak{z}^\nu = \sum_{\nu=0}^{\infty} (a_\nu + b_\nu) \mathfrak{z}^\nu,$$

$$\left(\sum_{\nu=0}^{\infty} a_\nu \mathfrak{z}^\nu\right) \circ \left(\sum_{\mu=0}^{\infty} b_\mu \mathfrak{z}^\mu\right) = \sum_{\lambda=0}^{\infty} \left(\sum_{\nu+\mu=\lambda} a_\nu b_\mu\right) \mathfrak{z}^\lambda.$$

Straight-forward calculation shows that $\|\cdot\cdot\|_{\mathfrak{t}}$ is a norm with properties (2) and (3).

Now if $c \in \mathbb{C}$, $f \in B_{\mathfrak{t}}$, then $c \cdot f \in B_{\mathfrak{t}}$ because of (2a). If f and g are in $B_{\mathfrak{t}}$, then $f + g \in B_{\mathfrak{t}}$ because of (2b) and $f \circ g \in B_{\mathfrak{t}}$ because of (3).

It is clear that 1 lies in $B_{\mathfrak{t}}$. All that remains is to show completeness:

Let (f_λ) be a Cauchy sequence in $B_{\mathfrak{t}}$ with $f_\lambda(\mathfrak{z}) = \sum_{\nu=0}^{\infty} a_\nu^{(\lambda)}\mathfrak{z}^\nu$. Then for every $\varepsilon > 0$ there is an $n = n(\varepsilon) \in \mathbb{N}$ such that for all $\lambda, \mu \geqslant n$

$$\sum_{\nu=0}^{\infty} |a_\nu^{(\lambda)} - a_\nu^{(\mu)}|\mathfrak{t}^\nu = \|f_\lambda - f_\mu\|_{\mathfrak{t}} < \varepsilon.$$

Because $\mathfrak{t}^\nu \neq 0$ it follows from this that

$$|a_\nu^{(\lambda)} - a_\nu^{(\mu)}| < \frac{\varepsilon}{\mathfrak{t}^\nu} \qquad \text{for every} \quad \nu \in \mathbb{N}^n.$$

For fixed ν $(a_\nu^{(\lambda)})$ is therefore a Cauchy sequence in $\mathbb{C}$ which converges to the complex number a_ν.

Let $f(\mathfrak{z}) := \sum_{\nu=0}^{\infty} a_\nu \mathfrak{z}^\nu$. Let $\delta > 0$ be given. Then there exists an $n = n(\delta)$ such that

$$\sum_{\nu=0}^{\infty} |a_\nu^{(\lambda)} - a_\nu^{(\lambda+\mu)}|\mathfrak{t}^\nu < \frac{\delta}{2} \qquad \text{for } \lambda \geqslant n \quad \text{and} \quad \mu \in \mathbb{N}.$$

Let $I \subset \mathbb{N}_0^n$ be an arbitrary finite set. There always exists a $\mu \in \mathbb{N}$ for $\lambda \geqslant n$ such that $\sum_{\nu \in I} |a_\nu^{(\lambda+\mu)} - a_\nu|\mathfrak{t}^\nu < \delta/2$, and then

$$\sum_{\nu \in I} |a_\nu^{(\lambda)} - a_\nu|\mathfrak{t}^\nu \leqslant \sum_{\nu \in I} |a_\nu^{(\lambda)} - a_\nu^{(\lambda+\mu)}|\mathfrak{t}^\nu + \sum_{\nu \in I} |a_\nu^{(\lambda+\mu)} - a_\nu|\mathfrak{t}^\nu < \delta \qquad \text{for } \lambda \geqslant n.$$

In particular $\|f_\lambda - f\|_{\mathfrak{t}} = \sum_{\nu=0}^{\infty} |a_\nu^{(\lambda)} - a_\nu|\mathfrak{t}^\nu \leqslant \delta$. Thus (f_λ) converges to f. Because $\|f\|_{\mathfrak{t}} \leqslant \|f - f_\lambda\|_{\mathfrak{t}} + \|f_\lambda\|_{\mathfrak{t}}$, it follows that f lies in $B_{\mathfrak{t}}$. □

For the following we need some additional notation:
If $\nu = (\nu_1, \ldots, \nu_n) \in \mathbb{N}_0^n$, we set $\nu' := (\nu_2, \ldots, \nu_n)$; if $\mathfrak{t} = (t_1, \ldots, t_n) \in \mathbb{R}_+^n$, we set $\mathfrak{t}' := (t_2, \ldots, t_n)$; if $\mathfrak{z} = (z_1, \ldots, z_n) \in \mathbb{C}^n$, we set $\mathfrak{z}' := (z_2, \ldots, z_n)$. Then $\nu = (\nu_1, \nu')$, $\mathfrak{t} = (t_1, \mathfrak{t}')$, $\mathfrak{z} = (z_1, \mathfrak{z}')$, and we can write an element $f \in \mathbb{C}\{\mathfrak{z}\}$ in the form

$$f(\mathfrak{z}) = \sum_{\lambda=0}^{\infty} f_\lambda(\mathfrak{z}')z_1^\lambda \qquad \text{where } f_\lambda(\mathfrak{z}') = \sum_{\nu'=0}^{\infty} a_{\lambda,\nu'}(\mathfrak{z}')^{\nu'}.$$

This representation is called the expansion of f with respect to z_1. The following assertions hold:

1. $f = \sum_{\lambda=0}^{\infty} f_\lambda z_1^\lambda$ lies in $B_{\mathfrak{t}}$ if and only if every f_λ lies in $B'_{\mathfrak{t}} = B_{\mathfrak{t}} \cap \mathbb{C}\{\mathfrak{z}'\}$ and $\sum_{\lambda=0}^{\infty} \|f_\lambda\|_{\mathfrak{t}'}\, t_1^\lambda < \infty$.
2. For $s \in \mathbb{N}_0$, $\|z_1^s \cdot f\|_{\mathfrak{t}} = t_1^s \cdot \|f\|_{\mathfrak{t}}$.

3. If $f = \sum_{\nu=0}^{\infty} a_\nu \mathfrak{z}^\nu$ converges and $a_{0\ldots 0} = 0$, then for every $\varepsilon > 0$ there exists a $\mathfrak{t} \in \mathbb{R}_+^n$ with $\|f\|_\mathfrak{t} < \varepsilon$.

PROOF

(1) $$\|f\|_\mathfrak{t} = \sum_{\nu=0}^{\infty} |a_\nu|\mathfrak{t}^\nu = \sum_{\lambda=0}^{\infty}\left(\sum_{\nu'=0}^{\infty} |a_{\lambda,\nu'}|(\mathfrak{t}')^{\nu'}\right) t_1^\lambda = \sum_{\lambda=0}^{\infty} \|f_\lambda\|_{\mathfrak{t}'} t_1^\lambda.$$

(2) $$\|z_1^s \cdot f\|_\mathfrak{t} = \left\|\sum_{\lambda=0}^{\infty} f_\lambda z_1^{\lambda+s}\right\|_\mathfrak{t} = \sum_{\lambda=0}^{\infty} \|f_\lambda\|_{\mathfrak{t}'} t_1^{\lambda+s} = t_1^s \cdot \sum_{\lambda=0}^{\infty} \|f_\lambda\|_{\mathfrak{t}'} \cdot t_1^\lambda$$
$$= t_1^s \cdot \|f\|_\mathfrak{t}.$$

3. If one sets $f_i := \sum_{\nu_i > 0}^{\infty} a_{0\ldots 0,\nu_i\nu_{i+1}\ldots\nu_n} z_i^{\nu_i - 1} z_{i+1}^{\nu_{i+1}} \cdots z_n^{\nu_n}$, then $z_1 \cdot f_1 + \cdots + z_n \cdot f_n = f$ and $\|f\|_\mathfrak{t} = t_1 \cdot \|f_1\|_\mathfrak{t} + \cdots + t_n \cdot \|f_n\|_\mathfrak{t}$. If f is convergent, then there exists a $\mathfrak{t}_0 \in \mathbb{R}_+^n$ with $\|f\|_{\mathfrak{t}_0} < \infty$, and for $\mathfrak{t} \leqslant \mathfrak{t}_0$

$$\|f\|_\mathfrak{t} = \sum_{i=1}^{n} t_i \|f_i\|_\mathfrak{t} \leqslant \sum_{i=1}^{n} t_i \|f_i\|_{\mathfrak{t}_0} \leqslant \max(t_1, \ldots, t_n) \cdot \sum_{i=1}^{n} \|f_i\|_{\mathfrak{t}_0},$$

which becomes arbitrarily small. □

2. The Weierstrass Formula

Let a fixed element $\mathfrak{t} \in \mathbb{R}_+^n$ be chosen. When no confusion is possible we shall write B in place of $B_\mathfrak{t}$, B' in place of $B'_\mathfrak{t}$, and $\|f\|$ in place of $\|f\|_\mathfrak{t}$.

Theorem 2.1 (Weierstrass formula). *Let $g = \sum_{\lambda=0}^{\infty} g_\lambda z_1^\lambda \in B$, let there be a $s \in \mathbb{N}_0$ for which g_s is a unit in B', and let there be an ε with $0 < \varepsilon < 1$ such that $\|z_1^s - g \cdot g_s^{-1}\| < \varepsilon \cdot t_1^s$. Then for every $f \in B$ there exists exactly one $q \in B$ and one $r \in B'[z_1]$ with $\deg(r) < s$ such that $f = q \cdot g + r$ ("Division with remainder"). Furthermore,*

(1) $$\|g_s \cdot q\| \leqslant t_1^{-1} \cdot \|f\| \cdot \frac{1}{1 - \varepsilon}$$

(2) $$\|r\| \leqslant \|f\| \cdot \frac{1}{1 - \varepsilon}.$$

PROOF. Let $h := -(z_1^s - g \cdot g_s^{-1})$. Then $\|h\| < \varepsilon \cdot t_1^s$ and $g \cdot g_s^{-1} = z_1^s + h$.

Let us start with an arbitrary $f \in B$ and inductively construct sequences (f_λ), (q_λ), and (r_λ). We set $f_0 := f$.

Suppose $f_0, \ldots, f_\lambda$ have been constructed. There exists a representation $f_\lambda = \sum_{\kappa=0}^{\infty} f_{\lambda,\kappa} z_1^\kappa$, and we define

$$q_\lambda := \sum_{\kappa=s}^{\infty} f_{\lambda,\kappa} z_1^{\kappa-s}, \qquad r_\lambda := \sum_{\kappa=0}^{s-1} f_{\lambda,\kappa} z_1^\kappa \qquad \text{and} \qquad f_{\lambda+1} := (z_1^s - g \cdot g_s^{-1}) q_\lambda.$$

Then $f_\lambda = z_1^s \cdot q_\lambda + r_\lambda$ and $f_{\lambda+1} = -h \cdot q_\lambda = f_\lambda - r_\lambda - g g_s^{-1} \cdot q_\lambda$. Clearly the following estimates hold:

$$\begin{aligned} \|r_\lambda\| &\leqslant \|f_\lambda\|,\\ \|q_\lambda\| &\leqslant t_1^{-s}\|f_\lambda\|,\\ \|f_{\lambda+1}\| &\leqslant \|h\| \cdot \|q_\lambda\| < \varepsilon \cdot \|f_\lambda\|, \qquad \text{so} \qquad \|f_\lambda\| < \varepsilon^\lambda \cdot \|f\|. \end{aligned}$$

Let $q := \sum_{\lambda=0}^{\infty} g_s^{-1} \cdot q_\lambda$ and $r := \sum_{\lambda=0}^{\infty} r_\lambda$. Then

$$\|g_s^{-1} q_\lambda\| < \varepsilon^\lambda t_1^{-s} \|g_s^{-1}\| \cdot \|f\| \qquad \text{and} \qquad \|r_\lambda\| < \varepsilon^\lambda \|f\|.$$

By the comparison test the series converge. Since each r_λ is a polynomial with $\deg(r_\lambda) \leqslant s - 1$, it follows that r is a polynomial with $\deg(r) \leqslant s - 1$. Since the series $\sum_{\lambda=0}^{\infty} f_\lambda$ also converges,

$$\begin{aligned} f = f_0 &= \sum_{\lambda=0}^{\infty} f_\lambda - \sum_{\lambda=0}^{\infty} f_{\lambda+1} = \sum_{\lambda=0}^{\infty} (f_\lambda - f_{\lambda+1}) = \sum_{\lambda=0}^{\infty} (r_\lambda + g g_s^{-1} q_\lambda)\\ &= g \cdot \sum_{\lambda=0}^{\infty} g_s^{-1} \cdot q_\lambda + \sum_{\lambda=0}^{\infty} r_\lambda = g \cdot q + r. \end{aligned}$$

The estimates now follow readily.

$$\text{(1)} \quad \|g_s q\| = \left\| \sum_{\lambda=0}^{\infty} q_\lambda \right\| \leqslant \sum_{\lambda=0}^{\infty} \|q_\lambda\| \leqslant t_1^{-s}\|f\| \cdot \sum_{\lambda=0}^{\infty} \varepsilon^\lambda = t_1^{-s} \cdot \|f\| \cdot \frac{1}{1-\varepsilon}.$$

$$\text{(2)} \quad \|r\| \leqslant \sum_{\lambda=0}^{\infty} \|r_\lambda\| \leqslant \|f\| \cdot \sum_{\lambda=0}^{\infty} \varepsilon^\lambda = \|f\| \cdot \frac{1}{1-\varepsilon}.$$

It still remains to show uniqueness.

Let there be two expressions of the form

$$f = q_1 g + r_1 = q_2 g + r_2.$$

Then $0 = (q_1 - q_2) \cdot g + (r_1 - r_2)$. From the representation $g = g_s(z_1^s + h)$ with $\|h\| < \varepsilon \cdot t_1^s$ we obtain

$$0 = (q_1 - q_2) g_s z_1^s + (q_1 - q_2) g_s h + (r_1 - r_2)$$

and

$$\begin{aligned} \|(q_1 - q_2) g_s z_1^s\| &\leqslant \|(q_1 - q_2) g_s z_1^s + (r_1 - r_2)\|\\ &= \|(q_1 - q_2) g_s \cdot h\| \leqslant \varepsilon \cdot t_1^s \cdot \|(q_1 - q_2) g_s\|\\ &= \varepsilon \cdot \|(q_1 - q_2) g_s z_1^s\|. \end{aligned}$$

Because $\varepsilon < 1$, $(q_1 - q_2) g_s z_1^s = 0$. Therefore $q_1 = q_2$ and $r_1 = r_2$. □

Corollary. *If the assumptions of Theorem 2.1 are satisfied and if in addition* $f \in B'[z_1]$, $g \in B'[z_1]$ *and* $\deg(g) = s$, *then* $q \in B'[z_1]$ *and* $\deg(q) = \max(-1, \deg(f) - s)$ [*with* $\deg(0) := -1$].

PROOF. Let $d := \deg(f)$. For $d < s$ one has the decomposition $f = 0 \cdot g + f$, let therefore $d \geqslant s$. Now $-1 \leqslant \deg(q_\lambda) \leqslant \max(-1, \deg(f_\lambda) - s)$ and $\deg(f_0) \leqslant d$. If we assume that $d_\nu := \deg(f_\nu) \leqslant d$ for $\nu = 0, \ldots, \lambda$, then $\deg(q_\lambda) \leqslant d - s$, therefore

$$\begin{aligned}\deg(f_{\lambda+1}) &= \deg(f_\lambda - r_\lambda - g g_s^{-1} q_\lambda)\\ &\leqslant \max(\deg(f_\lambda), \deg(r_\lambda), \deg(q_\lambda) + s)\\ &\leqslant \max(d, s - 1, (d - s) + s) \leqslant d.\end{aligned}$$

Hence $\deg(q_\lambda) \leqslant d - s$ for all λ, and $\deg(q) \leqslant d - s$. On the other hand, the representation $f = q \cdot g + r$ gives

$$\deg(f) \leqslant \max(\deg(q) + s, s - 1) = \deg(q) + s,$$

therefore $d - s \leqslant \deg(q)$. All together one obtains: $\deg(q) = \max(-1, d - s)$. □

Theorem 2.2. *If B is a Banach algebra, $f \in B$ and $\|1 - f\| < 1$, then f is a unit in B and $\|f^{-1}\| \leqslant 1/(1 - \|1 - f\|)$.*

PROOF. Let $g := \sum_{\lambda=0}^{\infty} (1 - f)^\lambda$, $\varepsilon := \|1 - f\|$. Then $0 \leqslant \varepsilon < 1$ and $\sum_{\lambda=0}^{\infty} \varepsilon^\lambda$ dominates g. Therefore the series $\sum_{\lambda=0}^{\infty} (1 - f)^\lambda$ converges and g is an element of B. Moreover $f \cdot g = (1 - (1 - f)) \cdot g = \sum_{\lambda=0}^{\infty} (1 - f)^\lambda - \sum_{\lambda=0}^{\infty} (1 - f)^{\lambda+1} = (1 - f)^0 = 1$, and $\|g\| \leqslant \sum_{\lambda=0}^{\infty} \varepsilon^\lambda = 1/(1 - \varepsilon)$. □

Def. 2.1. Let $s \in \mathbb{N}_0$. An element $g = \sum_{\lambda=0}^{\infty} g_\lambda z_1^\lambda \in B$ satisfies the Weierstrass condition (W-condition) at s if:

a. g_s is a unit in B'.
b. $\|z_1^s - g g_s^{-1}\| < \frac{1}{2} t_1^s$.

Theorem 2.3 (Weierstrass preparation theorem). *If $g \in B$ satisfies the W-condition at s, then there exists exactly one normalized polynomial $\omega \in B'[z_1]$ with $\deg(\omega) = s$ and one unit $e \in B$ such that $g = e \cdot \omega$.*

PROOF. We apply the Weierstrass formula to $f = z_1^s$: There are uniquely determined elements $q \in B$ and $r \in B'[z_1]$ with $z_1^s = q \cdot g + r$ and $\deg(r) < s$ (we take an $\varepsilon < \frac{1}{2}$ which satisfies the conditions of Theorem 2.1). But then $z_1^s - g g_s^{-1} = (q - g_s^{-1}) g + r$ is a decomposition of $z_1^s - g g_s^{-1}$ in the sense of Theorem 2.1; therefore we can employ formula (1):

$$\|q g_s - 1\| = \|(q - g_s^{-1}) \cdot g_s\| \leqslant t_1^{-s} \|z_1^s - g g_s^{-1}\| \frac{1}{1 - \varepsilon} < \frac{\varepsilon}{1 - \varepsilon} < 1.$$

That means that $q \cdot g_s$ and hence q is unit in B.

Let $e := q^{-1}$ and $\omega := z_1^s - r$. Then ω is a normalized polynomial with $\deg(\omega) = s$, and $e \cdot \omega = q^{-1}(z_1^s - r) = g$. If $g = e_1(z_1^s - r_1) = e_2(z_1^s - r_2)$, then

$$g \cdot e_1^{-1} + r_1 = z_1^s = g \cdot e_2^{-1} + r_2;$$

but on the other hand, in the decomposition $z_1^s = q \cdot g + r$, the elements q and r are uniquely determined. Therefore $e_1 = e_2$ and $r_1 = r_2$. □

Corollary. *If g is a polynomial in z_1, then e is also a polynomial in z_1.*

PROOF. If we use formula (2) in the decomposition $z_1^s - gg_s^{-1} = (q - g_s^{-1}) \cdot g + r$, we get

$$\|r\| \leqslant \|z_1^s - gg_s^{-1}\| \cdot \frac{1}{1-\varepsilon} < t_1^s \cdot \frac{\varepsilon}{1-\varepsilon} < t_1^s.$$

Because $\omega_s = 1$ it is also true that

$$\|z_1^s - \omega\omega_s^{-1}\| = \|z_1^s - \omega\| = \|r\| < t_1^s;$$

that is, ω satisfies the conditions of Theorem 2.1. Since $g = e \cdot \omega$ is a decomposition in the sense of the Weierstrass formula, the proposition follows from the corollary of that theorem. □

Comment. The Weierstrass preparation theorem serves as a "preparation of the examination of the zeroes of holomorphic functions".

A function holomorphic in a polycylinder will be represented by a convergent power series g. If there exists a decomposition $g = e \cdot \omega$ with a unit e and "pseudo polynomial" $\omega = z_1^s + A_1(\mathfrak{z}')z_1^{s-1} + \cdots + A_s(\mathfrak{z}')$, then g and ω have the same zeroes. However, the examination of ω is simpler than that of g.

3. Convergent Power Series

Def. 3.1. $g \in \mathbb{C}\{\mathfrak{z}\}$ is said to be *regular in z_1* if $g(z_1, 0, \ldots, 0)$ does not vanish identically.

If $g = \sum_{\lambda=0}^{\infty} g_\lambda z_1^\lambda$ is regular in z_1, then $\operatorname{ord}(g)$ is that number $s \in \mathbb{N}_0$ for which $g_0(0) = \cdots = g_{s-1}(0) = 0$, $g_s(0) \neq 0$.

We then say that g is *regular of order s in z_1*.

Theorem 3.1. For $g_1, g_2 \in \mathbb{C}\{\mathfrak{z}\}$

1. *$g_1 \cdot g_2$ is regular in z_1 if and only if g_1 and g_2 are regular in z_1,*
2. $\operatorname{ord}(g_1 \cdot g_2) = \operatorname{ord}(g_1) + \operatorname{ord}(g_2)$.

PROOF. $(g_s \cdot g_2)(z_1, 0) = g_1(z_1, 0) \cdot g_2(z_1, 0)$. Since $\mathbb{C}\{z_1\}$ is an integral domain, (1) holds; (2) is obtained by multiplying out. □

Theorem 3.2. *Let $g \in \mathbb{C}\{\mathfrak{z}\}$ be convergent and regular of order s in z_1. Then for every $\varepsilon > 0$ and every $\mathfrak{t}_0 \in \mathbb{R}_+^n$ there exists a $\mathfrak{t} \leqslant \mathfrak{t}_0$ such that g lies in $B_\mathfrak{t}$, g_s is a unit in $B'_\mathfrak{t}$, and $\|z_1^s - g g_s^{-1}\|_\mathfrak{t} \leqslant \varepsilon \cdot t_1^s$.*

Proof. Let $g = \sum_{\lambda=0}^{\infty} g_\lambda z_1^\lambda$ be the expansion of g with respect to z_1. Then $g_\lambda(0) = 0$ for $\lambda = 0, 1, \ldots, s-1$ and $g_s(0) \neq 0$.

1. Since g is convergent, there exists a $\mathfrak{t}_1 = (t_1^{(1)}, \ldots, t_n^{(1)}) \in \mathbb{R}_+^n$ with $\|g\|_{\mathfrak{t}_1} = \sum_{\lambda=0}^{\infty} \|g_\lambda\|_{\mathfrak{t}_1} \cdot t_1^{(1)\lambda} < \infty$; therefore $g_\lambda \in B'_{\mathfrak{t}_1}$. In particular then

$$\frac{g_s(\mathfrak{z}')}{g_s(0)} - 1 = :f(\mathfrak{z}') \in B'_{\mathfrak{t}_1},$$

and since $f(0) = 0$, there exists a $\mathfrak{t}_2 \leqslant \mathfrak{t}_1$ such that, for all $\mathfrak{t} \leqslant \mathfrak{t}_2$, $\|f\|_\mathfrak{t} < 1$.

$g_s/g_s(0)$ (and hence g_s also) is therefore a unit in $B'_\mathfrak{t}$. Moreover, it is clear that g lies in $B_\mathfrak{t}$.

2. Let $h := z_1^s - g \cdot g_s^{-1}$. Then $h \in B_\mathfrak{t}$ for all $\mathfrak{t} \leqslant \mathfrak{t}_2$, and we can write $h = \sum_{\lambda=0}^{\infty} d_\lambda z_1^\lambda$ with $d_s = 0$, $d_\lambda = -g_\lambda g_s^{-1}$ for $\lambda \neq s$ and $d_\lambda(0) = 0$ for $\lambda = 0, 1, \ldots, s-1$. For $\mathfrak{t} \leqslant \mathfrak{t}_2$

$$\left\| \sum_{\lambda=s+1}^{\infty} d_\lambda z_1^\lambda \right\|_\mathfrak{t} = \left\| z_1^{s+1} \cdot \sum_{\lambda=s+1}^{\infty} d_\lambda z_1^{\lambda-s-1} \right\|_\mathfrak{t}$$

$$= t_1^{s+1} \cdot \left\| \sum_{\lambda=s+1}^{\infty} d_\lambda z_1^{\lambda-s-1} \right\|_\mathfrak{t} \leqslant t_1^{s+1} \left\| \sum_{\lambda=s+1}^{\infty} d_\lambda z_1^{\lambda-s-1} \right\|_{\mathfrak{t}_2}.$$

3. If t_1 is sufficiently small, then

$$t_1 \cdot \left\| \sum_{\lambda=s+1}^{\infty} d_\lambda z_1^{\lambda-s-1} \right\|_{\mathfrak{t}_2} < \tfrac{1}{2}\varepsilon;$$

therefore

$$\left\| \sum_{\lambda=s+1}^{\infty} d_\lambda z_1^\lambda \right\|_\mathfrak{t} \leqslant \tfrac{1}{2}\varepsilon \cdot t_1^s.$$

Because $d_\lambda(0) = 0$ for $\lambda = 0, 1, \ldots, s-1$ we can choose for t_1 a suitable $\mathfrak{t}'$ so small that

$$\sum_{\lambda=0}^{s-1} \|d_\lambda\|_{\mathfrak{t}'} t_1^\lambda < \tfrac{1}{2}\varepsilon \cdot t_1^s.$$

For $\mathfrak{t} = (t_1, \mathfrak{t}')$ it then follows that

$$\|h\|_\mathfrak{t} \leqslant \sum_{\lambda=0}^{s-1} \|d_\lambda\|_{\mathfrak{t}'} \cdot t_1^\lambda + \left\| \sum_{\lambda=s+1}^{\infty} d_\lambda z_1^\lambda \right\|_\mathfrak{t} \leqslant \varepsilon \cdot t_1^s. \qquad \square$$

Remark. In a similar manner one can show that if $g_1, \ldots, g_N \in \mathbb{C}\{\mathfrak{z}\}$ are convergent power series and each g_i is regular of order s_i in z_1, $i = 1, \ldots, N$, then for every $\varepsilon > 0$ there is an arbitrarily small $\mathfrak{t} \in \mathbb{R}^n_+$ for which

$$g_i \in B_{\mathfrak{t}}, (g_i)_{s_i} \text{ is a unit in } B'_{\mathfrak{t}}$$

and

$$\|z_1^{s_i} - g_i (g_i)_{s_i}^{-1}\| \leqslant \varepsilon \cdot t_1^{s_i}.$$

The problem of what to do if g is not regular in z_1 now arises. We shall show that if g does not vanish identically one can always find a biholomorphic mapping which takes g into a power series g' regular in z_1.

Let $A(0)$ be the set of all holomorphic functions defined in a (not fixed) neighborhood of $0 \in \mathbb{C}^n$, let $\Phi: A(0) \to H_n$ with $\Phi(f) = (f)_0$ be the mapping which associates each local holomorphic function f with its Taylor series expansion about the origin. Φ is clearly surjective and commutes with addition and multiplication in $A(0)$. If U_1, U_2 are open neighborhoods of $0 \in \mathbb{C}^n$, $\sigma: U_1 \to U_2$ a biholomorphic mapping with $\sigma(0) = 0$, then for f, $g \in A(0)$ with $(f)_0 = (g)_0$ we have

$$(f \circ \sigma)_0 = (g \circ \sigma)_0.$$

Therefore the mapping $\sigma^*: H_n \to H_n$ with $\sigma^*((f)_0) = (f \circ \sigma)_0$ is well defined and moreover

1. $\sigma^*((f_1)_0 + (f_2)_0) = \sigma^*((f_1)_0) + \sigma^*((f_2)_0)$
2. $\sigma^*((f_1)_0 \cdot (f_2)_0) = \sigma^*((f_1)_0) \cdot \sigma^*((f_2)_0)$
3. $\mathrm{id}^*((f)_0) = (f)_0$
4. $(\sigma \circ \rho)^*((f)_0) = (\rho^* \circ \sigma^*)((f)_0)$
5. σ^* is bijective, and $(\sigma^*)^{-1} = (\sigma^{-1})^*$.

σ^* is therefore always a ring isomorphism. It is customary to write $(f)_0 \circ \sigma$ in place of $\sigma^*((f)_0)$.

Def. 3.2. Let $\mathfrak{c} = (c_2, \ldots, c_n) \in \mathbb{C}^{n-1}$. Then $\sigma_{\mathfrak{c}}: \mathbb{C}^n \to \mathbb{C}^n$ with $\sigma_{\mathfrak{c}}(w_1, \ldots, w_n) := (w_1, w_2 + c_2 w_1, \ldots, w_n + c_n w_1)$ is called a *shearing*. Let the set of all shearings be denoted by $\sum$.

Theorem 3.3. *$\sum$ is an abelian group of biholomorphic mappings of $\mathbb{C}^n$ onto itself.*

PROOF. Linear shearings are, of course, holomorphic. It follows from the equalities

$$\sigma_{\mathfrak{c}_1 + \mathfrak{c}_2} = \sigma_{\mathfrak{c}_1} \circ \sigma_{\mathfrak{c}_2}$$

and

$$\sigma_{\mathfrak{c}} \circ \sigma_{-\mathfrak{c}} = \sigma_0 = \mathrm{id}_{\mathbb{C}^n}$$

that $\sum$ is an abelian group and that shearings are biholomorphic. □

Theorem 3.4. *Let $g \in H_n$, $g \neq 0$. Then there exists a shearing $\sigma \in \sum$ such that $g \circ \sigma$ is regular in z_1.*

PROOF

1. Let $g = \sum_{\nu=0}^{\infty} a_\nu \mathfrak{z}^\nu = \sum_{\lambda=0}^{\infty} p_\lambda(\mathfrak{z})$ with $p_\lambda(\mathfrak{z}) = \sum_{|\nu|=\lambda} a_\nu \mathfrak{z}^\nu$ be the expansion of g into a series of homogeneous polynomials, $\lambda_0 := \min\{\lambda \in \mathbb{N}_0 : p_\lambda \neq 0\}$. Then for every shearing σ, $g \circ \sigma = \sum_{\lambda=\lambda_0}^{\infty} (p_\lambda \circ \sigma)$ is the expansion of $g \circ \sigma$ into a series of homogeneous polynomials.

2.

$$\begin{aligned} p_\lambda \circ \sigma(w_1, 0, \ldots, 0) &= \sum_{|\nu|=\lambda} a_\nu w_1^{\nu_1}(c_2 w_1)^{\nu_2} \cdots (c_n w_1)^{\nu_n} \\ &= \sum_{|\nu|=\lambda} a_\nu c_2^{\nu_2} \cdots c_n^{\nu_n} w_1^\lambda = \tilde{p}_\lambda(c_2, \ldots, c_n) \cdot w_1^\lambda \end{aligned}$$

with $\tilde{p}_\lambda$ a polynomial in $(n-1)$ variables. Since by definition not all the coefficients of $\tilde{p}_\lambda$ vanish, there are complex numbers $c_2^{(0)}, \ldots, c_n^{(0)}$ such that $\tilde{p}_{\lambda_0}(c_2^{(0)}, \ldots, c_n^{(0)}) \neq 0$. Let

$$\sigma_0 := \sigma_{(c_2^{(0)}, \ldots, c_n^{(0)})}.$$

Then

$$g \circ \sigma_0(w_1, 0, \ldots, 0) = \sum_{\lambda=\lambda_0}^{\infty} (p_\lambda \circ \sigma_0)(w_1, 0, \ldots, 0) = \sum_{\lambda=\lambda_0}^{\infty} \tilde{p}_\lambda(c_2^{(0)}, \ldots, c_n^{(0)}) w_1^\lambda,$$

and it is clear that $g \circ \sigma_0$ is regular of order λ_0 in w_1. □

Remark. One can show that if $g_1, \ldots, g_N$ are non-vanishing convergent power series, then there is a shearing $\sigma \in \sum$ such that all $g_i \circ \sigma$ are regular in z_1.

Theorem 3.5 (Weierstrass formula for convergent power series). *Let $g \in H_n$ be regular of order s in z_1. Then for every $f \in H_n$ there is exactly one $q \in H_n$ and one $r \in H_{n-1}[z_1]$ with $\deg(r) < s$ such that $f = q \cdot g + r$.*

PROOF

1. There is a $\mathfrak{t} \in \mathbb{R}_+^n$ such that f and g lie in $B_\mathfrak{t}$ and g_s is a unit in $B_\mathfrak{t}$ and $\|z_1^s - g g_s^{-1}\|_\mathfrak{t} \leqslant \varepsilon \cdot t_1^s$ for an ε with $0 < \varepsilon < 1$. The existence of q and r then follows from the earlier Weierstrass formula.

2. Let two decompositions of f be given:

$$f = q_1 \cdot g + r_1 = q_2 \cdot g + r_2.$$

We can find a $\mathfrak{t} \in \mathbb{R}_+^n$ such that f, q_1, q_2, r_1, r_2 lie in $B_\mathfrak{t}$ and g satisfies the W-condition in $B_\mathfrak{t}$. From the Weierstrass formula for $B_\mathfrak{t}$ it follows that $q_1 = q_2$ and $r_1 = r_2$. □

We also have the

Corollary. *If f and g are polynomials in z_1 with $\deg(g) = s$, then q is also a polynomial.*

Theorem 3.6 (Weierstrass preparation theorem for power series). *Let $g \in H_n$ be regular of order s in t_1. Then there exists a unit $e \in H_n$ and a normalized polynomial $\omega \in H_{n-1}[z_1]$ of degree s with*

$$g = e \cdot \omega$$

PROOF

1. There exists a $\mathfrak{t} \in \mathbb{R}_+^n$ such that g satisfies the W-condition in $B_\mathfrak{t}$. The existence of the decomposition "$g = e \cdot \omega$" therefore follows directly from the Weierstrass preparation theorem for $B_\mathfrak{t}$.

2. ω has the form $\omega = z_1^s - r$, where $r \in H_{n-1}[z_1]$ and $\deg(r) < s$. If there exist two representations $g = e_1(z_1^s - r_1) = e_2(z_1^s - r_2)$, then it follows that $e_1^{-1} \cdot g + r_1 = z_1^s = e_2^{-1} \cdot g + r_2$. But in this case the Weierstrass formula says that $e_1^{-1} = e_2^{-1}$ and $r_1 = r_2$. Therefore $e_1 = e_2$ and $\omega_1 = \omega_2$ (for $\omega_\lambda := z_1^s - r_\lambda$). □

Corollary. *If g is a polynomial in z_1, then e is also a polynomial in z_1.*

Theorem 3.7. *$f \in H_n$ is a unit if and only if $f(0) \neq 0$.*

PROOF

1. If $f \in H_n$ is a unit, then there exists a $g \in H_n$ with $f \cdot g = 1$. In particular $f(0) \cdot g(0) = 1$, so $f(0) \neq 0$.

2. If $f \in H_n$ and $f(0) \neq 0$, then $g := [f/f(0)] - 1$ lies in H_n and $g(0) = 0$. Therefore there is a $\mathfrak{t} \in \mathbb{R}_+^n$ such that $\|g\|_\mathfrak{t} < 1$, which means that $f/f(0)$ is a unit in $B_\mathfrak{t}$. But then f is a unit in H_n also. □

Remark. If the function $g \in A(0)$ does not vanish identically near 0, then there is a shearing σ such that $(g \circ \sigma)_0$ is regular in z_1. By the Weierstrass preparation theorem we can find a decomposition $(g \circ \sigma)_0 = e \cdot \omega$ with $e(0) \neq 0$ and $\omega = z_1^s + A_1(\mathfrak{z}')z_1^{s-1} + \cdots + A_s(\mathfrak{z}') \in H_{n-1}[z_1]$. g then has a zero at 0 if and only if $\omega(0) = 0$, and that is the case if and only if $A_s(0) = 0$. But A_s lies in H_{n-1}. The Weierstrass preparation theorem therefore allows an inductive examination of the zeroes of holomorphic functions.

4. Prime Factorization

In the following let I always be an arbitrary integral domain and $I^* := I - \{0\}$.

We quote some facts from elementary number theory (see for example, v.d. Waerden, Vol. I.).

Def. 4.1. Let $a \in I^*$, $b \in I$. We say that *a divides b* (symbolically $a|b$) if there exists a $c \in I$ such that $b = a \cdot c$.

Def. 4.2

1. Let $a \in I^*$, a not a unit. a is called *indecomposable* if it follows from $a = a_1 \cdot a_2$ with $a_1, a_2 \in I^*$ that a_1 is a unit or a_2 is a unit.

2. Let $a \in I^*$ not be a unit. a is called *prime* if $a|a_1 \cdot a_2$ implies that $a|a_1$ or $a|a_2$. It is true that a prime is always indecomposable. The converse is not always the case, but does hold in some important special cases, such as the ring of integers.

Def. 4.3. I is called *unique factorization domain* (or *UFD*) if every $a \in I^*$ which is not a unit can be written as a product of finitely many primes. This decomposition is then determined uniquely up to order and multiplication by units.

In unique factorization domains every indecomposable element is prime.

Theorem 4.1. *If k is a field, then $k[X]$ is a unique factorization domain.*

PROOF. The euclidean algorithm is valid in $k[X]$, hence $k[X]$ is a principal ideal domain. But every principal ideal domain is a unique factorization domain. (Details are found in van der Waerden.) □

Def. 4.4

1. Let I be an integral domain. Then *the quotient field of I*, denoted by $Q(I)$, is defined by

$$Q(I) := \left\{\frac{a}{b} : a, b \in I, b \neq 0\right\}.$$

2. If $I[X]$ is the polynomial ring over I, then we denote the set of normalized polynomials of $I[X]$ by $I^0[X]$.

Remark. $I^0[X]$ is closed with respect to multiplication but not with respect to addition. Therefore $I^0[X]$ is not a ring. One can, however, consider prime factorization in $I^0[X]$.

Theorem 4.2. *Let I be a unique factorization domain, $Q := Q(I)$ the quotient field. Furthermore, let $\omega_1, \omega_2 \in Q^0[X]$, $\omega \in I^0[X]$, and $\omega = \omega_1 \cdot \omega_2$. Then ω_1, ω_2 also lie in $I^0[X]$.*

PROOF. For $\lambda = 1, 2$, $\omega_\lambda = X^{s_\lambda} + A_{\lambda,1}X^{s_\lambda - 1} + \cdots + A_{\lambda,s}$ with $A_{\lambda,\mu} \in Q$. Therefore there exists a $d_\lambda \in I$ such that $d_\lambda \cdot \omega_\lambda \in I[X]$. In the coefficients of $d_\lambda \cdot \omega_\lambda$ any common divisors are cancelled.

Now let $d := d_1 \cdot d_2$. We assume that there exists a prime element p which divides d. It follows that $d \nmid d_\lambda \cdot \omega_\lambda$ for $\lambda = 1, 2$. Let μ_λ be minimal with the property that $p \nmid d_\lambda A_{\lambda,\mu_\lambda}$. Now $(d_1\omega_1) \cdot (d_2\omega_2) = \cdots + X^{\mu_1+\mu_2} \cdot [(d_1 \cdot A_{1,\mu_1}) \cdot (d_2 A_{2,\mu_2}) + \text{terms divisible by } p] + \cdots$. Therefore the coefficient of $X^{\mu_1+\mu_2}$ is not divisible by p, hence $(d_1 \cdot \omega_1)(d_2 \cdot \omega_2)$ is not divisible by p, which clearly is a contradiction of the fact that $(d_1 \cdot \omega_1)(d_2 \cdot \omega_2) = d \cdot \omega$ with $\omega \in I^0[X]$ and $p|d$.

Therefore d has no prime divisors, that is $d = d_1 \cdot d_2$ is a unit. Hence d_λ, $\lambda = 1, 2$ are units in I. It follows that $\omega_\lambda = d_\lambda^{-1} \cdot d_\lambda \cdot \omega_\lambda \in I[X]$ and hence $\omega_\lambda \in I^0[X]$. □

Theorem 4.3 (Gauss' lemma). *If I is a unique factorization domain, then so is $I^0[X]$; that is, every element of $I^0[X]$ is a product of finitely many prime elements of $I^0[X]$. (Only the multiplicative structure plays a role, so one can employ the notion of "factorization" in $I^0[X]$.)*

Proof

1. Let $\omega \in I^0[X] \subset Q[X]$. Then $\omega = \omega_1 \cdot \omega_2 \cdots \omega_\ell$ with $\omega_\lambda \in Q[X]$ prime (Theorem 4.1). In each case let a_λ be the coefficient of the term of highest degree in ω_λ. Then clearly $1 = a_1 \cdots a_\ell$. Therefore

$$\omega = \frac{\omega}{a_1 \cdots a_1} = \left(\frac{\omega_1}{a_1}\right) \cdots \left(\frac{\omega_1}{a_1}\right).$$

Without loss of generality we may assume, then, that the ω_λ are normalized.

2. By induction on ℓ it follows from Theorem 4.2 that all ω_λ lie in $I^0[X]$. It still remains to be shown that the ω_λ are also prime in $I^0[X]$. Let $\omega_\lambda | \omega' \cdot \omega''$ with $\omega', \omega'' \in I^0[X]$.

This relation also holds in $Q[X]$ and there either $\omega_\lambda|\omega'$ or $\omega_\lambda|\omega''$. Say $\omega_\lambda|\omega'$. Then $\omega' = \omega_\lambda \cdot \omega'_\lambda$ with $\omega'_\lambda \in Q[X]$ and hence $Q^0[X]$. By Theorem 4.2 it further follows that $\omega'_\lambda \in I^0[X]$. Therefore ω_λ is prime in $I^0[X]$. □

We now apply these results to the special case $I = H_n$.

Def. 4.5. Let $f \in H_n$, $f = \sum_{\lambda=0}^{\infty} p_\lambda$ be the expansion of f as a series of homogeneous polynomials. Then one defines the *order of f* by the number $\operatorname{ord}(f) := \min\{\lambda \in \mathbb{N}_0 : p_\lambda \neq 0\}$, $\operatorname{ord}(0) := \infty$.

Then:

1. $\operatorname{ord}(f) \geqslant 0$.
2. $\operatorname{ord}(f_1 \cdot f_2) = \operatorname{ord}(f_1) + \operatorname{ord}(f_2)$
3. $\operatorname{ord}(f_1 + f_2) \geqslant \min(\operatorname{ord}(f_1), \operatorname{ord}(f_2))$.
4. f is a unit if and only if $\operatorname{ord}(f) = 0$.

Theorem 4.4. *H_n is a unique factorization domain.*

Proof. We proceed by induction on n. For $n = 0$, $H_n = \mathbb{C}$ is a field, and the statement is trivial. Suppose the proposition has been proved for $n - 1$.

1. If $f \in H_n$ is not a unit, and $f = f_1 \cdot f_2$ a proper decomposition, then $\operatorname{ord}(f) = \operatorname{ord}(f_1) + \operatorname{ord}(f_2)$; therefore the orders of the factors are strictly smaller. Consequently we can decompose f into a finite number of indecomposable terms: $f = f_1 \cdots f_\ell$.

2. Now let f be indecomposable, f_1, f_2 arbitrary and $\neq 0$, and $f|f_1 \cdot f_2$. A shearing makes $f_1 \circ \sigma$, $f_2 \circ \sigma$, and $f \circ \sigma$ regular in z_1. Thus it follows that there exists a decomposition $f \circ \sigma = e \circ \omega$ and $f_\nu \circ \sigma = e_\nu \cdot \omega_\nu$, $\nu = 1, 2$, in the sense of Theorem 3.7. Since $f|f_1 \cdot f_2$ we have $(f \circ \sigma)|(f_1 \circ \sigma) \cdot (f_2 \circ \sigma)$; therefore $\omega|\omega_1 \cdot \omega_2$ in H_n. There exists a $q \in H_n$ with $q \cdot \omega = \omega_1 \cdot \omega_2$. By the Weierstrass formula (Theorem 3.6) q is uniquely determined, and by the corollary $q \in H_{n-1}^0[z_1]$.

Since f is indecomposable, so is $f \circ \sigma$ and thus ω is indecomposable (in $H_{n-1}^0[z_1]$). By the induction hypothesis H_{n-1} is a unique factorization domain, and by Gauss' lemma so is $H_{n-1}^0[z_1]$. Thus ω is prime in $H_{n-1}^0[z_1]$. Suppose $\omega|\omega_1$; then $f \circ \sigma|f_1 \circ \sigma$, so $f|f_1$ in H_n. Every indecomposable element in H_n is prime. □

5. Further Consequences (Hensel Rings, Noetherian Rings)

Hensel Rings

Let R be a commutative $\mathbb{C}$-algebra with 1 in which the set $\mathfrak{m}$ of all non-units is an ideal. Let $\pi: R \to R/\mathfrak{m}$ and $\imath: \mathbb{C} \to R$ be the canonical mappings.

Proposition

1. $\mathfrak{m}$ *is the only maximal ideal in* R.
2. $R/\mathfrak{m}$ *is a field.*
3. $\pi \circ \imath: \mathbb{C} \to R/\mathfrak{m}$ *is an injective ring homorphism.*

PROOF

1. Let $\mathfrak{a} \subset R$ be an arbitrary maximal ideal. If $\mathfrak{a}$ contains unit, then $\mathfrak{a} = R$, and that cannot be. Therefore $\mathfrak{a} \subset \mathfrak{m}$; that is, $\mathfrak{a} = \mathfrak{m}$.

2. If $\pi(a) \neq 0$, then $a \notin \mathfrak{m}$, and therefore is a unit in R. There exists an $a' \in R$ with $aa' = 1$, and then $\pi(a) \cdot \pi(a') = \pi(a \cdot a') = \pi(1) = 1 \in R/\mathfrak{m}$.

3. It is clear that $\pi \circ \imath$ is a ring homomorphism. If $\pi \circ \imath(c) = 0$, then $\imath(c) = c \cdot 1$ must lie in $\mathfrak{m}$, and that is possible only if $c = 0$. Therefore $\pi \circ \imath$ is injective. □

Def. 5.1. Let R be a commutative $\mathbb{C}$-algebra with 1. R is called a *local $\mathbb{C}$-algebra* if:

1. The set $\mathfrak{m}$ of all non-units of R is an ideal in R.
2. The canonical ring monomorphism $\pi \circ \imath: \mathbb{C} \to R/\mathfrak{m}$ is surjective.

Theorem 5.1. *H_n is a local $\mathbb{C}$-algebra.*

PROOF

1. $\mathfrak{m} = \{f \in H_n : f(0) = 0\}$ is clearly an ideal in H_n.

2. For $f \in H_n$, $f = \imath(f(0)) + (f - \imath(f(0)))$ with $f - \imath(f(0)) \in \mathfrak{m}$; therefore $\pi(f) = \pi \circ \imath(f(0))$. Hence $\pi \circ \imath$ is surjective. Moreover, $(\pi \circ \imath)^{-1} \circ \pi(f) = f(0)$. □

Let R be a local $\mathbb{C}$-algebra with maximal ideal $\mathfrak{m}$ and the canonical mappings $\pi: R \to R/\mathfrak{m}$, $\iota: \mathbb{C} \to R$. Then there is a mapping $\rho: R[X] \to \mathbb{C}[X]$ with $\rho\left(\sum_{\nu=0}^{n} r_\nu X^\nu\right) = \sum_{\nu=0}^{n} (\pi \circ \iota)^{-1} \cdot \pi(r_\nu) X^\nu$ which is clearly surjective.

Def. 5.2. Let R be a local $\mathbb{C}$-algebra, $\rho: R[X] \to \mathbb{C}[X]$ the mapping given above. R is called *henselian* if there exist normalized polynomials $f_1, f_2 \in R[X]$ with $\rho(f_1) = g_1$, $\rho(f_2) = g_2$ and $f = f_1 \cdot f_2$, whenever $f \in R[X]$ is a normalized polynomial and $\rho(f) = g_1 \cdot g_2$ is a decomposition of $\rho(f)$ into two relatively prime normalized polynomials $g_1, g_2 \in \mathbb{C}[X]$.

Theorem 5.2. *H_n is a henselian ring.*

This theorem follows directly from Hensel's lemma:

Theorem 5.3 (Hensel's lemma). *Let $\omega(u, \mathfrak{z}) \in H_n^0[u]$ have the decomposition $\omega(u, 0) = \prod_{\lambda=1}^{\ell} (u - c_\lambda)^{s_\lambda}$ into linear factors (with $c_\nu \neq c_\mu$ for $\nu \neq \mu$ and $s_1 + \cdots + s_\ell = : s = \deg(\omega)$). Then there are uniquely determined polynomials $\omega_1, \ldots, \omega_\ell \in H_n^0[u]$ with $\deg(\omega_\lambda) = s_\lambda$ and $\omega_\lambda(u, 0) = (u - c_\lambda)^{s_\lambda}$ for $\lambda = 1, \ldots, \ell$ such that $\omega = \omega_1 \cdots \omega_\ell$.*

PROOF. We proceed by induction on ℓ. The case $\ell = 1$ is trivial; we assume that the theorem has been proved for $\ell - 1$.

1. First assume that $\omega(0, 0) = 0$. Without loss of generality we can assume that $c_1 = 0$; thus $\omega(u, 0) = u^{s_1} \cdot h(u)$ with $\deg(h) = s - s_1$ and $h(0) \neq 0$. This means that ω is regular of order s_1 in u and we can apply the Weierstrass preparation theorem:

There is a unit $e \in H_{n+1}$ and a polynomial $\omega_1 \in H_n^0[u]$ with $\deg(\omega_1) = s_1$ such that $\omega = e \cdot \omega_1$. From the corollary it follows that e lies in $H_n^0[u]$. $\omega_1(0, 0) = 0$, since $\omega(0, 0) = 0$ and $e(0, 0) \neq 0$; so $\omega_1(u, 0) = u^{s_1}$. Therefore $e(u, 0) = h(u) = \prod_{\lambda=2}^{\ell} (u - c_\lambda)^{s_\lambda}$. By induction there are elements $\omega_2, \ldots, \omega_\ell \in H_n^0[u]$ with $\deg(\omega_\lambda) = s_\lambda$, $\omega_\lambda(u, 0) = (u - c_\lambda)^{s_\lambda}$ and $e = \omega_2 \cdots \omega_\ell$. $\omega = \omega_1 \omega_2 \cdots \omega_\ell$ is the desired decomposition.

2. If $\omega(0, 0) \neq 0$, then let $\omega'(u, \mathfrak{z}): = \omega(u + c_1, \mathfrak{z})$. As in (1) we find a decompsotion $\omega' = \omega_1' \cdots \omega_\ell'$ and with $\omega_\lambda(u, \mathfrak{z}): = \omega_\lambda'(u - c_1, \mathfrak{z})$ obtain a decomposition in the sense of the theorem.

The uniqueness of the decomposition is also proved by induction on ℓ. In Case 1 the induction step follows directly from the Weierstrass preparation theorem, and Case 2 reduces to Case 1. □

Noetherian Rings

Def. 5.3. Let R be a commutative ring with 1. An R-module M is called *finite* if there exists a $q \in \mathbb{N}$ and an R module epimorphism $\varphi: R^q \to M$.

This is equivalent to the existence of elements $e_1, \ldots, e_q \in M$ such that every element $x \in M$ can be written in the form $x = \sum_{\nu=1}^{q} r_\nu e_\nu$ with $r_\nu \in R$.

Def. 5.4. Let R be a commutative ring with 1. R is called *noetherian* if every ideal $\mathscr{I} \subset R$ is finitely generated. An R-module M is called noetherian if every submodule $M' \subset M$ is finite.

Theorem 5.4. *If R is a noetherian ring and $q \in \mathbb{N}$, then R^q is a noetherian R-module.*

PROOF. We proceed by induction on q.

The case $q = 1$ is trivial. Assume the theorem is proved for $q - 1$. Let $M \subset R^q$ be an R-submodule. Then $\mathscr{I}: = \{r_1 \in R$: There exist $r_2, \ldots, r_q \in R$ with $(r_1, r_2, \ldots, r_q) \in M\}$ is an ideal in R and as such is finitely generated by elements $r_1^{(\lambda)}, \lambda = 1, \ldots, \ell$. For every $r_1^{(\lambda)}$ there are elements $r_2^{(\lambda)}, \ldots, r_q^{(\lambda)} \in R$ such that $\mathfrak{r}_\lambda: = (r_1^{(\lambda)}, r_2^{(\lambda)}, \ldots, r_q^{(\lambda)})$ lies in M for $\lambda = 1, \ldots, \ell$. $M': = M \cap (\{0\} \times R^{q-1})$ can be identified with an R-submodule of R^{q-1} and is therefore, by the induction assumption, finite.

Let $\mathfrak{r}_\lambda = (0, r_2^{(\lambda)}, \ldots, r_q^{(\lambda)}), \lambda = \ell + 1, \ldots, p$ be generators of M'. If $\mathfrak{r} \in M$, we can write $\mathfrak{r} = (r_1, \mathfrak{r}')$ with $r_1 \in \mathscr{I}$, therefore $r_1 = \sum_{\lambda=1}^{\ell} a_\lambda r_1^{(\lambda)}, a_\lambda \in R$. But then

$$\mathfrak{r} - \sum_{\lambda=1}^{\ell} a_\lambda \mathfrak{r}_\lambda = \left(0, \mathfrak{r}' - \sum_{\lambda=1}^{\ell} a_\lambda (r_2^{(\lambda)}, \ldots, r_q^{(\lambda)})\right) \in M'.$$

That is, there are elements $a_{\ell+1}, \ldots, a_p \in R$ such that

$$\mathfrak{r} - \sum_{\lambda=1}^{\ell} a_\lambda \mathfrak{r}_\lambda = \sum_{\lambda=\ell+1}^{p} a_\lambda \mathfrak{r}_\lambda,$$

Hence

$$\mathfrak{r} = \sum_{\lambda=1}^{p} a_\lambda \mathfrak{r}_\lambda.$$

$\{\mathfrak{r}_1, \ldots, \mathfrak{r}_p\}$ is a system of generators for M. □

Theorem 5.5 (Rückert basis theorem). *H_n is a noetherian ring.*

PROOF. We proceed by induction on n. For $n = 0$, $H_n = \mathbb{C}$ and the statement is trivial. We now assume that the theorem is proved for $n - 1$. Let $\mathscr{I} \subset H_n$ be an ideal. We may assume that we are not dealing with the zero ideal, so there exists an element $g \neq 0$ in $\mathscr{I}$. By application of a suitable shearing σ, $g': = g \circ \sigma$ is regular of order s in z_1. σ induces an isomorphism $\sigma^*: H_n \to H_n$ with $\sigma^*(g) = g'$. $\sigma^*(\mathscr{I})$ is an ideal in H_n along with $\mathscr{I}$, and if $\sigma^*(\mathscr{I})$ is finitely generated, then $\mathscr{I} = (\sigma^*)^{-1} \sigma^*(\mathscr{I})$ is also finitely generated. Without loss of generality we can then assume that g is already regular of order s in z_1. Let $\Phi_g: H_n \to (H_{n-1})^s$ be the *Weierstrass homomorphism*, which will be defined in the following manner: For every $f \in H_n$ there are uniquely defined elements $q \in H_n$ and $r = r_0 + r_1 z_1 + \cdots + r_{s-1} z_1^{s-1} \in H_{n-1}[z_1]$, such that $f = q \cdot g + r$. Let $\Phi_g(f) := (r_0, \ldots, r_{s-1})$. Φ_g is an H_{n-1}-module homomorphism. By the induction hypothesis H_{n-1} is noetherian and so by Theorem 5.4, $(H_{n-1})^s$ is a noetherian H_{n-1}-module. $M := \Phi_g(\mathscr{I})$ is an H_{n-1}-submodule, and therefore finitely generated. Let $\mathfrak{r}^{(\lambda)} = (r_0^{(\lambda)}, \ldots, r_{s-1}^{(\lambda)})$, $\lambda = 1, \ldots, \ell$, be generators of M. If $f \in \mathscr{I}$ is arbitrary, then $f = q \cdot g + (r_0 + r_1 z_1 + \cdots + r_{s-1} z_1^{s-1})$, and there are elements $a_1, \ldots, a_\ell \in H_{n-1}$ such that $(r_0, r_1, \ldots, r_{s-1}) = \sum_{\lambda=1}^{\ell} a_\lambda \mathfrak{r}^{(\lambda)}$. Hence we obtain the representation

$$f = q \cdot g + \sum_{\lambda=1}^{\ell} a_\lambda (r_0^{(\lambda)} + r_1^{(\lambda)} z_1 + \cdots + r_{s-1}^{(\lambda)} z_1^{s-1}),$$

i.e.,

$$\{g, r_0^{(1)} + r_1^{(1)} z_1 + \cdots + r_{s-1}^{(1)} z_1^{s-1}, \ldots, r_0^{(\ell)} + r_1^{(\ell)} z_1 + \cdots + r_{s-1}^{(\ell)} z_1^{s-1}\}$$

is a system of generators of $\mathscr{I}$. □

Remark. We have up to now shown that H_n has a unique factorization, and is a henselian and noetherian local $\mathbb{C}$-algebra. If $\mathscr{I} \subset H_n$ is an arbitrary ideal (with $\mathscr{I} \neq H_n$), then $A := H_n/\mathscr{I}$ is called an *analytic algebra*. A is likewise noetherian and henselian. Analytic algebras play a decisive role in the local theory of *complex spaces*, a generalization of the *theory of analytic sets* sketched in the following section.

6. Analytic Sets

Def. 6.1. Let $B \subset \mathbb{C}^n$ be a region, $M \subset B$ a subset and $\mathfrak{z}_0 \in B$ a point. M is called *analytic at* $\mathfrak{z}_0$ if there exists an open neighborhood $U = U(\mathfrak{z}_0) \in B$ and functions $f_1, \ldots, f_\ell$ holomorphic in U such that

$$U \cap M = \{\mathfrak{z} \in U : f_1(\mathfrak{z}) = \cdots = f_\ell(\mathfrak{z}) = 0\}.$$

M is called analytic in B if M is analytic at every point of B.

Remark. If $B \subset \mathbb{C}^n$ is a region and $f_1, \ldots, f_\ell$ are elements of $A(B)$, then we call the set

$$N(f_1, \ldots, f_\ell) := \{\mathfrak{z} \in B : f_1(\mathfrak{z}) = \cdots f_\ell(\mathfrak{z}) = 0\}$$

the *zero set* of the functions $f_1, \ldots, f_\ell$.

Theorem 6.1. *If $B \subset \mathbb{C}^n$ is a region and $M \subset B$ is an analytic set in B, then M is closed in B.*

PROOF. We will show that $B - M$ is open. If $\mathfrak{z}_0 \in B - M$, then there exists an open neighborhood $U = U(\mathfrak{z}_0) \subset B$ and functions $f_1, \ldots, f_\ell \in A(U)$ with $N(f_1, \ldots, f_\ell) = U \cap M$ such that, say, $f_1(\mathfrak{z}_0) \neq 0$. Then there is an entire neighborhood $V = V(\mathfrak{z}_0) \subset U$ such that $f_1|V$ vanishes nowhere, and hence V is contained in $B - M$. Therefore $\mathfrak{z}_0$ is an interior point, and $B - M$ is open. □

Theorem 6.2. *Let $G \subset \mathbb{C}^n$ be a domain. Then the ring $A(G)$ of functions holomorphic on G is an integral domain.*

PROOF. We need only to show that $A(G)$ has no zero divisors: Suppose f_1, f_2 are two elements of $A(G)$ with $f_1 \neq 0$ and $f_1 \cdot f_2 = 0$. Then there is a $\mathfrak{z}_0 \in G$ with $f_1(\mathfrak{z}_0) \neq 0$, and hence an entire neighborhood $V = V(\mathfrak{z}_0) \subset G$ such that f_1 never vanishes on V. But then we must have $f_2|V = 0$, and, by the identity theorem, $f_2 = 0$. □

We cannot conclude from this that $A = A(G)$ is a unique factorization domain. But we shall show that $A^0[u]$ is a unique factorization domain, as a consequence of the following theorem:

Theorem 6.3. *Let I be an integral domain, $Q = Q(I)$ the quotient field of I. $I^0[X]$ is a unique factorization domain if I satisfies the condition:*

$$\omega_1, \omega_2 \in Q^0[X] \quad \text{and} \quad \omega_1 \cdot \omega_2 \in I^0[X] \quad \text{imply} \quad \omega_1, \omega_2 \in I^0[X].$$

PROOF. Although Gauss' lemma assumed that I was a unique factorization domain, the proof only used the above property of I, which is satisfied for every unique factorization domain. □

We now show that $I = A$ satisfies the hypothesis of Theorem 6.3. The quotient field $Q := Q(A)$ is the field of "meromorphic functions" on G. The elements $h = f/g$ can naturally be interpreted as functions only in a very broad sense. Poles may occur, and more besides! If f and g vanish independently at a point, then in general one cannot assign any reasonable value to h at that point. Such *indeterminate points* only occur for meromorphic functions of several variables. In what follows we confine ourselves to the algebraic properties of Q.

For $\mathfrak{z} \in G$ let $I_\mathfrak{z} = (H_n)_\mathfrak{z}$ be the ring of convergent power series at $\mathfrak{z}$ and $Q_\mathfrak{z} = Q(I_\mathfrak{z})$ the quotient field of $I_\mathfrak{z}$. Moreover, let $A(\mathfrak{z})$ be the set of all functions defined and holomorphic on a neighborhood of $\mathfrak{z}$. For $f \in A(\mathfrak{z})$ let $(f)_\mathfrak{z}$ denote the power series of f at the point $\mathfrak{z}$. Then for every $\mathfrak{z} \in G$ there exists a ring homomorphism

$$\theta_\mathfrak{z}: Q \to Q_\mathfrak{z} \quad \text{with} \quad \theta_\mathfrak{z}\left(\frac{f}{g}\right) = \frac{(f)_\mathfrak{z}}{(g)_\mathfrak{z}}.$$

By the identity theorem, $(g)_{\mathfrak{z}} \neq 0$ and furthermore $\theta_{\mathfrak{z}}$ is injective. Now if

$$h = \frac{f}{g} \in Q \qquad \text{and} \qquad (h)_{\mathfrak{z}_0} := \theta_{\mathfrak{z}_0}(h) = \frac{(f)_{\mathfrak{z}_0}}{(g)_{\mathfrak{z}_0}} \in I_{\mathfrak{z}_0},$$

then $(g)_{\mathfrak{z}_0}$ must be a unit in $I_{\mathfrak{z}_0}$ and therefore $g(\mathfrak{z}_0) \neq 0$. But then there is an open neighborhood $V = V(\mathfrak{z}_0) \subset G$ such that g is nowhere vanishing on V, and on V, h represents a holomorphic function. If $(h)_{\mathfrak{z}_0} \in I_{\mathfrak{z}_0}$ for every point $\mathfrak{z}_0 \in G$, then h is a holomorphic function on G.

Theorem 6.4. *If ω_1, ω_2 are elements of $Q^0[u]$ with $\omega_1 \cdot \omega_2 \in A^0[u]$ then $\omega_1, \omega_2 \in A^0[u]$.*

PROOF

1. If $\omega \in Q^0[u]$, then ω has the form $\omega = u^s + A_1 u^{s-1} + \cdots + A_s$ with $A_i \in Q$ for $i = 1, \ldots, s$. Let $(\omega)_{\mathfrak{z}} := u^s + (A_1)_{\mathfrak{z}} u^{s-1} + \cdots + (A_s)_{\mathfrak{z}} \in Q_{\mathfrak{z}}^0[u]$. If $(\omega)_{\mathfrak{z}}$ lies in $I_{\mathfrak{z}}^0[u]$ for all $\mathfrak{z} \in G$, then it follows from the above considerations that $A_1, \ldots, A_s$ are holomorphic functions; that is, $\omega \in A^0[u]$.

2. If ω_1, ω_2 are elements of $Q^0[u]$ with $\omega_1 \cdot \omega_2 \in A^0[u]$, then for all $\mathfrak{z} \in G$ $(\omega_1)_{\mathfrak{z}}, (\omega_2)_{\mathfrak{z}} \in Q_{\mathfrak{z}}^0[u]$ and $(\omega_1)_{\mathfrak{z}} \cdot (\omega_2)_{\mathfrak{z}} \in I_{\mathfrak{z}}^0[u]$. Since $I_{\mathfrak{z}} = (H_n)_{\mathfrak{z}}$ is a UFD, it follows that $(\omega_1)_{\mathfrak{z}}, (\omega_2)_{\mathfrak{z}} \in I_{\mathfrak{z}}^0[u]$. By (1) this means that $\omega_1, \omega_2 \in A^0[u]$. □

Theorem 6.5. *Let $G \subset \mathbb{C}^n$ be a domain, $A = A(G)$. Then $A^0[u]$ is a unique factorization domain.*

The proof follows directly from Theorems 6.3 and 6.4.

Def. 6.2. Let I be an integral domain. I is called a *euclidean ring* if there exists a mapping $N: I \to \mathbb{N}_0$ with the following properties:

1. $N(a \cdot b) = N(a) \cdot N(b)$.
2. $a = 0 \Leftrightarrow N(a) = 0$.
3. For all $a, b \in I$ with $a \neq 0$ there exists a $q \in I$ with $N(b - q \cdot a) < N(a)$.

EXAMPLES

a. $\mathbb{Z}$ is a euclidean ring, with $N: \mathbb{Z} \to \mathbb{N}_0$ with $N(a) := |a|$.

b. If k is a field, then $k[X]$ is a euclidean ring, by virtue of the mapping $N: k[X] \to \mathbb{N}_0$ with

$$N(f) := 2^{\deg(f)} \qquad (\text{and } N(0) := 0).$$

Every euclidean ring is a principal ideal domain (and thus a unique factorization domain). If a_1, a_2 are elements of a euclidean ring, then their greatest common divisor can be written as a linear combination,

$$\gcd(a_1, a_2) = r_1 \cdot a_1 + r_2 \cdot a_2,$$

where $N(r_1 \cdot a_1 + r_2 \cdot a_2)$ is minimal. Of course, the greatest common divisor is uniquely determined up to units only.

Again let $G \subset \mathbb{C}^n$ be a domain, $A = A(G)$, $Q = Q(A)$ the field of meromorphic functions on G. $Q[u]$ is a euclidean ring. If ω_1, ω_2 are elements of $Q[u]$, consider all linear combinations $\omega = p_1\omega_1 + p_2\omega_2$ with $p_1, p_2 \in Q[u]$ and $\omega \neq 0$. If ω has minimal degree, then ω is a greatest common divisor of ω_1 and ω_2. Let $h \in A$ be the product of the denominator of p_1 and p_2. The polynomials $h \cdot p_i$ lie in $A[u]$ and $(h \cdot p_1)\omega_1 + (h \cdot p_2)\omega_2 = h \cdot \omega$. But since h is a unit in $Q[u]$, we have

Theorem 6.6. *If ω_1, ω_2 are elements of $Q[u]$, then there exists a greatest common divisor of ω_1 and ω_2 which can be written as a linear combination of ω_1 and ω_2 over $A[u]$.*

Def. 6.3. An element $\omega \in A^0[u]$ is called a *pseudopolynomial without multiple factors* if the factors ω_i (by Theorem 6.5 uniquely determined) of the prime decomposition $\omega = \omega_1 \cdots \omega_\ell$ are pairwise distinct.

Def. 6.4. Let a mapping $D: A[u] \to A[u]$ be defined by

$$D\left(\sum_{\nu=0}^{s} A_\nu(\mathfrak{z})u^\nu\right) := \sum_{\nu=1}^{s} \nu \cdot A_\nu(\mathfrak{z})u^{\nu-1}.$$

If $\omega \in A[u]$, then one calls $D(\omega) \in A[u]$ the derivative of ω.

Remark. The following formulas are readily verified:

1. $D(\omega_1 + \omega_2) = D(\omega_1) + D(\omega_2)$.
2. $D(\omega_1 \cdot \omega_2) = \omega_1 \cdot D(\omega_2) + \omega_2 \cdot D(\omega_1)$.
3. $D(\omega_1 \cdots \omega_\ell) = \sum_{\nu=1}^{\ell} \omega_1 \cdots \hat{\omega}_\nu \cdots \omega_\ell \cdot D(\omega_\nu)$. (Here, the hat on ω_ν indicates that this term is to be omitted.)

Now let $\omega = \omega_1 \cdots \omega_\ell = u^s + A_1(\mathfrak{z})u^{s-1} + \cdots + A_s(\mathfrak{z})$ be a pseudopolynomial without multiple factors (in $A^0[u]$). Then

$$D(\omega) = \omega_2 \cdots \omega_\ell \cdot D(\omega_1) + \sum_{\nu=2}^{\ell} \omega_1 \cdots \hat{\omega}_\nu \cdots \omega_\ell \cdot D(\omega_\nu).$$

Clearly ω_1 can only be divided by $D(\omega)$ if ω_1 is a divisor of $D(\omega_1)$. However, since $\deg(D(\omega_1)) < \deg(\omega_1)$, a $\omega_1' \in Q[u]$ with $\omega_1 \cdot \omega_1' = D(\omega_1)$ cannot exist. Therefore ω_1 is not a divisor of $D(\omega)$, and the same holds for $\omega_2, \ldots, \omega_\ell$. Hence ω and $D(\omega)$ have no common divisor.

Theorem 6.7. *Let $\omega \in A^0[u]$ be a pseudopolynomial without multiple factors. Then there are elements $q_1, q_2 \in A[u]$ such that $h := q_1 \cdot \omega + q_2 \cdot D(\omega)$ lies in A and does not vanish identically.*

PROOF. We have shown above that $\gcd(\omega, D(\omega)) = 1$, so there exist elements $p_1, p_2 \in Q[u]$ with $p_1\omega + p_2 \cdot D(\omega) = 1$. If we multiply the equation by an appropriate factor $h \in A$ (with $h \neq 0$), we obtain $(p_1 \cdot h) \cdot \omega + (p_2 \cdot h) \cdot D(\omega) = h$, with $p_1 \cdot h, p_2 \cdot h \in A[u]$. □

In the same way one proves:

Theorem 6.8. *If $\omega_1, \omega_2 \in A[u]$ are relatively prime, then there exist elements $q_1, q_2 \in A[u]$ such that $q_1 \cdot \omega_1 + q_2 \cdot \omega_2$ lies in A and does not vanish identically.*

We must briefly entertain the notion of a symmetric polynomial.

Def. 6.5. A polynomial $p \in \mathbb{Z}[X_1, \ldots, X_s]$ is called *symmetric* if for all ν, μ

$$p(X_1, \ldots, X_\nu, \ldots, X_\mu, \ldots, X_s) = p(X_1, \ldots, X_\mu, \ldots, X_\nu, \ldots, X_s).$$

The most important example are the elementary symmetric polynomials $\sigma_1, \ldots, \sigma_s$ where

$$\begin{aligned}
\sigma_1(X_1, \ldots, X_s) &= X_1 + \cdots + X_s,\\
\sigma_2(X_1, \ldots, X_s) &= (X_1 \cdot X_2 + \cdots + X_1 \cdot X_s)\\
&\quad + (X_2 \cdot X_3 + \cdots + X_2 \cdot X_s) + \cdots + X_{s-1} \cdot X_s\\
&\ \vdots\\
\sigma_s(X_1, \ldots, X_s) &= X_1 \cdots X_s
\end{aligned}$$

(so in general

$$\left.\sigma_\nu(X_1, \ldots, X_s) := \sum_{1 \leqslant i_1 < \cdots < i\nu \leqslant s} X_{i_1} \cdots X_{i_\nu}\right).$$

In algebra (see van der Waerden I) one proves:

Theorem 6.9. *Let $p(X_1, \ldots, X_s)$ be a symmetric polynomial with integer coefficients. Then there is exactly one polynomial $Q(Y_1, \ldots, Y_s)$ with integer coefficients such that*

$$p(X_1, \ldots, X_s) = Q(\sigma_1(X_1, \ldots, X_s), \ldots, \sigma_s(X_1, \ldots, X_s)).$$

Another important example of a symmetric polynomial is the square of the Vandermonde determinant:

$$D(X_1, \ldots, X_s) := \det{}^2 \begin{bmatrix} 1, X_1, X_1^2, \ldots, X_1^{s-1} \\ \cdot \qquad\qquad \cdot \\ \cdot \qquad\qquad \cdot \\ \cdot \qquad\qquad \cdot \\ 1, X_s, X_s^2, \ldots, X_s^{s-1} \end{bmatrix} = \prod_{\nu < \mu} (X_\nu - X_\mu)^2.$$

Clearly $D(X_1, \ldots, X_s) = 0$ if and only if there exists a pair (ν, μ) with $\nu \neq \mu$ and $X_\nu = X_\mu$.

Def. 6.6. Let $f(X) = X^s - a_1 X^{s-1} + a_2 X^{s-2} + \cdots + (-1)^s a_s \in \mathbb{C}[X]$ be a polynomial and let $Q \in \mathbb{Z}[X_1, \ldots, X_s]$ be that polynomial for which the equation

$$D(X_1, \dots, X_s) = Q(\sigma_1(X_1, \dots, X_s), \dots, \sigma_s(X_1, \dots, X_s))$$

holds. Then $\Delta(f) := Q(a_1, \dots, a_s)$ is called the *discriminant* of $f(X)$.

EXAMPLE. Let $f(X) = X^2 - aX + b$. For $s = 2$ we have

$$D(X_1, X_2) = (X_1 - X_2)^2 = X_1^2 - 2X_1 \cdot X_2 + X_2^2,$$
$$\sigma_1(X_1, X_2) = X_1 + X_2, \sigma_2(X_1, X_2) = X_1 \cdot X_2.$$

If we set $Q(Y_1, Y_2) := Y_1^2 - 4Y_2$, then

$$Q(\sigma_1(X_1, X_2), \sigma_2(X_1, X_2)) = (X_1 + X_2)^2 - 4 \cdot X_1 \cdot X_2 = D(X_1, X_2).$$

Therefore $\Delta(f) = Q(a, b) = a^2 - 4b$. If c_1, c_2 are both zeroes of $f(X)$, then:

$$\begin{aligned} f(X) &= (X - c_1) \cdot (X - c_2) = X^2 - (c_1 + c_2)X + c_1 \cdot c_2 \\ &= X^2 - \sigma_1(c_1, c_2)X + \sigma_2(c_1, c_2), \end{aligned}$$

and therefore $\Delta(f) = D(c_1, c_2)$.

Thus $\Delta(f)$ vanishes if and only if $c_1 = c_2$.

Theorem 6.10. *Let* $f(X) = \prod_{\rho=1}^{s} (X - X_\rho) \in \mathbb{C}[X]$. *$f$ has a multiple root if and only if* $\Delta(f) = 0$.

PROOF

$$\begin{aligned} f(X) &= (X - X_1)(X - X_2) \cdots (X - X_s) \\ &= X^s - (X_1 + \cdots + X_s)X^{s-1} + (X_1X_2 + \cdots)X^{s-2} + \cdots \\ &\quad + (-1)^s X_1 \cdot X_2 \cdots X_s, \end{aligned}$$

i.e.: $f(X) = X^s - a_1X^{s-1} + a_2X^{s-2} + \cdots + (-1)^s a_s$

with

$$a_\nu = \sigma_\nu(X_1, \dots, X_s) \text{ for } \nu = 1, \dots, s.$$

Therefore

$$\begin{aligned} \Delta(f) &= Q(a_1, \dots, a_s) = Q(\sigma_1(X_1, \dots, X_s), \dots, \sigma_s(X_1, \dots, X_s)) \\ &= D(X_1, \dots, X_s) = \prod_{\nu<\mu} (X_\nu - X_\mu)^2. \end{aligned}$$ □

Now let $\omega(u, \mathfrak{z}) = u^s - A_1(\mathfrak{z})u^{s-1} + \cdots + (-1)^s A_s(\mathfrak{z})$ be a pseudopolynomial over G. A holomorphic function on G is defined by $\Delta_\omega(\mathfrak{z}) := \Delta(\omega(u, \mathfrak{z})) = Q(A_1(\mathfrak{z}), \dots, A_s(\mathfrak{z}))$. Clearly $\Delta_\omega(\mathfrak{z}) \neq 0$ if and only if $\omega(u, \mathfrak{z})$ has s distinct roots. But more is true:

Theorem 6.11. *Let $G \subset \mathbb{C}^n$ be a domain, $\omega(u, \mathfrak{z}) \in A^0[u]$ a pseudopolynomial. Δ_ω does not vanish identically if and only if ω has no multiple factors.*

PROOF

1. Let $\omega = \omega_1^2 \cdot \tilde{\omega}$ with $\deg(\omega_1) > 0$. If $\mathfrak{z} \in G$, then we can decompose $\omega_1(u, \mathfrak{z})$ into linear factors,

$$\omega_1(u, \mathfrak{z}) = (u - c_1) \cdots (u - c_t).$$

For $\omega(u, \mathfrak{z})$ we obtain a decomposition of the form

$$\omega(u, \mathfrak{z}) = (u - c_1)^2 \cdots (u - c_t)^2 (u - c_{t+1}) \cdots (u - c_p).$$

Hence

$$\Delta_\omega(\mathfrak{z}) = D(c_1, \ldots, c_t, c_1, \ldots, c_t, c_{t+1}, \ldots, c_p) = 0.$$

Since $\mathfrak{z}$ was arbitrary, $\Delta_\omega = 0$.

2. Let ω be a polynomial without multiple factors. Then there are elements $q_1, q_2 \in A[u]$ such that $h: = q_1 \cdot \omega + q_2 \cdot D(\omega) \in A$ does not vanish identically. We can find a $\mathfrak{z}_0 \in G$ with $h(\mathfrak{z}_0) \neq 0$. Let $a_i(u): = q_i(u, \mathfrak{z}_0) \in \mathbb{C}[u]$ for $i = 1, 2$. Then

$$a: = a_1(u) \cdot \omega(u, \mathfrak{z}_0) + a_2(u) \cdot D(\omega)(u, \mathfrak{z}_0) \neq 0 \text{ (independent of } u).$$

If $\omega(u, \mathfrak{z}_0) = (u - c_1)^2 \cdot \tilde{\omega}(u)$, then

$$\begin{aligned} D(\omega)(u, \mathfrak{z}_0) &= D(\omega(u, \mathfrak{z}_0)) = 2 \cdot (u - c_1) \cdot \tilde{\omega}(u) + (u - c_1)^2 \cdot D(\tilde{\omega}(u)) \\ &= (u - c_1) \cdot (2\tilde{\omega}(u) + (u - c_1) \cdot D(\tilde{\omega}(u))) = (u - c_1) \cdot \omega_1(u), \end{aligned}$$

and therefore

$$a = a_1(c_1) \cdot \omega(c_1, \mathfrak{z}_0) + a_2(c_1) \cdot D(\omega)(c_1, \mathfrak{z}_0) = 0,$$

which cannot be. Hence all the roots $c_1, \ldots, c_s$ of $\omega(u, \mathfrak{z}_0)$ must be distinct, and $\Delta_\omega(\mathfrak{z}_0) = D(c_1, \ldots, c_s) \neq 0$. □

Theorem 6.12. *Let $G \subset \mathbb{C}^n$ be a domain, $A = A(G)$,*

$$\omega(u, \mathfrak{z}) = u^s - A_1(\mathfrak{z})u^{s-1} + \cdots + (-1)^s A_s(\mathfrak{z}) \in A^0[u]$$

a pseudopolynomial without multiple factors,

$$M_\omega: = \{(u, \mathfrak{z}) \in \mathbb{C} \times G: \omega(u, \mathfrak{z}) = 0\}, D_\omega: = \{\mathfrak{z} \in G: \Delta_\omega(\mathfrak{z}) = 0\}.$$

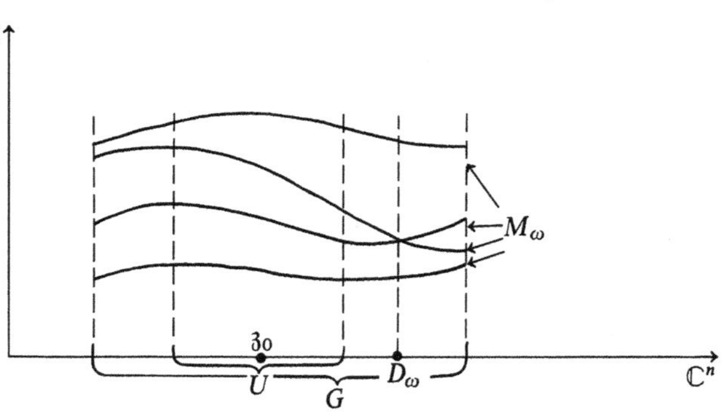

Figure 16. Illustration for Theorem 6.12.

Then M_ω and D_ω are analytic sets and:

1. *For $\mathfrak{z}_0 \in G - D_\omega$ there exists an open neighborhood $U(\mathfrak{z}_0) \subset G - D_\omega$ and holomorphic functions $f_1, \ldots, f_s$ on U with $f_\nu(\mathfrak{z}) \neq f_\mu(\mathfrak{z})$ for $\nu \neq \mu$ and $\mathfrak{z} \in U$, such that $\omega(u, \mathfrak{z}) = (u - f_1(\mathfrak{z})) \cdots (u - f_s(\mathfrak{z}))$ for all $\mathfrak{z} \in U$.*
2. *The points of D_ω are "branch points," that is, above a point $\mathfrak{z} \in D_\omega$ there always lie fewer than s points of the set M_ω.*

PROOF. $\omega(u, \mathfrak{z})$ always has exactly s distinct roots above $G - D_\omega$; above D_ω multiple roots appear. Now let $\mathfrak{z}_0 \in G - D_\omega$, $\omega(u, \mathfrak{z}_0) = (u - c_1) \cdots (u - c_s)$. $\omega_{\mathfrak{z}_0}$ is a polynomial over the ring $(H_n)_{\mathfrak{z}_0}$, and by the Hensel lemma there are polynomials $(\omega_i)_{\mathfrak{z}_0}$, $i = 1, \ldots, s$, with the following properties:

1. $(\omega_i)_{\mathfrak{z}_0}(u, \mathfrak{z}_0) = u - c_i$ for $i = 1, \ldots, s$
2. $(\omega_1)_{\mathfrak{z}_0} \cdots (\omega_s)_{\mathfrak{z}_0} = \omega_{\mathfrak{z}_0}$
3. $\deg((\omega_i)_{\mathfrak{z}_0}) = 1$ for $i = 1, \ldots, s$.

In particular we can write

$$(\omega_i)_{\mathfrak{z}_0} = u - r_i \text{ with } r_i \in (H_n)_{\mathfrak{z}_0} \text{ for } i = 1, \ldots, s.$$

Then there exist a connected open neighborhood $U(\mathfrak{z}_0) \subset G - D_\omega$ and holomorphic functions $f_1, \ldots, f_s$ on U such that the power series r_i converge to f_i. If we set $\tilde{\omega}(u, \mathfrak{z}) := (u - f_1(\mathfrak{z})) \cdots (u - f_s(\mathfrak{z}))$, we obtain

$$\tilde{\omega}_{\mathfrak{z}_0} = (u - (f_1)_{\mathfrak{z}_0}) \cdots (u - (f_s)_{\mathfrak{z}_0}) = (u - r_1) \cdots (u - r_s) = \omega_{\mathfrak{z}_0}.$$

Therefore, near $\mathfrak{z}_0$—and by the identity theorem in all of $U - \omega$ and $\tilde{\omega}$ must coincide. Hence $\omega(u, \mathfrak{z}) = (u - f_1(\mathfrak{z})) \cdots (u - f_s(\mathfrak{z}))$ on U, and because $U \subset G - D_\omega$, $f_\nu(\mathfrak{z}) \neq f_\mu(\mathfrak{z})$ for $\nu \neq \mu$. □

We now can continue with the study of analytic sets. We begin with *hypersurfaces*:

Let $G \subset \mathbb{C}^n$ be a domain, f be holomorphic and not identically zero on G and $N := \{\mathfrak{z} \in G : f(\mathfrak{z}) = 0\}$. Let $\mathfrak{z}_0 \in N$ be a fixed point. Since a shearing does not change an analytic set essentially, we can assume without loss of generality that $(f)_{\mathfrak{z}_0}$ is regular in z_1. By the Weierstrass preparation theorem there exists a unit $(e)_{\mathfrak{z}_0}$ and a pseudopolynomial $(\omega)_{\mathfrak{z}_0}$ such that $(f)_{\mathfrak{z}_0} = (e)_{\mathfrak{z}_0} \cdot (\omega)_{\mathfrak{z}_0}$. We can find a neighborhood $U(\mathfrak{z}_0) \subset G$ on which $(e)_{\mathfrak{z}_0}$ resp. $(\omega)_{\mathfrak{z}_0}$ converge to a holomorphic function e and a pseudopolynomial ω such that $f|U = e \cdot \omega$. If we choose U sufficiently small then $e(\mathfrak{z}) \neq 0$ for all $\mathfrak{z} \in U$, and therefore

$$\{\mathfrak{z} \in U : f(\mathfrak{z}) = 0\} = \{\mathfrak{z} \in U : \omega(z_1, \mathfrak{z}') = 0\}.$$

Now let $\omega = \omega_1 \cdots \omega_\ell$ be the prime decomposition of ω. Then

$$\{\mathfrak{z} \in U : f(\mathfrak{z}) = 0\} = \bigcup_{i=1}^{\ell} \{\mathfrak{z} \in U : \omega_i(\mathfrak{z}) = 0\}.$$

If multiple factors appear, then the corresponding components of the analytic set are equal; it is sufficient therefore to restrict our attention to pseudopolynomials without multiple factors. Let $\mathfrak{z}_0 = (z_1^{(0)}, \mathfrak{z}_0')$, G_1 be an open neighborhood of $z_1^{(0)} \in \mathbb{C}$ and G' be a connected open neighborhood of $\mathfrak{z}_0' \in \mathbb{C}^{n-1}$ such that

$$G_1 \times G' \subset U \qquad \text{and} \qquad \{(z_1, \mathfrak{z}') \in \mathbb{C} \times G' : \omega(z_1, \mathfrak{z}') = 0\} \subset G_1 \times G'.$$

Moreover, let

$$D_\omega = \{\mathfrak{z}' \in G' : \Delta_\omega(\mathfrak{z}') = 0\}.$$

$N \cap (G_1 \times G')$ represents a branched covering of G_1 whose branch points lie over D_ω (see Theorem 6.12); over $G_1 - D_\omega$ the covering is unbranched. One knows the analytic set $N \subset \mathbb{C}^n$ once we know the analytic set $D_\omega \subset \mathbb{C}^{n-1}$

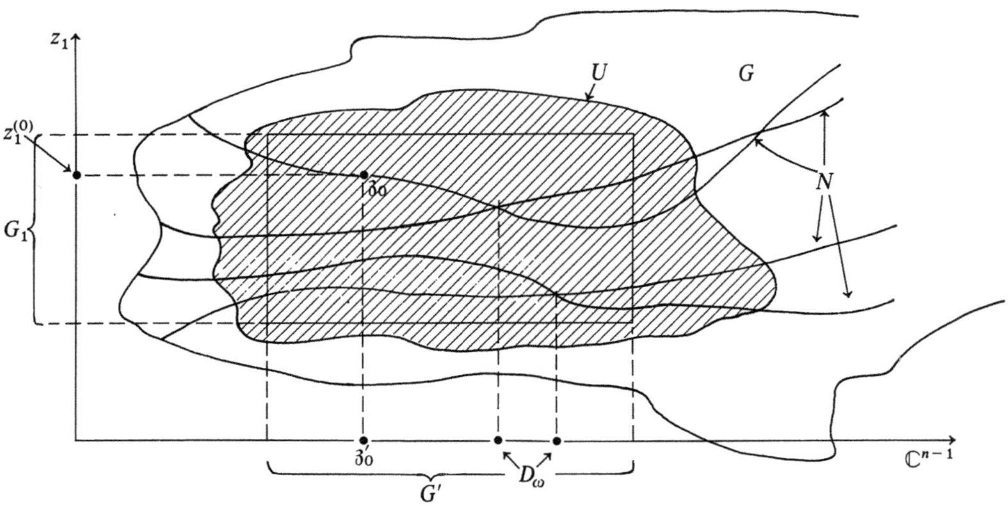

Figure 17. Representation of an analytic set as a branched covering.

and the branching behavior of N. Inductively one obtains such an overview of the construction of N. We will consider special cases:

(A) $n = 1$. Let $G \subset \mathbb{C}$ be a domain, $f: G \to \mathbb{C}$ a holomorphic function which vanishes identically nowhere. The local pseudopolynomials corresponding to f are polynomials over $\mathbb{C}$, each having finitely many zeroes. The analytic set $N = \{z \in G : f(z) = 0\}$ therefore consists of isolated points which may cluster at the boundary of G.

(B) $n = 2$. It suffices to consider pseudopolynomials.

1. Let $G = \mathbb{C}^2$, $\omega(u, z) := u^2 - z$, $N := \{(u, z) \in G : \omega(u, z) = 0\}$. $N - \{(0, 0)\}$ is the Riemann surface of $\sqrt{z}$. The discriminant is $\Delta_\omega(z) = 4z$. Clearly $D_\omega = \{z \in \mathbb{C} : \Delta_\omega(z) = 0\} = \{0\}$. For $z_0 \in \mathbb{C} - D_\omega$ there is a neighborhood $V(z_0) \subset \mathbb{C} - D_\omega$ and above V there is a decomposition $\omega(u, z) = (u - \sqrt{z})(u + \sqrt{z})$. This yields a 2-sheeted covering above $\mathbb{C} - D_\omega$ and a branch point above $D_\omega = \{0\}$. N as well as $N - \{(0, 0)\}$ are connected topological spaces.

2. Let $G = \mathbb{C}^2$, $\omega(u, z) := u^2 - z^2 = (u - z) \cdot (u + z)$. Then

$$\begin{aligned} N &:= \{(u, z) \in \mathbb{C}^2 : \omega(u, z) = 0\} \\ &= \{(u, z) \in \mathbb{C}^2 : u = z\} \cup \{(u, z) \in \mathbb{C}^2 : u = -z\}. \end{aligned}$$

The discriminant is $\Delta_\omega(z) = 4z^2$ with zero set

$$D_\omega = \{z \in \mathbb{C}: \Delta_\omega(z) = 0\} = \{0\}.$$

In this case globally N consists of two distinct schlicht sheets which intersect only above the origin. N is connected but $N - \{(0, 0)\}$ is no longer. In such a case one speaks of *pseudo-branching.*

(1) and (2) are the two characteristic cases which can occur. One inductively reduces cases of higher dimension—as described above—to cases A and B. There still remains the question how to proceed in the case of analytic sets which are described by several equations.

Let there be given a domain $G \subset \mathbb{C}^n$ and holomorphic functions f_1, f_2 on G. Both f_1 and f_2 vanish nowhere identically. Then let $M: = \{\mathfrak{z} \in G: f_1(\mathfrak{z}) = f_2(\mathfrak{z}) = 0\}$ and $\mathfrak{z}_0 \in M$. A shearing makes $(f_1)_{\mathfrak{z}_0}$ and $(f_2)_{\mathfrak{z}_0}$ simultaneously regular in z_1, and then there are a connected neighborhood $U = U_1 \times U'$ of $\mathfrak{z}_0$ and pseudopolynomials $\omega_1, \omega_2 \in A(U')^0[z_1]$ with

$$\omega_i(z_1, \mathfrak{z}') = z_1^{s_i} + A_1^{(i)}(\mathfrak{z}')z_1^{s_i - 1} + \cdots + A_{s_i}^{(i)}(\mathfrak{z}') \qquad \text{for} \qquad i = 1, 2$$

and

$$M \cap U = \{(z_1, \mathfrak{z}') \in \mathbb{C} \times U': \omega_1(z_1, \mathfrak{z}') = \omega_2(z_1, \mathfrak{z}') = 0\}.$$

We can assume that the polynomials ω_i contain no multiple factors; but in general they are not relatively prime. There are polynomials $\tilde{\omega}, \omega_1', \omega_2' \in A(U')^0[z_1]$ with

$$\omega_1 = \tilde{\omega} \cdot \omega_1', \omega_2 = \tilde{\omega} \cdot \omega_2' \qquad \text{and} \qquad \gcd(\omega_1', \omega_2') = 1.$$

Hence $M \cap U = M_1 \cup M_2$ with

$$M_1 = \{(z_1, \mathfrak{z}') \in U: \tilde{\omega}(z_1, \mathfrak{z}') = 0\}$$

and

$$M_2 = \{(z_1, \mathfrak{z}') \in U: \omega_1'(z_1, \mathfrak{z}') = \omega_2'(z_1, \mathfrak{z}') = 0\}.$$

M_1 is a "hypersurface" such as we have already considered. M_2 is given by two relatively prime pseudopolynomials. By Theorem 6.8 there exist polynomials $q_1, q_2 \in A(u')[z_1]$ such that $h: = q_1 \cdot \omega_1' + q_2 \cdot \omega_2'$ is a nowhere identically vanishing holomorphic function on U'. Let

$$M': = \{\mathfrak{z} \in U': h(\mathfrak{z}') = 0\}.$$

If $\pi: U \to U'$ is the projection with $\pi(z_1, \mathfrak{z}') = \mathfrak{z}'$, then it is clear that $\pi(M_2)$ lies in M'. Naturally above each point $\mathfrak{z}' \in M'$ there lie only finitely many points of M_2. "M_2 lies discretely over M'."

By means of sheaf theory one can show that $\pi(M_2)$ is itself an analytic hypersurface in U' and that there exists a nowhere dense analytic subset N in $\pi(M_2)$ such that $M_2 - \pi^{-1}(N)$ is a smooth several sheeted covering of $\pi(M_2) - N$.

Similar considerations apply to analytic sets which are given by several functions. At this point we want to give one example showing that in general analytic sets cannot be defined by global equations. Let

$$Q_1 := \{\mathfrak{z} = (z_1, z_2) \in \mathbb{C}^2 : |z_1| < \tfrac{1}{2}, |z_2| < 1\},$$
$$Q_2 := \{\mathfrak{z} = (z_1, z_2) \in \mathbb{C}^2 : |z_1| < 1, \tfrac{1}{2} < |z_2| < 1\},$$
$$Q := Q_1 \cup Q_2.$$

Furthermore let

$$M := \{(z_1, z_2) \in Q_2 : z_1 = z_2\}.$$

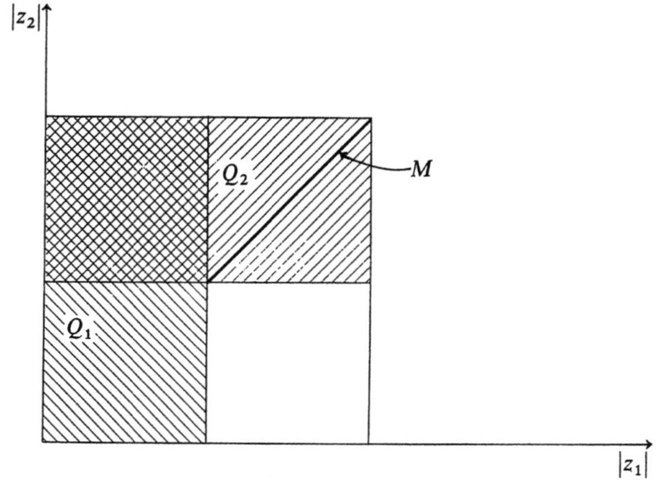

Figure 18. An analytic set which cannot be defined globally.

1. Q_1, Q_2 are open subsets of Q and $M \cap Q_1 = \varnothing$, $M \cap Q_2 = \{(z_1, z_2) \in Q_2 : z_1 - z_2 = 0\}$. M is therefore an analytic subset of Q.

2. We assume that there exist holomorphic functions $f_1, \ldots, f_\ell$ on Q such that $M = \{\mathfrak{z} \in Q : f_1(\mathfrak{z}) = \cdots = f_\ell(\mathfrak{z}) = 0\}$. But then there exist holomorphic extensions $F_1, \ldots, F_\ell$ on P (with $F_i|Q = f_i$ for $i = 1, \ldots, \ell$), and holomorphic functions $F_i^* : \{z \in \mathbb{C} : |z| < 1\} \to \mathbb{C}$ are defined by $F_i^*(z) := F_i(z, z)$. For $\frac{1}{2} < |z| < 1$ (z, z) lies in M and therefore $F_i^*(z) = F_i(z, z) = f_i(z, z) = 0$. By the identity theorem it then follows that $F_i^* = 0$ for $i = 1, \ldots, \ell$. Therefore $f_i(0, 0) = F_i(0, 0) = F_i^*(0) = 0$ for $i = 1, \ldots, \ell$; that is, (0. 0) lies in M. That is a contradiction, and the analytic set M cannot be defined globally.

Nevertheless, by means of sheaf theory one can prove the following theorem:

Theorem 6.13. *Let $G \subset \mathbb{C}^n$ be a domain of holomorphy, $M \subset G$ analytic. Then there exist holomorphic functions $f_1, \ldots, f_{n+1}$ on G such that $M = \{\mathfrak{z} \in G : f_1(\mathfrak{z}) = \cdots = f_{n+1}(\mathfrak{z}) = 0\}$.*

Next we present a short survey of further results from the theory of analytic sets.

Theorem 6.14. *Let $G \subset \mathbb{C}^n$ be a domain. Then:*

1. *$\varnothing$ and G are analytic subsets of G.*
2. *If $M_1, \ldots, M_\ell$ are analytic in G, so is $\bigcup_{i=1}^{\ell} M_i$.*
3. *If $M_1, \ldots, M_\ell$ are analytic in G, so is $\bigcap_{i=1}^{\ell} M_i$.*

3′. *If $(M_\iota)_{\iota \in I}$ is a system of analytic sets, $\bigcap_{\iota \in I} M_\iota$ is analytic in G.*

PROOF

1. $\varnothing = \{\mathfrak{z} \in G : 1 = 0\}$, $G = \{\mathfrak{z} \in G : 0 = 0\}$.
2. Let $\mathfrak{z}_0 \in M := \bigcup_{i=1}^{\ell} M_i$. Then there exists an open neighborhood $U(\mathfrak{z}_0) \subset G$ and holomorphic functions $f_{i,j}$, $j = 1, \ldots, d_i$ such that

$$M_i \cap U = \{\mathfrak{z} \in U : f_{i,1}(\mathfrak{z}) = \cdots = f_{i,d_i}(\mathfrak{z}) = 0\}.$$

Let $f_{(j_1,\ldots,j_\ell)} := f_{1,j_1} \cdots f_{\ell,j_\ell}$. Then

$$M \cap U = \{\mathfrak{z} \in U : f_{(j_1,\ldots,j_\ell)}(\mathfrak{z}) = 0 \text{ for all indices } (j_1, \ldots, j_\ell)\}.$$

3. Let $\mathfrak{z}_0 \in M' := \bigcap_{i=1}^{\ell} M_i$. Then

$$U \cap M' = \{\mathfrak{z} \in U : f_{i,j}(\mathfrak{z}) = 0 \text{ for } i = 1, \ldots, \ell, j = 1, \ldots, d_i\}.$$

3′. is more difficult to prove. The proof will be omitted here. □

Comment. (1), (2), and (3′) are the axiomatic properties of closed sets in a topology. In fact, we get the so-called *Zariski topology* on G by defining $U \subset G$ to be open if and only if there exists an analytic set M in G with $U = G - M$.

Def. 6.7. Let $G \subset \mathbb{C}^n$ be a domain, M analytic in G. A point $\mathfrak{z}_0 \in M$ is called a *regular point* (*ordinary smooth point*) *of M* (*of dimension* $2k$) if there exists an open neighborhood $U(\mathfrak{z}_0) \subset G$ and functions $f_1, \ldots, f_{n-k}$ holomorphic on U such that

(1) $$U \cap M = \{\mathfrak{z} \in U : f_1(\mathfrak{z}) = \cdots = f_{n-k}(\mathfrak{z}) = 0\}.$$

(2) $$\operatorname{rk}\left(\left(\frac{\partial f_i}{\partial z_j}(\mathfrak{z}_0)\right)\begin{matrix} i = 1, \ldots, n-k \\ j = 1, \ldots, n \end{matrix}\right) = n - k.$$

A point $\mathfrak{z}_0 \in M$ is called *singular* (a *singularity of M*) if it is not regular. One denotes the set of singular points of M by $S(M)$. Let $\mathfrak{z}_0$ be a regular point of M. Without loss of generality we can assume that

$$\det\left(\left(\frac{\partial f_i}{\partial z_j}(\mathfrak{z}_0)\right)\begin{matrix} i = 1, \ldots, n-k \\ j = 1, \ldots, n-k \end{matrix}\right) \neq 0.$$

Now let $F: U \to \mathbb{C}^n$ be defined by

$$F(z_1, \ldots, z_n) := (f_1(z_1, \ldots, z_n), \ldots, f_{n-k}(z_1, \ldots, z_n), z_{n-k+1} - z_{n-k+1}^{(0)}, \ldots, z_n - z_n^{(0)}).$$

Let

$$\mathcal{F} := \left(\begin{array}{c|c} \left(\left(\dfrac{\partial f_i}{\partial z_j}(\mathfrak{z}_0)\right)\begin{matrix} i = 1, \ldots, n-k \\ j = 1, \ldots, n-k \end{matrix}\right) & * \\ \hline 0 & \begin{matrix} 1 & & \\ & \ddots & \\ & & 1 \end{matrix} \end{array}\right)$$

be the functional matrix of F at the point $\mathfrak{z}_0$. Then clearly $\det \mathcal{F} \neq 0$ and there exist open neighborhoods $V(\mathfrak{z}_0) \subset U$, $W(0) \subset \mathbb{C}^n$ such that $F|V: V \to W$ is biholomorphic. But $F(V \cap M) = W \cap \{(w_1, \ldots, w_n) \in \mathbb{C}^n: w_1 = \cdots = w_{n-k} = 0\}$ is a real $2k$-dimensional plane segment.

Theorem 6.15. *Let $G \subset \mathbb{C}^n$ be a domain, M analytic in G and $\mathfrak{z}_0 \in M$ a regular point of M of dimension $2k$. Then there exists an open neighborhood $V(\mathfrak{z}_0) \subset G$ such that $M \cap V$ is biholomorphically equivalent to a plane segment of real dimension $2k$.*

Theorem 6.16. *Let $G \subset \mathbb{C}^n$ be a domain, M analytic in G. Then the set $S(M)$ of singular points of M is a nowhere dense analytic subset of G.*

Def. 6.8. An analytic set M is called *reducible* if there exist analytic subsets $M_i \subset G$, $i = 1, 2$, such that:

1. $M = M_1 \cup M_2$.
2. $M_i \neq M$, $i = 1, 2$.

If M is not reducible, it is called *irreducible.*

Theorem 6.17. *Let $G \subset \mathbb{C}^n$ be a domain, M analytic. Then there is a countable system (M_i) of irreducible analytic subsets of G such that*

1. $\bigcup_{i \in \mathbb{N}} M_i = M$.
2. The system $(M_i)_{i \in \mathbb{N}}$ is locally finite in G.
3. If $M_{i_1} \neq M_{i_2}$, then $M_{i_1} \not\subset M_{i_2}$.

We speak of a decomposition of M into irreducible components. This decomposition is unique up to the order in which the components appear.

The proof is lengthy and requires the help of sheaf theory.

Remark. Let M be irreducible. Then:
1. $M - S(M)$ is connected. (This condition is equivalent to irreducibility.)
2. The dimension $\dim_{\mathfrak{z}}(M)$ of the point $\mathfrak{z} \in M - S(M)$ is independent of

$\mathfrak{z}$. The number thus obtained is denoted by $\dim_{\mathbb{R}}(M)$. The *complex dimension* of M is $\dim_{\mathbb{C}}(M) := \frac{1}{2}\dim_{\mathbb{R}}(M)$.

If $M = \bigcup_{i \in \mathbb{N}} M_i$ is the decomposition of an arbitrary analytic set into irreducible components, then we define

$$\dim_{\mathbb{C}}(M) := \max_{i \in \mathbb{N}} \dim_{\mathbb{C}}(M_i).$$

$\dim_{\mathbb{C}}(M) \leqslant n$ always. In particular, if $\dim_{\mathbb{C}}(M_i) = k$ for all $i \in \mathbb{N}$, then we say that M is of *pure dimension* k.

Theorem 6.18. *If M is an irreducible analytic set in G and f a holomorphic function on G with $f|M \neq 0$, then* $\dim_{\mathbb{C}}(M \cap \{\mathfrak{z}: f(\mathfrak{z}) = 0\}) = \dim_{\mathbb{C}}(M) - 1$. *For every irreducible component $N \subset M \cap \{\mathfrak{z}: f(\mathfrak{z}) = 0\}$ we have* $\dim_{\mathbb{C}}(N) = \dim_{\mathbb{C}}(M) - 1$, *hence*:

Theorem 6.19. *Let $G \subset \mathbb{C}^n$ be a domain, let $f_1, \ldots, f_{n-k}$ be holomorphic functions in G, $M := \{\mathfrak{z} \in G: f_1(\mathfrak{z}) = \cdots = f_{n-k}(\mathfrak{z}) = 0\}$, $M' \subset M$ an irreducible component. Then* $\dim_{\mathbb{C}}(M') \geqslant k$.

Proof. G itself is an irreducible analytic set. Then, by Theorem 6.18, $\dim_{\mathbb{C}}(\{\mathfrak{z} \in G: f_1(\mathfrak{z}) = 0\}) \geqslant n - 1$, and the set $M_1 = \{\mathfrak{z} \in G: f_1(\mathfrak{z}) = 0\}$ is pure dimensional. Let $M_1 = \bigcup_{i \in \mathbb{N}} N_i^{(1)}$ be the decomposition of M_1 into irreducible components.

Then $\dim_{\mathbb{C}}(N_i^{(1)} \cap \{\mathfrak{z} \in G: f_2(\mathfrak{z}) = 0\}) \geqslant n - 2$ and we obtain the same value for each $i \in \mathbb{N}$. Therefore $\dim_{\mathbb{C}}(\{\mathfrak{z} \in G: f_1(\mathfrak{z}) = f_2(\mathfrak{z}) = 0\}) = \dim_{\mathbb{C}}\left(\bigcup_{i \in \mathbb{N}} N_i^{(1)} \cap \{\mathfrak{z} \in G: f_2(\mathfrak{z}) = 0\}\right) \geqslant n - 2$. We are finished after finitely many steps. □

In conclusion we consider one more example of an analytic set: Let $f: \mathbb{C}^n \to \mathbb{C}$ be defined by

$$f(z_1, \ldots, z_n) := z_1^{s_1} + \cdots + z_n^{s_n} \qquad \text{with} \qquad s_i \in \mathbb{N}, s_i \geqslant 2.$$

Let $M := \{\mathfrak{z} \in \mathbb{C}^n: f(\mathfrak{z}) = 0\}$.

$$0 = f_{z_i}(z_1, \ldots, z_n) = s_i \cdot z_i^{s_i - 1}$$

if and only if $z_i = 0$. Thus only the origin could be a singularity. It can be shown that $S(M) = \{0\}$. In this case we say that M has an *isolated singularity at the origin*.

Clearly M belongs to the family $(M_t)_{t \in \mathbb{C}}$ of analytic sets which are given by

$$M_t = \{(z_1, \ldots, z_n) \in \mathbb{C}^n: z_1^{s_1} + \cdots + z_n^{s_n} = t\}.$$

$M = M_0$ is an analytic set with an isolated singularity at the origin, while all sets M_t with $t \neq 0$ are regular. The family $(M_t)_{t \in \mathbb{C}}$ is called a *deformation of M*.

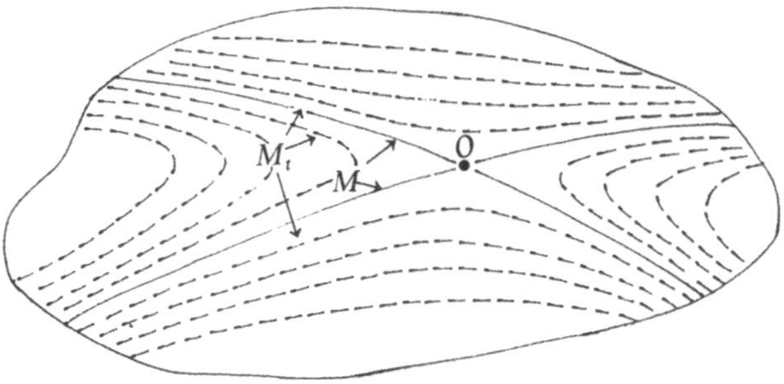

Figure 19. Deformation of an analytic set.

One can consider corresponding situations in the real analytic case. Suppose a, b are real numbers with $a < 0 < b$ and let $(M_t)_{t \in [a, b]}$ be a family of real analytic sets which are free of singularities for $t \neq 0$ and which have a singularity at the origin. It can then occur that for $t = 0$ the topological structure *jumps*, that is:

All sets M_{t_1}, M_{t_2} with $t_1, t_2 < 0$ are homeomorphic, all sets M_{t_1}, M_{t_2} with $t_1, t_2 > 0$ are homeomorphic, but for $t_1 < 0$ and $t_2 > 0$, M_{t_1} and M_{t_2} are not homeomorphic.

R. Thom recently applied this theory to the developmental processes in biology for example. One can call the jumping of the structure a revolution. Thom speaks instead of a *catastrophe*!

CHAPTER IV
Sheaf Theory

If $\mathfrak{z}_0 \in \mathbb{C}^n$ is a point, then $\mathcal{O}_{\mathfrak{z}_0} = (H_n)$ denotes the $\mathbb{C}$-algebra of convergent power series convergent at $\mathfrak{z}_0$. An arbitrary element of $\mathcal{O}_{\mathfrak{z}_0}$ has the form $f_{\mathfrak{z}_0} = \sum_{\nu=0}^{\infty} a_\nu(\mathfrak{z} - \mathfrak{z}_0)^\nu$.

Therefore there is a $\mathbb{C}$-algebra $\mathcal{O}_{\mathfrak{z}}$ for each point $\mathfrak{z} \in \mathbb{C}^n$. The disjoint union $\mathcal{O} := \dot{\bigcup}_{\mathfrak{z} \in \mathbb{C}^n} \mathcal{O}_{\mathfrak{z}}$ of these algebras is a set over $\mathbb{C}^n$ with a natural projection $\pi: \mathcal{O} \to \mathbb{C}^n$ taking a power series $f_{\mathfrak{z}}$ onto the point of expansion $\mathfrak{z}$. There exists a natural topology on $\mathcal{O}$ which makes π a continuous mapping and induces the discrete topology on every *stalk* $\mathcal{O}_{\mathfrak{z}}$, derived as follows.

If $f_{\mathfrak{z}_0} \in \mathcal{O}$, then there exists an open neighborhood $U(\mathfrak{z}_0) \subset \mathbb{C}^n$ and a holomorphic function f on U such that the series $f_{\mathfrak{z}_0}$ converges uniformly to f in U. Therefore, the function f can also be expanded in a convergent power series at each point $\mathfrak{z} \in U$. Hence f induces a mapping $s: U \to \mathcal{O}$ with the following properties:

1. $\pi \circ s = \mathrm{id}_U$
2. $s(\mathfrak{z}_0) = f_{\mathfrak{z}_0} \in s(U) \subset \mathcal{O}$.

All such sets $s(U)$ form a system of neighborhoods of $f_{\mathfrak{z}_0}$ in $\mathcal{O}$. If we give $\mathcal{O}$ the topology induced in this way, then the topological space $\mathcal{O}$ is called the sheaf of convergent power seies. The $\mathbb{C}$-algebras $\mathcal{O}_{\mathfrak{z}} = \pi^{-1}(\mathfrak{z})$ are called *stalks* of the sheaf. π is locally topological and the algebraic operations in $\mathcal{O}$ are continuous in this topology.

1. Sheaves of Sets

Def. 1.1. Let $B \subset \mathbb{C}^n$ be a region, $\mathcal{S}$ a topological space, and $\pi: \mathcal{S} \to B$ a locally topological mapping. Then $\mathfrak{S} = (\mathcal{S}, \pi)$ is called a *sheaf of sets over* B. If $\mathfrak{z} \in B$, then we call $\mathcal{S}_{\mathfrak{z}} := \pi^{-1}(\mathfrak{z})$ the stalk of $\mathfrak{S}$ over $\mathfrak{z}$.

Remark. In exactly the same manner we define sheaves over arbitrary topological spaces. If it is clear how the mapping π is defined, we shall also write $\mathscr{S}$ in place of $\mathfrak{S}$.

Def. 1.2. Let $(\mathscr{S}, \pi)$ be a sheaf over B, $\mathscr{S}^* \subset \mathscr{S}$ open and $\pi^* := \pi|\mathscr{S}^*$. Then $(\mathscr{S}^*, \pi^*)$ is called a *subsheaf of* $\mathscr{S}$.

Remark. Each subsheaf $(\mathscr{S}^*, \pi^*)$ of a sheaf $(\mathscr{S}, \pi)$ is a sheaf. We need only show that $\pi^*: \mathscr{S}^* \to B$ is locally topological.

For every element $\sigma \in \mathscr{S}^*$ there are open neighborhoods $U(\sigma) \subset \mathscr{S}$ and $V(\pi(\sigma)) \subset B$ such that $\pi|U: U \to V$ is topological. But then $U^* := U \cap \mathscr{S}^*$ is an open neighborhood of σ in $\mathscr{S}^*$, $V^* := \pi(U^*)$ is an open neighborhood of $\pi(\sigma)$ in B and $\pi^*|U^* = \pi|U^*: U^* \to V^*$ is a topological mapping. □

If $W \subset B$ is open, $\mathscr{S}|W := \pi^{-1}(W)$, then $(\mathscr{S}|W, \pi|(\mathscr{S}|W))$ is also a sheaf, *the restriction of* $\mathscr{S}$ *to* W.

Def. 1.3. Let $(\mathscr{S}, \pi)$ be a sheaf over B, $W \subset B$ open and $s: W \to \mathscr{S}$ a continuous mapping with $\pi \circ s = \mathrm{id}_W$. Then s is called a *section of* $\mathscr{S}$ over W. We denote the set of all sections of $\mathscr{S}$ over W by $\Gamma(W, \mathscr{S})$.

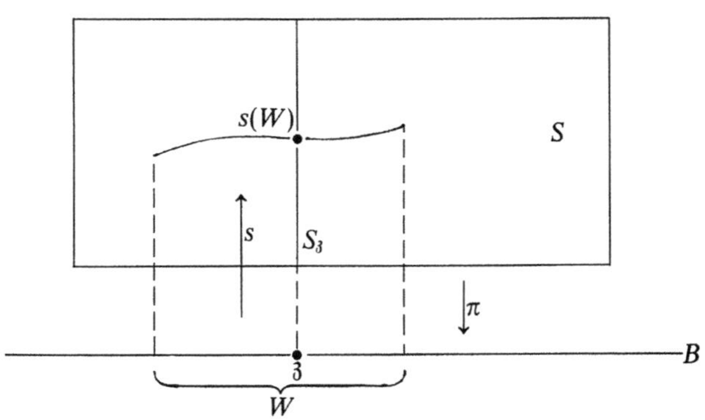

Figure 20. The definition of sheaves and sections.

Theorem 1.1. *Let $(\mathscr{S}, \pi)$ be a sheaf over B, $W \subset B$ open and $s \in \Gamma(W, \mathscr{S})$. Then $\pi: s(W) \to W$ is topological and $s = (\pi|s(W))^{-1}$.*

PROOF. By definition $\pi \circ s = \mathrm{id}_W$. For $\mathfrak{z} \in W$

$$s \circ (\pi|s(W))(s(\mathfrak{z})) = s \circ \pi \circ s(\mathfrak{z}) = s(\mathfrak{z}).$$

Therefore $s \circ (\pi|s(W)) = \mathrm{id}_{s(W)}$. □

Remark. The equation $s = (\pi|s(W))^{-1}$ holds even if s is not continuous.

Theorem 1.2. *Let $(\mathcal{S}, \pi)$ be a sheaf over B, $W \subset B$ open and $s: W \to \mathcal{S}$ a mapping with $\pi \circ s = \mathrm{id}_W$. Then $s \in \Gamma(W, \mathcal{S})$ if and only if $s(W)$ is open in $\mathcal{S}$.*

PROOF

1. Let s be continuous, $\sigma_0 \in s(W)$, and $\mathfrak{z}_0 := \pi(\sigma_0)$. Then $s(\mathfrak{z}_0) = \sigma_0$ and there are open neighborhoods $V(\mathfrak{z}_0) \subset W$ and $U(\sigma_0) \subset \mathcal{S}$ such that $\pi|U: U \to V \cap W$ is topological. Moreover, there exists an open neighborhood $V'(\mathfrak{z}_0) \subset V$ with $s(V') \subset U$. Therefore $(\pi|U) \circ (s|V') = (\pi \circ s)|V' = \mathrm{id}_{V'}$. But then $(\pi|U)^{-1}(V') = s(V') \subset s(W)$ is an open neighborhood of σ_0; that is, σ_0 is an interior point of $s(W)$.

2. Let $s(W)$ be open, $\mathfrak{z}_0 \in W$, and $\sigma_0 := s(\mathfrak{z}_0)$. Then there are open neighborhoods $V(\mathfrak{z}_0) \subset W$, $U(\sigma_0) \subset s(W)$ such that $\pi|U: U \to V$ is topological. $s = (\pi|s(W))^{-1}$, so $s|V = (\pi|U)^{-1}$, and this mapping is continuous at $\mathfrak{z}_0$. □

Theorem 1.3. *Let $(\mathcal{S}, \pi)$ be a sheaf over B, $\sigma \in \mathcal{S}$. Then there exists an open set $V \subset B$ and a section $s \in \Gamma(V, \mathcal{S})$ with $\sigma \in s(V)$.*

PROOF. Let $\mathfrak{z} := \pi(\sigma)$. Let open neighborhoods $U(\sigma) \subset \mathcal{S}$ and $V(\mathfrak{z}) \subset B$ be chosen so that $\pi|U: U \to V$ is topological. Then V and $s := (\pi|U)^{-1}$ satisfy the conditions. □

Theorem 1.4. *Let $(\mathcal{S}, \pi)$ be a sheaf over B, $W \subset B$ open. If for two sections $s_1, s_2 \in \Gamma(W, \mathcal{S})$ there is a point $\mathfrak{z} \in W$ with $s_1(\mathfrak{z}) = s_2(\mathfrak{z})$, then there is an open neighborhood $V(\mathfrak{z}) \subset W$ with $s_1|V = s_2|V$.*

PROOF. Let $\sigma := s_1(\mathfrak{z}) = s_2(\mathfrak{z})$. Then $U := s_1(W) \cap s_2(W)$ is an open neighborhood of σ and $\pi|U: U \to V := \pi(U) \subset W$ is a topological mapping of U onto the (consequently) open set V. Hence $s_1|V = (\pi|U)^{-1} = s_2|V$. □

Def. 1.4. Let $(\mathcal{S}_1, \pi_1)$, $(\mathcal{S}_2, \pi_2)$ be sheaves over B.

1. A mapping $\varphi: \mathcal{S}_1 \to \mathcal{S}_2$ is called *stalk preserving* if $\pi_2 \circ \varphi = \pi_1$ (therefore $\varphi((\mathcal{S}_1)_\mathfrak{z}) \subset (\mathcal{S}_2)_\mathfrak{z}$ for all $\mathfrak{z} \in B$).

2. A *sheaf morphism* is a continuous stalk preserving mapping $\varphi: \mathcal{S}_1 \to \mathcal{S}_2$.

3. A *sheaf isomorphism* is a topological stalk preserving mapping $\varphi: \mathcal{S}_1 \to \mathcal{S}_2$. The sheaves $\mathcal{S}_1$, $\mathcal{S}_2$ are called isomorphic if there exists a sheaf isomorphism between them.

Theorem 1.5. *Let $(\mathcal{S}_1, \pi_1)$, $(\mathcal{S}_2, \pi_2)$ be sheaves over B, $\varphi: \mathcal{S}_1 \to \mathcal{S}_2$ a stalk preserving mapping. Then the following statements are equivalent:*

1. *φ is a sheaf morphism.*

2. *For every open set $W \subset B$ and every section $s \in \Gamma(W, \mathcal{S}_1)$ $\varphi \circ s \in \Gamma(W, \mathcal{S}_2)$.*

3. *For every element $\sigma \in \mathcal{S}_1$ there exists an open set $W \subset B$ and a section $s \in \Gamma(W, \mathcal{S}_1)$ with $\sigma \in s(W)$ and $\varphi \circ s \in \Gamma(W, \mathcal{S}_2)$.*

PROOF

a. If φ is continuous, $W \subset B$ open and $s \in \Gamma(W, \mathscr{S}_1)$ then $\varphi \circ s$ is also continuous. Moreover: $\pi_2 \circ (\varphi \circ s) = (\pi_2 \circ \varphi) \circ s = \pi_1 \circ s = \mathrm{id}_W$. Therefore $\varphi \circ s$ lies in $\Gamma(W, \mathscr{S}_2)$.

b. If $\sigma \in \mathscr{S}_1$, then there exists an open set $W \subset B$ and an $s \in \Gamma(W, \mathscr{S}_1)$ with $\sigma \in s(W)$. If the conditions of (2) are also satisfied, then $\varphi \circ s$ lies in $\Gamma(W, \mathscr{S}_2)$.

c. If for a given $\sigma \in \mathscr{S}_1$, a $W \subset B$ and a $s \in \Gamma(W, \mathscr{S}_1)$ with $\sigma \in s(W)$ and $\varphi \circ s \in \Gamma(W, \mathscr{S}_2)$ are chosen according to condition (3), then $s: W \to s(W)$ is topological. Therefore $\varphi|s(W) = (\varphi \circ s) \circ s^{-1}: s(W) \to \mathscr{S}_2$ is continuous, and therefore φ is continuous at σ. □

Remark. For every open subset $W \subset B$ a sheaf morphism $\varphi: \mathscr{S}_1 \to \mathscr{S}_2$ defines a mapping $\varphi_*: \Gamma(W, \mathscr{S}_1) \to \Gamma(W, \mathscr{S}_2)$ by $\varphi_*(s) := \varphi \circ s$.

Def. 1.5. Let $B \subset \mathbb{C}^n$ be a region. For every open set $W \subset B$ let there be given a set M_W and for every pair (V, W) of open subsets of B with $V \subset W$ let there be given a mapping $r_V^W: M_W \to M_V$ such that:

1. $r_W^W = \mathrm{id}_{M_W}$ for every open set $W \subset B$.
2. If $U \subset V \subset W$, then $r_U^V \circ r_V^W = r_U^W$.

Then the system $\{M_W, r_V^W\}$ is called a *pre-sheaf* (of sets) and the mappings r_V^W are called *restriction mappings*.

With every sheaf $(\mathscr{S}, \pi)$ over B a pre-sheaf is associated in a natural manner:

If V, W are open subsets of B, then we set $M_W := \Gamma(W, \mathscr{S})$ and $r_V^W(s) := s|V$ for $s \in M_W$. Clearly $\{\Gamma(W, \mathscr{S}), r_V^W\}$ is a pre-sheaf; it is called the *canonical pre-sheaf of the sheaf* $\mathscr{S}$.

Conversely a sheaf can be constructed for each pre-sheaf:

Let the system $\{M_W, r_V^W\}$ be given, $\mathfrak{z} \in B$ fixed. On the sets $\{(W, s): W$ is an open neighborhood of $\mathfrak{z}$, $s \in M_W\}$ the following equivalence relation is introduced: $(W_1, s_1) \underset{\mathfrak{z}}{\sim} (W_2, s_2)$ if and only if there exists an open neighborhood V of $\mathfrak{z}$ with $V \subset W_1 \cap W_2$ and $r_V^{W_1}(s_1) = r_V^{W_2}(s_2)$. Let the equivalence class of (W, s) be denoted by $(W, s)_{\mathfrak{z}}$, and let $\mathscr{S}_{\mathfrak{z}}$ be the set of all classes $(W, s)_{\mathfrak{z}}$. Finally, let $\mathscr{S} := \dot{\bigcup}_{\mathfrak{z} \in B} \mathscr{S}_{\mathfrak{z}}$ and $\pi: \mathscr{S} \to B$ be the canonical projection. $\mathscr{S}$ will now be provided with a topology such that π becomes locally topological:

If $W \subset B$ is open and $s \in M_W$, then define $rs: W \to \mathscr{S}$ by $rs(\mathfrak{z}) := (W, s)_{\mathfrak{z}}$. Let $\mathfrak{B} := \{rs(W): W \subset B$ is open, $s \in M_W\} \cup \{\mathscr{S}\}$. If $W_1, W_2 \subset B$ are open sets, $s_1 \in M_{W_1}$, $s_2 \in M_{W_2}$, then let $W := \{\mathfrak{z} \in W_1 \cap W_2 : rs_1(\mathfrak{z}) = rs_2(\mathfrak{z})\}$.

a. W is open: If $\mathfrak{z}_0 \in W$, then $(W_1, s_1)_{\mathfrak{z}_0} = (W_2, s_2)_{\mathfrak{z}_0}$; therefore there exists an open neighborhood $V(\mathfrak{z}_0) \subset W_1 \cap W_2$ with $r_V^{W_1}(s_1) = r_V^{W_2}(s_2)$. But then for every $\mathfrak{z} \in V$ also $(W_1, s_1)_{\mathfrak{z}} = (W_2, s_2)_{\mathfrak{z}}$, therefore $rs_1(\mathfrak{z}) = rs_2(\mathfrak{z})$. Hence V lies in W and $\mathfrak{z}_0$ is an interior point of W.

b. Let $s := r_W^{W_1}(s_1) \in M_W$. Then $rs_1(W_1) \cap rs_2(W_2) = rs(W)$: An element $\sigma \in \mathscr{S}$ lies in $rs_1(W_1) \cap rs_2(W_2)$ if and only if there exists a $\mathfrak{z} \in W_1 \cap W_2$ with $rs_1(\mathfrak{z}) = \sigma = rs_2(\mathfrak{z})$, that is, a $\mathfrak{z} \in W$ with $rs_1(\mathfrak{z}) = \sigma$. This holds if and only if $\sigma = (W_1, s_1)_{\mathfrak{z}} = (W, s)_{\mathfrak{z}} \in rs(W)$.

For two sets $rs_1(W_1), rs_2(W_2) \in \mathfrak{B}$ the intersection $rs_1(W_1) \cap rs_2(W_2)$ also lies in $\mathfrak{B}$. Hence $\mathfrak{B}$ is a basis for a topology on $\mathscr{S}$ whose open sets are arbitrary unions of elements of $\mathfrak{B}$. It remains to show that π is locally topological.

a. Let $\sigma \in \mathscr{S}$, $\mathfrak{z} = \pi(\sigma)$. Then there is an open set $W \subset B$ with $\mathfrak{z} \in W$ and an $s \in M_W$ with $\sigma = (W, s)_{\mathfrak{z}} = rs(\mathfrak{z})$. We set $U := rs(W)$. U is an open neighborhood of σ in $\mathscr{S}$ and $\pi \circ rs = \mathrm{id}_W$, $rs \circ (\pi|U) = \mathrm{id}_U$. Therefore $\pi|U : U \to W$ is bijective, $(\pi|U)^{-1} = rs$.

b. Ev y open set $U' \subset U$ is of the form $U' = \bigcup_{\iota \in I} rs_\iota(W_\iota)$ where in every case W_ι is open in W and $s_\iota \in M_{W_\iota}$. Therefore $\pi(U') = \bigcup_{\iota \in I} W_\iota$ is open in W. $\pi|U$ then takes open sets onto open sets and hence rs is continuous.

c. If $W' \subset W$ is open, then $(W, s) \underset{\mathfrak{z}}{\sim} (W', r_{W'}^W s)$ for every $\mathfrak{z} \in W'$, so that $rs|W' = r(r_{W'}^W s)$. Hence $rs(W') = r(r_{W'}^W s)(W')$, which is an open set. Also rs maps open sets onto open sets, and thus $\pi|U$ is continuous.

We now have proved the following theorem:

Theorem 1.6. *Every pre-sheaf $\{M_W, r_V^W\}$ defines a sheaf $\mathscr{S}$ over B in the above manner (forming the inductive limit). Every element $s \in M_W$ is associated with a section $rs \in \Gamma(W, \mathscr{S})$. If $\mathfrak{z} \in B$ and $\sigma \in \mathscr{S}_{\mathfrak{z}}$, then there is an open neighborhood $W(\mathfrak{z}) \subset B$ and an $s \in M_W$ such that $\sigma = rs(\mathfrak{z})$.*

Theorem 1.7. *If $\mathscr{S}$ is a sheaf over B, then the sheaf defined by the canonical pre-sheaf $\{\Gamma(W, \mathscr{S}), r_V^W\}$ is canonically isomorphic to $\mathscr{S}$.*

PROOF. Let $(\hat{\mathscr{S}}, \hat{\pi})$ be the sheaf defined by the canonical pre-sheaf.

a. If $(W_1, s_1) \underset{\mathfrak{z}}{\sim} (W_2, s_2)$ then $s_1(\mathfrak{z}) = s_2(\mathfrak{z})$ and the converse also holds. Therefore $\varphi : (W, s)_{\mathfrak{z}} \mapsto s(\mathfrak{z})$ defines an injective mapping $\varphi : \hat{\mathscr{S}} \to \mathscr{S}$ which is stalk preserving. If $\sigma \in \mathscr{S}_{\mathfrak{z}}$, then there exists a neighborhood $W(\mathfrak{z})$ and an $s \in \Gamma(W, \mathscr{S})$ with $s(\mathfrak{z}) = \sigma$. $rs(\mathfrak{z}) = (W, s)_{\mathfrak{z}}$ then lies in $\hat{\mathscr{S}}_{\mathfrak{z}}$, and $\varphi(rs(\mathfrak{z})) = \sigma$. Hence φ is also surjective.

b. If $\sigma \in \hat{\mathscr{S}}_{\mathfrak{z}}$, then there exists an open set $W \subset B$ and an $s \in \Gamma(W, \mathscr{S})$ with $\sigma = (W, s)_{\mathfrak{z}} = rs(\mathfrak{z})$. Then $rs \in \Gamma(W, \hat{\mathscr{S}}_{\mathfrak{z}})$, $\sigma \in rs(W)$ and $\varphi \circ (rs) = s \in \Gamma(W, \mathscr{S})$. Therefore φ is continuous at σ.

c. If $W \subset B$ is open and $s \in \Gamma(W, \mathscr{S})$, then $\varphi^{-1}(s) = rs \in \Gamma(W, \hat{\mathscr{S}})$. Therefore φ^{-1} is also continuous. □

Theorem 1.8. *Every sheaf morphism is an open mapping.*

PROOF. Let $\varphi : \mathscr{S}_1 \to \mathscr{S}_2$ be a sheaf morphism. Since $\mathscr{S}_1$ is canonically isomorphic to the sheaf $\hat{\mathscr{S}}_1$ defined by the canonical pre-sheaf $\{\Gamma(W, \mathscr{S}_1), r_V^W\}$, the sets $s(W)$ with $s \in \Gamma(W, \mathscr{S}_1)$ form a basis of the topology of $\mathscr{S}_1$. If s

lies in $\Gamma(W, \mathscr{S}_1)$, then $\varphi \circ s$ lies in $\Gamma(W, \mathscr{S}_2)$ and hence $\varphi(s(W)) = (\varphi \circ s)(W)$ is open in $\mathscr{S}_2$. The proposition follows. □

Def. 1.6. Let $(\mathscr{S}_1, \pi_1), \ldots, (\mathscr{S}_\ell, \pi_\ell)$ be sheaves over B. For open sets $W \subset B$ let $M_W := \Gamma(W, \mathscr{S}_1) \times \cdots \times \Gamma(W, \mathscr{S}_\ell)$, for $s = (s_1, \ldots, s_\ell) \in M_W$ and open subsets $V \subset W$ let $r_V^W s := (s_1|V, \ldots, s_\ell|V) \in M_V$. Then $\{M_W, r_V^W\}$ is a pre-sheaf and the corresponding sheaf $\mathscr{S} = \mathscr{S}_1 \oplus \cdots \oplus \mathscr{S}_\ell$ is called the *Whitney sum of the sheaves* $\mathscr{S}_1, \ldots, \mathscr{S}_\ell$.

Theorem 1.9. *Let* $(\mathscr{S}_1, \pi_1), \ldots, (\mathscr{S}_\ell, \pi_\ell)$ *be sheaves over B, and let* $\mathscr{S} = \mathscr{S}_1 \oplus \cdots \oplus \mathscr{S}_\ell$ *be their Whitney sum. Then for every* $\mathfrak{z} \in B$ *there is a bijection* $\varphi: \mathscr{S}_\mathfrak{z} \to (\mathscr{S}_1)_\mathfrak{z} \times \cdots \times (\mathscr{S}_\ell)_\mathfrak{z}$ *defined by* $(W, (s_1, \ldots, s_\ell))_\mathfrak{z} \mapsto (s_1(\mathfrak{z}), \ldots, s_\ell(\mathfrak{z}))$.

Proof

a. Let $s_\lambda = (s_1^{(\lambda)}, \ldots, s_\ell^{(\lambda)}) \in \Gamma(W_\lambda, \mathscr{S})$ for $\lambda = 1, 2, \mathfrak{z} \in W_1 \cap W_2$.

$(W_1, s_1) \underset{\mathfrak{z}}{\sim} (W_2, s_2)$ if and only if there exists a neighborhood $V(\mathfrak{z}) \subset W_1 \cap W_2$ such that

$$(s_1^{(1)}|V, \ldots, s_\ell^{(1)}|V) = (s_1^{(2)}|V, \ldots, s_\ell^{(2)}|V).$$

This is equivalent to $s_i^{(1)}(\mathfrak{z}) = s_i^{(2)}(\mathfrak{z})$ for $i = 1, \ldots, \ell$. Therefore an injective mapping is defined by $(W, (s_1, \ldots, s_\ell))_\mathfrak{z} \mapsto (s_1(\mathfrak{z}), \ldots, s_\ell(\mathfrak{z}))$.

b. If $\sigma = (\sigma_1, \ldots, \sigma_\ell) \in (\mathscr{S}_1)_\mathfrak{z} \times \cdots \times (\mathscr{S}_\ell)_\mathfrak{z}$ and, say, $\sigma_i = s_i^*(\mathfrak{z})$ with $s_i^* \in \Gamma(W_i, \mathscr{S}_i)$, then $W := \bigcap_{i=1}^{\ell} W_i$ is an open neighborhood of $\mathfrak{z}$ and $s_i := s_i^*|W \in \Gamma(W, \mathscr{S}_i)$. Consequently $s := (s_1, \ldots, s_\ell)$ lies in M_W, and rs is a section of the sheaf $\mathscr{S}$ with $rs(\mathfrak{z}) = (W, s)_\mathfrak{z} \mapsto s(\mathfrak{z}) = \sigma$. The mapping defined above is therefore also surjective. □

Henceforth we identify $(\mathscr{S}_1 \oplus \cdots \oplus \mathscr{S}_\ell)_\mathfrak{z}$ with $(\mathscr{S}_1)_\mathfrak{z} \times \cdots \times (\mathscr{S}_\ell)_\mathfrak{z}$.

Theorem 1.10. *Let* $(\mathscr{S}_1, \pi_1), \ldots, (\mathscr{S}_\ell, \pi_\ell)$ *be sheaves over B. Then the canonical projections* $p_i: \mathscr{S}_1 \oplus \cdots \oplus \mathscr{S}_\ell \mapsto \mathscr{S}_i$ *(with* $p_i(\sigma_1, \ldots, \sigma_\ell) := \sigma_i$*) are sheaf morphisms.*

Proof. The mappings p_i are stalk preserving, by definition. If $\sigma \in (\mathscr{S}_1 \oplus \cdots \oplus \mathscr{S}_\ell)_\mathfrak{z} = (\mathscr{S}_1)_\mathfrak{z} \times \cdots \times (\mathscr{S}_\ell)_\mathfrak{z}$, then there exists sections s_i in $\mathscr{S}_i$ with $s_i(\mathfrak{z}) = p_i(\sigma)$ and $rs(\mathfrak{z}) = \sigma$ for $s := (s_1, \ldots, s_\ell)$. Therefore $p_i \circ rs = s_i$ is continuous, p_i a sheaf morphism. □

For $s_i \in \Gamma(W, \mathscr{S}_i)$, $i = 1, \ldots, \ell$, defined a mapping $s_1 \oplus \cdots \oplus s_\ell: W \to \mathscr{S}_1 \oplus \cdots \oplus \mathscr{S}_\ell$ by $(s_1 \oplus \cdots \oplus s_\ell)(\mathfrak{z}) := (s_1(\mathfrak{z}), \ldots, s(\mathfrak{z}))$. Clearly, $(s_1, \ldots, s_\ell)$ lies in M_W, and $r(s_1, \ldots, s_\ell)(\mathfrak{z}) = (W, (s_1, \ldots, s_\ell))_\mathfrak{z} = (s_1(\mathfrak{z}), \ldots, s_\ell(\mathfrak{z})) = (s_1 \oplus \cdots \oplus s_\ell)(\mathfrak{z})$; therefore $s_1 \oplus \cdots \oplus s_\ell = r(s_1, \ldots, s_\ell) \in \Gamma(W, \mathscr{S}_1 \oplus \cdots \oplus \mathscr{S}_\ell)$. Hence we can identify the sets $\Gamma(W, \mathscr{S}_1 \oplus \cdots \oplus \mathscr{S}_\ell)$ and $\Gamma(W, \mathscr{S}_1) \times \cdots \times \Gamma(W, \mathscr{S}_\ell)$.

If global sections $s_i \in \Gamma(B, \mathscr{S}_i)$ for $i = 1, \dots, \ell$ are given, then we can define corresponding injections $j_i := j_i(s_1, \dots, \hat{s}_i, \dots, s_\ell): \mathscr{S}_i \to \mathscr{S}_1 \oplus \cdots \oplus \mathscr{S}_\ell$ where

$$j_i(\sigma) := (s_1(\mathfrak{z}), \dots, s_{i-1}(\mathfrak{z}), \sigma, s_{i+1}(\mathfrak{z}), \dots, s_\ell(\mathfrak{z})), \qquad \text{for} \qquad \sigma \in (\mathscr{S}_i).$$

Clearly j_i is stalk preserving and for $s \in \Gamma(W, \mathscr{S}_i)$

$$j_i \circ s = (s_1|W, \dots, s_{i-1}|W, s, s_{i+1}|W, \dots, s|W)$$

lies in $\Gamma(W, \mathscr{S}_1) \times \cdots \times \Gamma(W, \mathscr{S}_\ell) = \Gamma(W, \mathscr{S}_1 \oplus \cdots \oplus \mathscr{S}_\ell)$ that is, j_i is continuous. $p_i \circ j_i = \mathrm{id}_{S_i}$ holds for $i = 1, \dots, \ell$.

2. Sheaves with Algebraic Structure

Let $B \subset \mathbb{C}^n$ be an open set.

Def. 2.1. A sheaf $(\mathscr{S}, \pi)$ over B is called a *$\mathbb{C}$-algebra sheaf* if:

1. Every stalk $\mathscr{S}_\mathfrak{z}$ is a commutative $\mathbb{C}$ algebra with 1.
2. $\mathscr{S} \oplus \mathscr{S} \xrightarrow{+} \mathscr{S}$ (with $(\sigma_1, \sigma_2) \mapsto \sigma_1 + \sigma_2$) is continuous.
3. $\mathscr{S} \oplus \mathscr{S} \xrightarrow{\cdot} \mathscr{S}$ (with $(\sigma_1, \sigma_2) \mapsto \sigma_1 \cdot \sigma_2$) is continuous.
4. For every $c \in \mathbb{C}$, $\mathscr{S} \xrightarrow{c\cdot} \mathscr{S}$ (with $\sigma \mapsto c \cdot \sigma$) is continuous.
5. The mapping $\mathbf{1}: \mathfrak{z} \mapsto \mathbf{1}_\mathfrak{z} \in \mathscr{S}_\mathfrak{z}$ lies in $\Gamma(B, \mathscr{S})$.

Consequences

1. $\mathbf{O}: \mathfrak{z} \mapsto \mathbf{O}_\mathfrak{z} \in \mathscr{S}_\mathfrak{z}$ lies in $\Gamma(B, \mathscr{S})$.
2. $\mathscr{S} \xrightarrow{-} \mathscr{S}$ (with $\sigma \mapsto -\sigma$) is continuous.
3. If $W \subset B$ is open, then $\Gamma(W, \mathscr{S})$ is also a $\mathbb{C}$-algebra.

PROOF

1. Because $0 \cdot \mathbf{1}_\mathfrak{z} = \mathbf{O}_\mathfrak{z}$, $0 \cdot \mathbf{1} = \mathbf{O}$, and the zero section $\mathbf{O}$ is continuous.

2. It follows from the definition that the mapping $\sigma \mapsto -\sigma = (-1) \cdot \sigma$ is continuous.

3. Addition, multiplication, and multiplication by a complex number are defined pointwise, so the axioms of a $\mathbb{C}$-algebra are satisfied since they hold in every stalk. Continuous sections go into continuous sections. □

Theorem 2.1. *Let $\mathscr{S}_1, \dots, \mathscr{S}_\ell, \mathscr{S}$ be sheaves over B given by pre-sheaves $\{M_W^{(i)}, r_{iV}^W\}$, $i = 1, \dots, \ell$ and $\{M_W, r_V^W\}$. Suppose that for every open set $W \subset B$ there exists a mapping $\varphi_W: M_W^{(1)} \times \cdots \times M_W^{(\ell)} \to M_W$ (for example, an algebraic operation) with $r_V^W \varphi_W(s_1, \dots, s_\ell) = \varphi_V(r_{1V}^W s_1, \dots, r_{\ell V}^W s_\ell)$ for arbitrary elements $s_i \in M_W^{(i)}$, $i = 1, \dots, \ell$, and open sets $V \subset W$. Then there exists exactly one sheaf morphism $\varphi: \mathscr{S}_1 \oplus \cdots \oplus \mathscr{S}_\ell \to \mathscr{S}$ with $\varphi(rs_1, \dots, rs_\ell) = r\varphi_W(s_1, \dots, s_\ell)$.*

PROOF

1. Let $W, \tilde{W}$ be open in B, $\mathfrak{z} \in W \cap \tilde{W}$ and $(W, s_i) \underset{\mathfrak{z}}{\sim} (\tilde{W}, \tilde{s}_i)$ for $i = 1, \dots, \ell$. Then there exists a neighborhood $V(\mathfrak{z}) \subset W \cap \tilde{W}$ with $r_{iV}^W s_i = r_{iV}^{\tilde{W}} \tilde{s}_i$ for

$i = 1, \ldots, \ell$; and then

$$r_V^W \varphi_W(s_1, \ldots, s_\ell) = \varphi_V(r_{1V}^W s_1, \ldots, r_{\ell V}^W s_\ell) = \varphi_V(r_{1V}^{\tilde{W}} \tilde{s}_1, \ldots, r_{\ell V}^{\tilde{W}} \tilde{s}_\ell) = r_V^{\tilde{W}} \varphi_{\tilde{W}}(\tilde{s}_1, \ldots, \tilde{s}_\ell),$$

therefore

$$(W, \varphi_W(s_1, \ldots, s_\ell)) \underset{\mathfrak{z}}{\sim} (\tilde{W}, \varphi_{\tilde{W}}(\tilde{s}_1, \ldots, \tilde{s}_\ell)).$$

Hence a mapping $\varphi: \mathscr{S}_1 \oplus \cdots \oplus \mathscr{S}_\ell \to \mathscr{S}$ is defined by

$$(rs_1(\mathfrak{z}), \ldots, rs_\ell(\mathfrak{z})) \mapsto (W, \varphi_W(s_1, \ldots, s_\ell))_{\mathfrak{z}} = r\varphi_W(s_1, \ldots, s_\ell)(\mathfrak{z});$$

it is stalk preserving and $\varphi(rs_1, \ldots, rs_\ell) = r\varphi_W(s_1, \ldots, s_\ell)$. Hence φ is uniquely determined.

2. For $\sigma = (\sigma_1, \ldots, \sigma_\ell) \in (\mathscr{S}_1)_{\mathfrak{z}} \times \cdots \times (\mathscr{S}_\ell)$ there is a neighborhood $W(\mathfrak{z})$ and elements $s_i \in M_W^{(i)}$ for $i = 1, \ldots, \ell$ such that $\sigma_i = (W, s_i)_{\mathfrak{z}}$. Then $s := (rs_1, \ldots, rs_\ell) \in \Gamma(W, \mathscr{S}_1 \oplus \cdots \oplus \mathscr{S}_\ell)$, $\sigma \in s(W)$ and $\varphi \circ s = r\varphi_W(s_1, \ldots, s_\ell) \in \Gamma(W, \mathscr{S})$. Therefore φ is continuous. □

Def. 2.2. Let $\{M_W, r_V^W\}$ be a pre-sheaf with the following properties:

1. Every M_W is a $\mathbb{C}$-algebra.
2. $r_V^W: M_W \to M_V$ is always a homomorphism of $\mathbb{C}$-algebras. Then $\{M_W, r_V^W\}$ is called a *pre-sheaf of* $\mathbb{C}$*-algebras.*

Theorem 2.2. *Let $\{M_W, r_V^W\}$ be a pre-sheaf of $\mathbb{C}$-algebras, $\mathscr{S}$ the corresponding sheaf. Then $\mathscr{S}$ is a sheaf of $\mathbb{C}$-algebras and for every open set $W \subset B$ $r: M_W \to \Gamma(W, \mathscr{S})$ is a homomorphism of $\mathbb{C}$-algebras.*

PROOF. For $W \subset B$ let $\varphi_W: M_W \times M_W \to M_W$ be defined by $\varphi_W(s_1, s_2) := s_1 + s_2$. Then

$$r_V^W \varphi_W(s_1, s_2) = r_V^W(s_1 + s_2) = r_V^W s_1 + r_V^W s_2 = \varphi_V(r_V^W s_1, r_V^W s_2).$$

By Theorem 2.1 there is exactly one sheaf morphism $\varphi: \mathscr{S} \oplus \mathscr{S} \to \mathscr{S}$ with

$$\varphi(rs_1, rs_2) = r\varphi_W(s_1, s_2) = r(s_1 + s_2).$$

An addition $\mathscr{S} \oplus \mathscr{S} \xrightarrow{+} \mathscr{S}$ is defined by $\sigma_1 + \sigma_2 := \varphi(\sigma_1, \sigma_2)$, so

$$rs_1(\mathfrak{z}) + rs_2(\mathfrak{z}) = \varphi(rs_1(\mathfrak{z}), rs_2(\mathfrak{z})) = [\varphi(rs_1, rs_2)](\mathfrak{z}) = r(s_1 + s_2)(\mathfrak{z});$$

therefore

$$rs_1 + rs_2 = r(s_1 + s_2).$$

The remaining operations are defined analogously; r transfers them to the stalks, and it is clear that r is a homomorphism of $\mathbb{C}$-algebras. □

Def. 2.3. Let $\mathscr{A}$ be a sheaf of $\mathbb{C}$-algebras over B and $\mathscr{S}$ some sheaf over B. $\mathscr{S}$ is called a *sheaf of* $\mathscr{A}$*-modules* if:

1. For every $\mathfrak{z} \in B$, $\mathscr{S}_{\mathfrak{z}}$ is a unitary $\mathscr{A}_{\mathfrak{z}}$-module.
2. $\mathscr{S} \oplus \mathscr{S} \xrightarrow{+} \mathscr{S}$ is continuous.
3. $\mathscr{A} \oplus \mathscr{S} \xrightarrow{\cdot} \mathscr{S}$ is continuous.

Remarks

1. Let $\mathbf{O}_\mathfrak{z}$ be the zero element of $\mathscr{S}_\mathfrak{z}$. Then $\mathbf{O}:\mathfrak{z} \mapsto \mathbf{O}_\mathfrak{z}$ defines the zero section $\mathbf{O} \in \Gamma(B, \mathscr{S})$.

2. For every W, $\Gamma(W, \mathscr{S})$ is a $\Gamma(W, \mathscr{A})$-module.

Def. 2.4. Let $\{M_W, r_V^W\}$ be a pre-sheaf of $\mathbb{C}$-algebras, $\{\tilde{M}_W, \tilde{r}_V^W\}$ a pre-sheaf of abelian groups, and $\mathscr{A}$ resp. $\mathscr{S}$ the corresponding sheaves. If for every open set $W \subset B$, $\tilde{M}_W$ is a (unitary) M_W-module and for every $s \in M_W$ and every $\tilde{s} \in \tilde{M}_W$, $\tilde{r}_V^W(s \cdot \tilde{s}) = r_V^W(s) \cdot \tilde{r}_V^W(\tilde{s})$, then $(\{M_W, r_V^W\}, \{\tilde{M}_W, \tilde{r}_V^W\})$ is called a *pre-sheaf of modules.*

Analogous to Theorem 2.2 it can be shown that every pre-sheaf of modules defines a sheaf of $\mathscr{A}$-modules. $r:\tilde{M}_W \to \Gamma(W, \mathscr{S})$ is then a homomorphism of abelian groups with $r(s \cdot \tilde{s}) = rs \cdot r\tilde{s}$.

Example. Let M_W be the set of holomorphic functions in W and let r_V^W: $M_W \to M_V$ be defined by $r_V^W(f) := f|V$. Clearly $\{M_W, r_V^W\}$ is a pre-sheaf of $\mathbb{C}$-algebras. The corresponding sheaf $\mathcal{O}$ is a sheaf of $\mathbb{C}$-algebras and called the *sheaf of germs of holomorphic functions on B.*

An element $(W, f)_\mathfrak{z}$ of the stalk $\mathcal{O}_\mathfrak{z}$ is an equivalence class of pairs (W_ι, F_ι), where W_ι is an open neighborhood of $\mathfrak{z}$ and f_ι a holomorphic function on W_ι. Two pairs (W_1, f_1) and (W_2, f_2) are equivalent if there exists a neighborhood $V(\mathfrak{z}) \subset W_1 \cap W_2$ with $f_1|V = f_2|V$, that is, if and only if f_1 and f_2 have the same power series expansion about $\mathfrak{z}$. Hence we can identify the stalk $\mathcal{O}_\mathfrak{z}$ with the $\mathbb{C}$-algebra of convergent power series, so that nothing new has been added to $\mathcal{O}_\mathfrak{z}$ as introduced above. In particular the power series $f_\mathfrak{z}$ and the equivalence class $(W, f)_\mathfrak{z}$ coincide.

For every open set $W \subset B$ $r:M_W \to \Gamma(W, \mathcal{O})$ is a homomorphism of $\mathbb{C}$-algebras and $r(f|V) = rf|V$.

Proposition. *r is bijective.*

Proof

1. If $rf = \mathbf{O}$, then for every $\mathfrak{z} \in W$ we have $rf(\mathfrak{z}) = \mathbf{O}_\mathfrak{z}$, therefore $(W, f)_\mathfrak{z} = \mathbf{O}_\mathfrak{z}$; that is, there exists a neighborhood $V(\mathfrak{z}) \subset W$ with $f|V = 0$, in particular $f(\mathfrak{z}) = 0$. Therefore $f = 0$.

2. If $s \in \Gamma(W, \mathcal{O})$ then for every $\mathfrak{z} \in W$ there exists a neighborhood $U(\mathfrak{z}) \subset W$ and a holomorphic function f on U with $(U, f)_\mathfrak{z} = s(\mathfrak{z})$. Then there is a neighborhood $V(\mathfrak{z}) \subset U$ with $rf|V = s|V$.

Now let $(U_\iota)_{\iota \in I}$ be an open covering of W such that there is a holomorphic function f_ι on each U_ι with $rf_\iota = s|U_\iota$. Then a holomorphic function f on W is given by $f|U_\iota := f_\iota$, for which

$$rf|U_\iota = r(f|U_\iota) = rf_\iota = s|U_\iota.$$

Therefore $f \in M_W$ and $rf = s$. □

Hence the following theorem holds:

Theorem 2.3. *$r: M_W \to \Gamma(W, \mathcal{O})$ is an isomorphism of $\mathbb{C}$-algebras.*

Henceforth we shall identify the functions holomorphic on W with the elements of $\Gamma(W, \mathcal{O})$.

Example. Let $M_W = \mathbb{C}$ and $r_V^W = \mathrm{id}_{\mathbb{C}}$ for all V, W. Then $\{M_W, r_V^W\}$ is a pre-sheaf of $\mathbb{C}$-algebras, indeed, of fields. Let $\mathscr{A}$ be the corresponding sheaf. $(W_1, c_1) \underset{\mathfrak{z}}{\sim} (W_2, c_2)$ if and only if $c_1 = c_2$, that is $\mathscr{A}_\mathfrak{z} = \mathbb{C}$ for all $\mathfrak{z} \in B$.

If $s \in \Gamma(W, A)$ and $\mathfrak{z} \in W$, then $c := s(\mathfrak{z})$ lies in $\mathscr{A}_\mathfrak{z} = \mathbb{C} = M_W$, and $rc(\mathfrak{z}) = c = s(\mathfrak{z})$. Then there exists a neighborhood $V(\mathfrak{z}) \subset W$ with $s|V = rc|V$, that is $s(\mathfrak{z}) = c$ for $\mathfrak{z} \in V$. One can thus regard s as a locally constant complex function.

We call $\mathscr{A}$ the constant sheaf of the complex numbers. Clearly $\mathscr{A}$ is a subsheaf of $\mathcal{O}$.

Def. 2.5. An *analytic sheaf* over B is a sheaf $\mathscr{S}$ of $\mathcal{O}$-modules over B.

Examples

1. $\mathcal{O}$ is an analytic sheaf.
2. Let $\mathscr{S}$ be an analytic sheaf, $\mathscr{S}^* \subset \mathscr{S}$ a subsheaf. If for every $\mathfrak{z} \in B$, $\mathscr{S}_\mathfrak{z}^* \subset \mathscr{S}_\mathfrak{z}$ is a submodule, then $\mathscr{S}^*$ is likewise an analytic sheaf: If, say, $(s_1, s_2) \in \Gamma(W, \mathscr{S}^* \oplus \mathscr{S}^*) \subset \Gamma(W, \mathscr{S} \oplus \mathscr{S})$, then $s_1 + s_2$ belongs to $\Gamma(W, \mathscr{S})$, and therefore to $\Gamma(W, \mathscr{S}^*)$. This shows addition is continuous. Multiplication by scalars is treated similarly. Note that if $\mathscr{S}^* \subset \mathscr{S}$ is an analytic subsheaf, then $\Gamma(W, \mathscr{S}^*) \subset \Gamma(W, \mathscr{S})$ is a $\Gamma(W, \mathcal{O})$-submodule.
3. If $\mathscr{I} \subset \mathcal{O}$ is an analytic subsheaf, then $\mathscr{I}_\mathfrak{z} \subset \mathcal{O}_\mathfrak{z}$ is an ideal. Hence we also call $\mathscr{I}$ an ideal sheaf.

Def. 2.6. Let $\mathscr{I} \subset \mathcal{O}$ be an ideal sheaf. Then we call $N(\mathscr{I}) := \{\mathfrak{z} \in B : \mathcal{O}_\mathfrak{z} \neq \mathscr{I}_\mathfrak{z}\}$ the zero set of $\mathscr{I}$.

For $f_\mathfrak{z} \in \mathcal{O}_\mathfrak{z}$, $f_\mathfrak{z}$ converges near $\mathfrak{z}$ to a holomorphic function which we denote by f.

Theorem 2.4. *Let $\mathscr{I} \subset \mathcal{O}$ be an ideal sheaf over B. Then $N(\mathscr{I}) = \{\mathfrak{z} \in B$: For all $f_\mathfrak{z} \in \mathscr{I}_\mathfrak{z}$, $f(\mathfrak{z}) = 0\}$.*

Proof

1. Let $\mathfrak{z} \in N(\mathscr{I})$, $f_\mathfrak{z} \in \mathscr{I}_\mathfrak{z}$, but $f(\mathfrak{z}) \neq 0$. Then on a neighborhood of $W(\mathfrak{z})$, $1/f$ is holomorphic and $\mathbf{1}_\mathfrak{z} = r1(\mathfrak{z}) = r(1/f)r(f)(\mathfrak{z}) \in \mathscr{I}_\mathfrak{z}$; therefore $\mathscr{I}_\mathfrak{z} = \mathcal{O}_\mathfrak{z}$. That is a contradiction, so $f(\mathfrak{z})$ must be zero.
2. If $\mathfrak{z} \notin N(\mathscr{I})$, then $\mathscr{I}_\mathfrak{z} = \mathcal{O}_\mathfrak{z}$, therefore $\mathbf{1}_\mathfrak{z} \in \mathscr{I}_\mathfrak{z}$; on the other hand, $1(\mathfrak{z}) \neq 0$.

□

EXAMPLE. Let $\mathsf{O}: = \dot{\bigcup}_{\mathfrak{z} \in B} \{\mathbf{O}_\mathfrak{z}\}$ and let $\pi: \mathsf{O} \to B$ be the canonical mapping. If we give O the topology of B, then π is a topological mapping. In this way O becomes an analytic sheaf, the *zero sheaf*.

Theorem 2.5. *Let $\mathscr{S}_1, \ldots, \mathscr{S}_\ell$ be analytic sheaves over B. Then $\mathscr{S}: = \mathscr{S}_1 \oplus \cdots \oplus \mathscr{S}_\ell$ is analytic.*

PROOF. Clearly $\mathscr{S}_\mathfrak{z} = (\mathscr{S}_1)_\mathfrak{z} \times \cdots \times (\mathscr{S}_\ell)_\mathfrak{z}$ is always an $\mathcal{O}_\mathfrak{z}$-module. It remains to show that the operations are continuous. We only carry out the proof for addition:

Let

$$
\begin{aligned}
(s, \tilde{s}) \in \Gamma(W, \mathscr{S} \oplus \mathscr{S}) &= \Gamma(W, \mathscr{S}) \times \Gamma(W, \mathscr{S}) \\
(s_i, \tilde{s}_i): &= (p_i \circ s, p_i \circ \tilde{s}) \in \Gamma(W, \mathscr{S}_i) \times \Gamma(W, \mathscr{S}_i) \\
&= \Gamma(W, \mathscr{S}_i \oplus \mathscr{S}_i) \qquad \text{for} \qquad i = 1, \ldots, \ell.
\end{aligned}
$$

Then

$$s_i + \tilde{s}_i \in \Gamma(W, \mathscr{S}_i) \qquad \text{for} \qquad i = 1, \ldots, \ell,$$

therefore

$$s + \tilde{s} = (s_1 + \tilde{s}_1, \ldots, s_\ell + \tilde{s}_\ell) \in \Gamma(W, \mathscr{S}_1) \times \cdots \times \Gamma(W, \mathscr{S}_\ell) = \Gamma(W, \mathscr{S}). \quad \square$$

Def. 2.7. For $q \in \mathbb{N}$ let $q\mathcal{O}: = \underbrace{\mathcal{O} \oplus \cdots \oplus \mathcal{O}}_{q \text{ times}}$. (In the literature $\mathcal{O}^q$ is the most common notation for this.) $q\mathcal{O}$ is always an analytic sheaf.

To conclude this section, we consider quotient sheaves. Let $\mathscr{S}$ be an analytic sheaf over B, $\mathscr{S}^* \subset \mathscr{S}$ an analytic subsheaf. For open sets $W \subset B$ we define $N_W: = \Gamma(W, \mathcal{O})$ and $M_W: = \Gamma(W, \mathscr{S})/\Gamma(W, \mathscr{S}^*)$ interpreted as N_W-module. There is a canonical projection $\mathfrak{q}: \Gamma(W, \mathscr{S}) \to M_W$. For $s \in \Gamma(W, S)$ let $\langle s \rangle: = \mathfrak{q}(s)$. Then (for $V \subset W \subset B$) we can define

$$r_V^W(\langle s \rangle): = \langle s | V \rangle \qquad \text{for} \qquad \langle s \rangle \in M_W.$$

Clearly r_V^W is well-defined: $\langle s_1 \rangle = \langle s_2 \rangle$ if and only if $s_1 - s_2$ lies in $\Gamma(W, \mathscr{S}^*)$. But then $(s_1 - s_2)|V \in \Gamma(V, \mathscr{S}^*)$, so $\langle s_1 | V \rangle = \langle s_2 | V \rangle$. Hence $\{M_W, r_V^W\}$ is a pre-sheaf of abelian groups and for $\langle s \rangle \in M_W$ and $f \in N_W$,

$$
\begin{aligned}
r_V^W(f \cdot \langle s \rangle) &= r_V^W(\langle f \cdot s \rangle) = \langle (f \cdot s) | V \rangle = \langle (f|V) \cdot (s|V) \rangle \\
&= (f|V) \cdot \langle s | V \rangle = (f|V) \cdot r_V^W(\langle s \rangle).
\end{aligned}
$$

$(\{N_W, r_V^W\}, \{M_W, r_V^W\})$ is a pre-sheaf of modules whose associated sheaf $\mathscr{Q}$ is an analytic sheaf. We call $\mathscr{Q}$ the *quotient sheaf of $\mathscr{S}$ by $\mathscr{S}^*$* and write $\mathscr{Q} = \mathscr{S}/\mathscr{S}^*$.

Theorem 2.6. *Let $\mathscr{S}$ be an analytic sheaf over B, $\mathscr{S}^* \subset \mathscr{S}$ an analytic subsheaf, $\mathscr{Q} = \mathscr{S}/\mathscr{S}^*$ the quotient sheaf. Then for every $\mathfrak{z} \in B$ there is an isomorphism $\psi: \mathscr{Q}_\mathfrak{z} \to \mathscr{S}_\mathfrak{z}/\mathscr{S}_\mathfrak{z}^*$ (of $\mathcal{O}_\mathfrak{z}$-modules) defined by $(W, \langle s \rangle) \mapsto \overline{s(\mathfrak{z})}$. ($\bar{\sigma}$ denotes the image of $\sigma \in \mathscr{S}_\mathfrak{z}$ in $\mathscr{S}_\mathfrak{z}/\mathscr{S}_\mathfrak{z}^*$.)*

PROOF

1. $(W_1, \langle s_1\rangle) \underset{\mathfrak{z}}{\sim} (W_2, \langle s_2\rangle)$ if and only if there is a neighborhood $V(\mathfrak{z}) \subset W_1 \cap W_2$ such that

$$\langle s_1|V\rangle = r_V^{W_1}(\langle s_1\rangle) = r_V^{W_2}(\langle s_2\rangle) = \langle s_2|V\rangle$$

and that is exactly the case when $(s_1 - s_2)|V$ lies in $\Gamma(V, \mathscr{S}^*)$. But by the continuity of $s_1 - s_2$, this is equivalent to having $s_1(\mathfrak{z}) - s_2(\mathfrak{z}) \in \mathscr{S}^*_{\mathfrak{z}}$; therefore $\overline{s_1(\mathfrak{z})} = \overline{s_2(\mathfrak{z})}$. Hence ψ is well-defined and injective.

2. Since

$$(W, \langle s_1\rangle)_{\mathfrak{z}} + (W, \langle s_2\rangle)_{\mathfrak{z}} = (W, \langle s_1 + s_2\rangle)_{\mathfrak{z}}, \psi(\sigma_1 + \sigma_2) = \psi(\sigma_1) + \psi(\sigma_2).$$

Moreover

$$\psi(f_{\mathfrak{z}} \cdot (W, \langle s\rangle)_{\mathfrak{z}}) = \psi((W, \langle f \cdot s\rangle)_{\mathfrak{z}}) = \overline{(f \cdot s)(\mathfrak{z})} = f_{\mathfrak{z}} \cdot \overline{s(\mathfrak{z})} = f_{\mathfrak{z}} \cdot \psi((W, \langle s\rangle)_{\mathfrak{z}}).$$

ψ is therefore an $\mathcal{O}_{\mathfrak{z}}$-module homomorphism.

3. If $\bar{\sigma} \in \mathscr{S}_{\mathfrak{z}}/\mathscr{S}^*_{\mathfrak{z}}$, then there exists a neighborhood $W(\mathfrak{z}) \subset B$ and an $s \in \Gamma(W, \mathscr{S})$ with $s(\mathfrak{z}) = \sigma$. But then $(W, \langle s\rangle)_{\mathfrak{z}}$ is in $\mathscr{Q}_{\mathfrak{z}}$, and $\psi((W, \langle s\rangle)_{\mathfrak{z}}) = \overline{s(\mathfrak{z})} = \bar{\sigma}$. Therefore ψ is also surjective. □

Henceforth we can identify $\mathscr{Q}_{\mathfrak{z}}$ with $\mathscr{S}_{\mathfrak{z}}/\mathscr{S}^*_{\mathfrak{z}}$.

Remark. If $s \in \Gamma(W, \mathscr{S})$, then $r\langle s\rangle$ lies in $\Gamma(W, \mathscr{Q})$. If we define $\bar{s}: W \to \mathscr{Q}$ by $\bar{s}(\mathfrak{z}) := \overline{s(\mathfrak{z})}$, then $\psi \circ r\langle s\rangle = \bar{s}$. Hence we can identify $r\langle s\rangle$ and $\bar{s}$ by means of ψ.

3. Analytic Sheaf Morphisms

Def. 3.1. Let $\mathscr{S}_1, \mathscr{S}_2$ be analytic sheaves over B, $\varphi: \mathscr{S}_1 \to \mathscr{S}_2$ a sheaf morphism. φ is called an *analytic sheaf morphism* (or a *sheaf homomorphism*) if for every $\mathfrak{z} \in B$, $\varphi: (\mathscr{S}_1)_{\mathfrak{z}} \to (\mathscr{S}_2)_{\mathfrak{z}}$ is an $\mathcal{O}_{\mathfrak{z}}$-module homomorphism.

EXAMPLES

1. Let $\mathscr{S}$ be an analytic sheaf over B and $\mathscr{S}^* \subset \mathscr{S}$ an analytic subsheaf. Let $\mathrm{q}: \mathscr{S} \to \mathscr{S}/\mathscr{S}^*$ be the canonical projection with $\mathrm{q}(\sigma) = \bar{\sigma}$. Then $\mathrm{q}: \mathscr{S}_{\mathfrak{z}} \to \mathscr{S}_{\mathfrak{z}}/\mathscr{S}^*_{\mathfrak{z}}$ is always an $\mathcal{O}_{\mathfrak{z}}$-module homomorphism and for $s \in \Gamma(W, \mathscr{S})$, $\mathrm{q} \circ s = \bar{s} = r\langle s\rangle \in \Gamma(W, \mathscr{S}/\mathscr{S}^*)$. Therefore q is a surjective sheaf homomorphism (a *sheaf epimorphism*).

2. If $\mathscr{S}$ is an analytic sheaf, then there is exactly one sheaf morphism $\mathscr{S} \to \mathsf{O}$, and it is clearly a sheaf epimorphism.

3. Conversely, though there can be several sheaf morphisms $\mathsf{O} \to \mathscr{S}$, there is only one analytic sheaf morphism (mapping $\mathbf{O}_{\mathfrak{z}}$ onto $\mathbf{O}_{\mathfrak{z}}$). This homomorphism is injective (a *sheaf monomorphism*).

4. If $\mathscr{S}^* \subset \mathscr{S}$ is an analytic subsheaf, then the canonical injection $\imath = \mathrm{id}_{\mathscr{S}}|\mathscr{S}^*: \mathscr{S}^* \hookrightarrow \mathscr{S}$ is a sheaf morphism.

Remark. (2) is a special case of (1), with $\mathscr{S}^* = \mathscr{S}$; and (3) is a special case of (4), with $\mathscr{S}^* = \mathsf{O}$.

5. If $\mathscr{S}_1, \ldots, \mathscr{S}_\ell$ are analytic sheaves, then the canonical projections $p_i:\mathscr{S}_1 \oplus \cdots \oplus \mathscr{S}_\ell \to \mathscr{S}_i$ are sheaf epimorphisms.

6. If $\mathbf{O}_i$ is the zero section in $\mathscr{S}_i$, then the canonical injection $j_i = j_i(\mathbf{O}_1, \ldots, \hat{\mathbf{O}}_i, \ldots, \mathbf{O}_\ell):\mathscr{S}_i \hookrightarrow \mathscr{S}_1 \oplus \cdots \oplus \mathscr{S}_\ell$ are sheaf monomorphisms.

7. Let $j_i:\mathcal{O} \hookrightarrow q\mathcal{O}$ be the canonical injections. If $\mathbf{l} \in \Gamma(B, \mathcal{O})$ is the "unit 1-section", then we define the unit sections in $q\mathcal{O}$ by

$$\mathfrak{e}_i: = j_i \circ \mathbf{l} = (\mathbf{O}, \ldots, \mathbf{l}, \ldots, \mathbf{O}).$$

Now let $\varphi:q\mathcal{O} \to \mathscr{S}$ be an analytic sheaf morphism and let $s_i: = \varphi \circ \mathfrak{e}_i \in \Gamma(B, \mathscr{S})$. Then for $(a_1, \ldots, a_q) \in q\mathcal{O}_\mathfrak{z}$,

$$\varphi(a_1, \ldots, a_q) = \varphi\left(\sum_{i=1}^{q} a_i\mathfrak{e}_i(\mathfrak{z})\right) = \sum_{i=1}^{q} a_i s_i(\mathfrak{z}).$$

So the sections $s_1, \ldots, s_q$ determine the homomorphism completely, and conversely we can define an analytic sheaf morphism $\varphi = \varphi_{(s_1, \ldots, s_q)}$ by the above equations for $s_1, \ldots, s_q$.

8. If $\varphi:\mathscr{S}_1 \to \mathscr{S}_2$ and $\psi:\mathscr{S}_2 \to \mathscr{S}_3$ are analytic morphisms, then so is $\psi \circ \varphi:\mathscr{S}_1 \to \mathscr{S}_3$.

Def. 3.2. Let $\varphi:\mathscr{S}_1 \to \mathscr{S}_2$ be an analytic sheaf morphism. Then we define $\operatorname{Im} \varphi: = \varphi(\mathscr{S}_1) \subset \mathscr{S}_2$; $\operatorname{Ker} \varphi: = \varphi^{-1}(\mathsf{O}) \subset \mathscr{S}_1$.

Theorem 3.1. *If $\varphi:\mathscr{S}_1 \to \mathscr{S}_2$ is an analytic sheaf morphism, then* Im φ *and* Ker φ *are analytic sheaves.*

PROOF

1. Since every sheaf morphism is an open mapping, $\operatorname{Im} \varphi = \varphi(\mathscr{S}_1) \subset \mathscr{S}_2$ is open in $\mathscr{S}_2$, and is therefore a subsheaf. Since $(\operatorname{Im} \varphi)_\mathfrak{z} = \varphi((\mathscr{S}_1)_\mathfrak{z})$, Im φ is analytic.

2. Because φ is continuous and $\mathsf{O} \subset \mathscr{S}_2$ is open, $\operatorname{Ker} \varphi = \varphi^{-1}(\mathsf{O}) \subset \mathscr{S}_1$ is open and therefore a subsheaf. Because $(\operatorname{Ker} \varphi)_\mathfrak{z} = \{\sigma \in (\mathscr{S}_1)_\mathfrak{z}:\varphi(\sigma) = \mathbf{O}_\mathfrak{z} \in (\mathscr{S}_2)_\mathfrak{z}\} = \operatorname{Ker}(\varphi|(\mathscr{S}_1)_\mathfrak{z})$, Ker φ is analytic. □

Def. 3.3. Let $\mathscr{S}_1, \mathscr{S}_2$ be analytic sheaves over B. A mapping $\varphi:\mathscr{S}_1 \to \mathscr{S}_2$ is called an *analytic sheaf isomorphism* if (1) φ is stalk preserving; (2) φ is topological; and (3) $\varphi|(\mathscr{S}_1)_\mathfrak{z}:(\mathscr{S}_1)_\mathfrak{z} \to (\mathscr{S}_2)_\mathfrak{z}$ is an $\mathcal{O}_\mathfrak{z}$-module isomorphism for every $\mathfrak{z} \in B$.

We write $\mathscr{S}_1 \simeq \mathscr{S}_2$ if there exists a sheaf isomorphism $\varphi:\mathscr{S}_1 \to \mathscr{S}_2$.

Remark. If a mapping $\varphi:\mathscr{S}_1 \to \mathscr{S}_2$ is a bijective sheaf homomorphism then it is an analytic sheaf isomorphism. Namely, φ is stalk preserving and continuous, and for every $\mathfrak{z} \in B$, $\varphi|(\mathscr{S}_1)_\mathfrak{z}:(\mathscr{S}_1)_\mathfrak{z} \to (\mathscr{S}_2)_\mathfrak{z}$ is an $\mathcal{O}_\mathfrak{z}$-module isomorphism. Since every sheaf morphism is open, it now also follows that φ^{-1} is continuous; φ is therefore topological.

Theorem 3.2. *If $\varphi:\mathscr{S}_1 \to \mathscr{S}_2$ is an analytic sheaf morphism, then $\mathscr{S}_1/\mathrm{Ker}\ \varphi \simeq \mathrm{Im}\ \varphi$.*

Proof. $\bar{\varphi}(\bar{\sigma}) := \varphi(\sigma)$ defines a stalk preserving bijective mapping $\bar{\varphi}:\mathscr{S}_1/\mathrm{Ker}\ \varphi \to \mathrm{Im}\ \varphi$ which induces a $\mathcal{O}_{\mathfrak{z}}$-module isomorphism in every stalk. If $\bar{\sigma} \in (\mathscr{S}_1/\mathrm{Ker}\ \varphi)_{\mathfrak{z}}$, then there exists a neighborhood $W(\mathfrak{z})$ and an $s \in \Gamma(W, \mathscr{S}_1)$ with $\bar{s}(\mathfrak{z}) = \bar{\sigma}$ and $\bar{\varphi} \circ \bar{s} = \varphi \circ s \in \Gamma(W, \mathrm{Im}\ \varphi)$. Therefore $\bar{\varphi}$ is also continuous. Hence by the above remark $\bar{\varphi}$ is a sheaf isomorphism. □

Remark. If $\varphi:\mathscr{S}_1 \to \mathscr{S}_2$ is an analytic sheaf morphism and if $\mathfrak{q}:\mathscr{S}_1 \to \mathscr{S}_1/\mathrm{Ker}\ \varphi$ and $\imath:\mathrm{Im}\ \varphi \hookrightarrow \mathscr{S}_2$ are the canonical mappings, then one has the *canonical decomposition of* φ:

$$\varphi = \imath \circ \bar{\varphi} \circ \mathfrak{q}:\mathscr{S}_1 \twoheadrightarrow \mathscr{S}_1/\mathrm{Ker}\ \varphi \simeq \mathrm{Im}\ \varphi \hookrightarrow \mathscr{S}_2.$$

Def. 3.4. Let $\mathscr{S}_1, \ldots, \mathscr{S}_\ell$ be analytic sheaves over B, and let $\varphi_i:\mathscr{S}_i \to \mathscr{S}_{i+1}$ be analytic sheaf morphisms for $i = 1, \ldots, \ell - 1$. Then we call the sequence

$$\mathscr{S}_1 \xrightarrow{\varphi_1} \mathscr{S}_2 \xrightarrow{\varphi_2} \mathscr{S}_3 \to \cdots \to \mathscr{S}_{\ell-1} \xrightarrow{\varphi_{\ell-1}} \mathscr{S}_\ell$$

an *analytic sequence of sheaves*.

The sequence is called *exact at* $\mathscr{S}_i$ if $\mathrm{Im}\ \varphi_{i-1} = \mathrm{Ker}\ \varphi_i$. The sequence is exact if it is exact at each $\mathscr{S}_i$.

Remarks. The sheaf homomorphism which maps every element stalkwise onto zero will be denoted by 0.

1. $\mathrm{Im}\ \varphi_{i-1} = \mathrm{Ker}\ \varphi_i$ means:
   a. $\varphi_i \circ \varphi_{i-1} = 0$
   b. If $\varphi_i(\sigma) = \mathbf{O}$, then there is a $\hat{\sigma}$ with $\varphi_{i-1}(\hat{\sigma}) = \sigma$.
2. $0 \to \mathscr{S}' \xrightarrow{\varphi} \mathscr{S}$ is exact if and only if φ is injective.
3. $\mathscr{S} \xrightarrow{\psi} \mathscr{S}'' \to 0$ is exact if and only if ψ is surjective.
4. If $\varphi:\mathscr{S}_1 \to \mathscr{S}_2$ is an analytic sheaf morphism, then we have a canonical exact sequence:

$$0 \to \mathrm{Ker}\ \varphi \to \mathscr{S}_1 \to \mathrm{Im}\ \varphi \to 0.$$

If φ is injective, then $\mathrm{Ker}\ \varphi = 0$ and $\mathscr{S}_1 \simeq \mathrm{Im}\ \varphi$; if φ is surjective, then $\mathscr{S}_1/\mathrm{Ker}\ \varphi \simeq \mathscr{S}_2$.

Def. 3.5. Let $(\mathscr{S}_1, \pi_1)$, $(\mathscr{S}_2, \pi_2)$ be analytic sheaves over B. $\mathrm{Hom}_{\mathcal{O}}(\mathscr{S}_1, \mathscr{S}_2)$ is the set of all analytic sheaf morphisms $\varphi:\mathscr{S}_1 \to \mathscr{S}_2$.

If we set $(\varphi_1 + \varphi_2)(\sigma) := \varphi_1(\sigma) + \varphi_2(\sigma)$ and $(f \cdot \varphi)(\sigma) := f_{\pi_1(\sigma)} \cdot \varphi(\sigma)$ for $\varphi, \varphi_1, \varphi_2 \in \mathscr{S}_1$, and $f \in \Gamma(B, \mathcal{O})$, then $\mathrm{Hom}_{\mathcal{O}}(\mathscr{S}_1, \mathscr{S}_2)$ becomes a $\Gamma(B, \mathcal{O})$-module.

($(\varphi_1 + \varphi_2) \circ s = \varphi_1 \circ s + \varphi_2 \circ s$ and $(f \cdot \varphi) \circ s = f \cdot (\varphi \circ s)$ are also sections, and hence $\varphi_1 + \varphi_2$ and $f \cdot \varphi$ are continuous.)

4. Coherent Sheaves

B will always be a region in $\mathbb{C}^n$.

Def. 4.1. An analytic sheaf $\mathscr{S}$ over B is called *finitely generated* if for every point $\mathfrak{z} \in B$ there exist an open neighborhood $W(\mathfrak{z}) \subset B$, a natural number q and a sheaf epimorphism $\varphi: q\mathcal{O}|W \twoheadrightarrow \mathscr{S}|W$.

Let $\mathfrak{e}_i$ be the i-th unit section of $q\mathcal{O}$, $s_i: = \varphi \circ (\mathfrak{e}_i|W)$ the images under φ. If $\sigma \in \mathscr{S}_\mathfrak{z}$ then σ comes from an element $(a_1, \ldots, a_q) \in q\mathcal{O}$; that is, $\sigma = \varphi(a_1, \ldots, a_q) = \sum_{i=1}^{q} a_i s_i(\mathfrak{z})$. The sections $s_1, \ldots, s_q$ therefore generate the $\mathcal{O}_\mathfrak{z}$-module $\mathscr{S}_\mathfrak{z}$ simultaneously over all of W.

Def. 4.2. If $\mathscr{S}$ is analytic over B, then one calls the set $\operatorname{Supp}(\mathscr{S}): = \{\mathfrak{z} \in B: \mathscr{S}_\mathfrak{z} \neq \mathbf{O}_\mathfrak{z}\}$ the *support of* $\mathscr{S}$.

Theorem 4.1. *If $\mathscr{S}$ is finitely generated, then* $\operatorname{Supp}(\mathscr{S})$ *is closed in B.*

Proof. We show that $B - \operatorname{Supp}(\mathscr{S})$ is open in $\mathbb{C}^n$. Let $\mathfrak{z}_0 \in B - \operatorname{Supp}(\mathscr{S})$ be chosen arbitrarily, and let $W(\mathfrak{z}_0) \subset B$ be an open neighborhood over which a sheaf epimorphism $\varphi: q\mathcal{O} \twoheadrightarrow \mathscr{S}|W$ exists. Let $s_1, \ldots, s_q$ be the images of the unit sections over W. Then $s_1(\mathfrak{z}_0) = \cdots = s_q(\mathfrak{z}_0) = \mathbf{O}_{\mathfrak{z}_0} = \mathbf{O}(\mathfrak{z}_0) \in \mathscr{S}_{\mathfrak{z}_0}$. Hence there exists a neighborhood $V(\mathfrak{z}_0) \subset W$ with $s_1|V = \cdots = s_q|V = \mathbf{O}|V$; therefore $\mathscr{S}|V = 0$, so $V \subset B\text{-}\operatorname{Supp}(\mathscr{S})$. □

Examples

1. $q\mathcal{O}$ is finitely generated, since $\mathrm{id}: q\mathcal{O} \to q\mathcal{O}$ is a sheaf epimorphism.

2. Let $\varepsilon: \mathscr{S}_1 \to \mathscr{S}_2$ be a sheaf epimorphism with $\mathscr{S}_1$ finitely generated. Then trivially $\mathscr{S}_2$ is also finitely generated.

3. Let $\mathscr{S}^* \subset \mathscr{S}$ be an analytic subsheaf with $\mathscr{S}$ finitely generated. Then 2, applied to the canonical projection $\mathfrak{q}: \mathscr{S} \to \mathscr{S}/\mathscr{S}^*$, shows that $\mathscr{S}/\mathscr{S}^*$ is finitely generated.

4. Let $A \subset B$ be analytic in B. The ideal sheaf $\mathscr{I}(A)$ is defined as follows: Let $\mathscr{I}_{\mathfrak{z}_0}: = \{\sigma \in \mathcal{O}_{\mathfrak{z}_0}$: There exists a $U(\mathfrak{z}_0) \subset B$ and a holomorphic f in U with $f|U \cap A = 0$ and $rf(\mathfrak{z}_0) = \sigma\}$ for $\mathfrak{z}_0 \in B$; then $\mathscr{I}(A): = \dot{\bigcup}_{\mathfrak{z} \in B} \mathscr{I}_\mathfrak{z}$.

a. $\mathscr{I}(A)$ is a subset of $\mathcal{O}$, and for $\sigma \in \mathscr{I}_\mathfrak{z}$ there exists a neighborhood $U(\mathfrak{z}) \subset B$ and an f such that $rf(\mathfrak{z}) = \sigma$. But then the set $rf(U)$, open in $\mathcal{O}$, lies in $\mathscr{I}$ and contains the element σ. Therefore σ is an interior point and $\mathscr{I}$ is open in $\mathcal{O}$.

b. That every stalk $\mathscr{I}_\mathfrak{z}$ is an ideal in the ring $\mathcal{O}_\mathfrak{z}$ follows immediately from the definition. Hence $\mathscr{I} \subset \mathcal{O}$ is an analytic subsheaf and an ideal sheaf.

By (3) the quotient sheaf $\mathscr{H} = \mathcal{O}/\mathscr{I}$ (a sheaf of $\mathbb{C}$-algebras!) is finitely generated. We show that $\operatorname{Supp}(\mathscr{H}) = A$. For $\mathfrak{z}_0 \in B - A$, $\mathscr{I}_{\mathfrak{z}_0} = \mathcal{O}_{\mathfrak{z}_0}$, therefore $\mathscr{H}_{\mathfrak{z}_0} = \mathbf{O}_{\mathfrak{z}_0}$. For $\mathfrak{z}_0 \in A$, $\mathscr{I}_{\mathfrak{z}_0} \neq \mathcal{O}_{\mathfrak{z}_0}$, since otherwise $\mathbf{1}_{\mathfrak{z}_0} \in \mathscr{I}_{\mathfrak{z}_0}$ and there would be a holomorphic function f on a neighborhood $U(\mathfrak{z}_0)$ with $f|U \cap A = 0$ and $rf(\mathfrak{z}_0) = \mathbf{1}_{\mathfrak{z}_0} = r1(\mathfrak{z}_0)$. But then rf and $r1$ would agree on a neighborhood $V(\mathfrak{z}_0) \subset U$. Since in this case r is bijective, it follows that $f|V = 1|V$, in particular $0 = f(\mathfrak{z}_0) = 1$.

Remark. Clearly

$$N(\mathscr{I}(A)) = \operatorname{Supp} \mathcal{O}/\mathscr{I}(A) = A.$$

Yet for an arbitrary ideal sheaf $\mathscr{J} \subset \mathcal{O}$, the equation $\mathscr{I}(N(\mathscr{J})) = \mathscr{J}$ is false.

5. Let B be connected, $B' \subset B$ open, $B' \neq \varnothing$ and $B' \neq B$. An open subset $\mathscr{S} = \pi^{-1}(B') \cup \mathbf{O}(B)$ of $\mathcal{O}$ is defined by $\mathscr{S}|B' := \mathcal{O}|B'$ and $\mathscr{S}|(B - B') = \mathsf{O}$. It is a subsheaf. Since $\mathscr{S}_{\mathfrak{z}} \subset \mathcal{O}_{\mathfrak{z}}$ is always an ideal, $\mathscr{S}$ is an ideal sheaf; but $\operatorname{Supp}(\mathscr{S}) = B'$ is not closed. Hence it follows that $\mathscr{S}$ is not finitely generated.

Def. 4.3. Let $\mathscr{S}$ be an analytic sheaf over B. $\mathscr{S}$ is called *coherent* if:

1. $\mathscr{S}$ is finite (that is, finitely generated).
2. $\mathscr{S}$ is relation finite (that is, if $U \subset B$ is open and $\varphi: q\mathcal{O}|U \to \mathscr{S}|U$ is an analytic sheaf morphism, then $\operatorname{Ker} \varphi$ is finitely generated).

Let $s_i \in \Gamma(U, \mathscr{S})$ be the images of the i-th unit section $\mathfrak{e}_i \in \Gamma(U, q\mathcal{O})$ under $\varphi: q\mathcal{O}|U \to \mathscr{S}|U$. Then an element $(a_1, \ldots, a_q) \in q\mathcal{O}_{\mathfrak{z}}$ is mapped onto $\mathbf{O}_{\mathfrak{z}}$ if and only if the "relation" $\sum_{i=1}^{q} a_i s_i(\mathfrak{z}) = \mathbf{O}$ is satisfied. We call $\operatorname{Ker} \varphi$ the *relation sheaf of* $s_1, \ldots, s_q$.

Consequences

1. *Coherence theorem of Oka*: $\mathcal{O}$ is coherent.
2. *Coherence theorem of Cartan*: The ideal sheaf $\mathscr{I}(A)$ of an analytic set is coherent.

These two results are very deep and cannot be proved here.

3. O is coherent. (This is trivial.)
4. If $\mathscr{S}$ is coherent and $\mathscr{S}^* \subset \mathscr{S}$ a finitely generated subsheaf, then $\mathscr{S}^*$ is also coherent.

PROOF. Let $W \subset B$ be open, $\varphi: q\mathcal{O}|W \to \mathscr{S}^*|W$ be an analytic sheaf morphism, $\imath: \mathscr{S}^*|W \hookrightarrow \mathscr{S}|W$ the canonical injection. Then $\imath \circ \varphi: q\mathcal{O}|W \to \mathscr{S}|W$ is also an analytic sheaf morphism, and $\operatorname{Ker} \varphi = \operatorname{Ker}(\imath \circ \varphi)$ is finitely generated. □

Theorem 4.2 (Existence of liftings). *Let $\varphi: \mathscr{S} \to \mathscr{S}^*$ be a sheaf epimorphism, $\varepsilon^*: q\mathcal{O} \to \mathscr{S}^*$ an arbitrary sheaf homomorphism. Then for every $\mathfrak{z}_0 \in B$ there is a neighborhood $U(\mathfrak{z}_0) \subset B$ and a (non-canonical) sheaf homomorphism $\varepsilon: q\mathcal{O}|U \to \mathscr{S}|U$ such that $\varphi \circ \varepsilon = \varepsilon^*$ (one calls any ε with these properties a lifting of ε^*).*

PROOF. Let $s_i^*: = \varepsilon^* \circ e_i \in \Gamma(B, \mathcal{S}^*)$ for $i = 1, \ldots, q$. Then for $\mathfrak{z}_0 \in B$ there are elements $\sigma_i \in \mathcal{S}_{\mathfrak{z}_0}$ with $\varphi(\sigma_i) = s_i^*(\mathfrak{z}_0)$. We can find a neighborhood $W(\mathfrak{z}_0) \subset B$ and sections $s_i \in \Gamma(W, \mathcal{S})$ with $s_i(\mathfrak{z}_0) = \sigma_i$; therefore $\varphi \circ s_i(\mathfrak{z}_0) = s_i^*(\mathfrak{z}_0)$. $\varphi \circ s_i$ and s_i^* coincide on a neighborhood $U(\mathfrak{z}_0) \subset W$. $\varepsilon: = \varphi_{(s_1, \ldots, s_q)}: q\mathcal{O}|U \to \mathcal{S}|U$ is an analytic sheaf morphism with

$$\varepsilon(a_1, \ldots, a_q) = \sum_{i=1}^{q} a_i s_i(\mathfrak{z}) \qquad \text{for} \qquad (a_1, \ldots, a_q) \in q\mathcal{O}_{\mathfrak{z}};$$

therefore

$$\varphi \circ \varepsilon(a_1, \ldots, a_q) = \sum_{i=1}^{q} a_i s_i^*(\mathfrak{z}) = \varepsilon^*(a_1, \ldots, a_q). \qquad \square$$

Theorem 4.3. *Let* $0 \to \mathcal{S}^* \xrightarrow{j} \mathcal{S} \xrightarrow{p} \mathcal{S}^{**} \to 0$ *be an exact sequence of analytic sheaves over B. If* $\mathcal{S}^*$ *and* $\mathcal{S}^{**}$ *are coherent,* $\mathcal{S}$ *is also coherent.*

PROOF

1. $\mathcal{S}$ is finitely generated: Since $\mathcal{S}^*$ and $\mathcal{S}^{**}$ are finitely generated, there are for every $\mathfrak{z}_0 \in B$ a neighborhood $W(\mathfrak{z}_0) \subset B$ and sheaf epimorphisms $\varepsilon^*: q^*\mathcal{O} \twoheadrightarrow \mathcal{S}^*$, $\varepsilon^{**}: q^{**}\mathcal{O} \twoheadrightarrow \mathcal{S}^{**}$ over W. Since $p: \mathcal{S} \to \mathcal{S}^{**}$ is surjective, there is (*w. l. o. g.* also over W) a lifting of ε^{**}

$$\varepsilon: q^{**}\mathcal{O} \to \mathcal{S} \qquad \text{with} \qquad p \circ \varepsilon = \varepsilon^{**}$$

If $\mathrm{pr}_1: q^*\mathcal{O} \oplus q^{**}\mathcal{O} \twoheadrightarrow q^*\mathcal{O}$ and $\mathrm{pr}_2: q^*\mathcal{O} \oplus q^{**}\mathcal{O} \twoheadrightarrow q^{**}\mathcal{O}$ are the canonical projections, then

$$\psi: (q^* + q^{**})\mathcal{O} \to \mathcal{S}$$

with

$$\psi(\sigma): = j \circ \varepsilon^* \circ \mathrm{pr}_1(\sigma) + \varepsilon \circ \mathrm{pr}_2(\sigma)$$

is an analytic sheaf morphism. It remains to show that ψ is surjective:

Let $\sigma \in \mathcal{S}_{\mathfrak{z}}$, $\mathfrak{z} \in W$. Then there is a $\sigma_1 \in q^{**}\mathcal{O}_{\mathfrak{z}}$ with $\varepsilon^{**}(\sigma_1) = p(\sigma)$. Clearly $\sigma - \varepsilon(\sigma_1)$ lies in $\mathrm{Ker}\, p = \mathrm{Im}\, j$, therefore there is a $\sigma_2 \in \mathcal{S}_{\mathfrak{z}}^*$ with $j(\sigma_2) = \sigma - \varepsilon(\sigma_1)$. Furthermore, we can find a $\sigma_3 \in q^*\mathcal{O}_{\mathfrak{z}}$ with $\varepsilon^*(\sigma_3) = \sigma_2$. Now

$$\psi(\sigma_3, \sigma_1) = j \circ \varepsilon^*(\sigma_3) + \varepsilon(\sigma_1) = j(\sigma_2) + \varepsilon(\sigma_1) = \sigma.$$

2. $\mathcal{S}$ is relation finite: Let $W \subset B$ be open, $\varphi: q\mathcal{O}|W \to \mathcal{S}|W$ an analytic sheaf morphism and $\mathfrak{z}_0 \in W$ an arbitrary point. Since $\mathcal{S}^{**}$ is relation finite there is a neighborhood $V(\mathfrak{z}_0) \subset W$ and, over V, a sheaf morphism $\psi_1: r\mathcal{O}|V \twoheadrightarrow \mathrm{Ker}(p \circ \varphi)|V$. This gives the exact sequence:

$$r\mathcal{O}|V \xrightarrow{\psi_1} q\mathcal{O}|V \xrightarrow{p \circ \varphi} \mathcal{S}^{**}.$$

Because $\mathrm{Ker}\, p = \mathrm{Im}\, j \simeq \mathcal{S}^*$, we can regard $\varphi \circ \psi_1: r\mathcal{O} \to \mathrm{Ker}\, p$ as a mapping $\varphi \circ \psi_1: r\mathcal{O} \to \mathcal{S}^*$, and since $\mathcal{S}^*$ is relation finite, there is a neighborhood $U(\mathfrak{z}_0) \subset V$ and a sheaf epimorphism $\psi_2: s\mathcal{O}|U \twoheadrightarrow \mathrm{Ker}(\varphi \circ \psi_1)|U$. This yields the following exact sequence:

$$s\mathcal{O}|U \xrightarrow{\psi_2} r\mathcal{O}|U \xrightarrow{\varphi \circ \psi_1} \mathcal{S}.$$

Hence we obtain (over U):

a. $\varphi \circ (\psi_1 \circ \psi_2) = (\varphi \circ \psi_1) \circ \psi_2 = 0$

b. If $\sigma \in q\mathcal{O}$ and $\varphi(\sigma) = \mathbf{O}$, then $p \circ \varphi(\sigma) = \mathbf{O}$ also and there is a $\sigma_1 \in r\mathcal{O}$ with $\psi_1(\sigma_1) = \sigma$. Then $\varphi \circ \psi_1(\sigma_1) = \mathbf{O}$ and there is a $\sigma_2 \in s\mathcal{O}$ with $\psi_2(\sigma_2) = \sigma_1$. Then $\psi_1 \circ \psi_2(\sigma_2) = \sigma$.

(a) and (b) imply that the following sequence is exact:

$$s\mathcal{O}|U \xrightarrow{\psi_1 \circ \psi_2} q\mathcal{O}|U \xrightarrow{\varphi} \mathcal{S}|U$$

Therefore Ker φ is finitely generated. □

Theorem 4.4. *Let $\mathcal{S}^* \xrightarrow{j} \mathcal{S} \xrightarrow{p} \mathcal{S}^{**} \to \mathrm{O}$ be an exact sequence of sheaves over B. If $\mathcal{S}^*$ and $\mathcal{S}$ are coherent, then $\mathcal{S}^{**}$ is also coherent.*

Proof

1. Since p is surjective, it follows immediately that $\mathcal{S}^{**}$ is finitely generated.

2. Let $\varepsilon^{**}: q^{**}\mathcal{O} \to \mathcal{S}^{**}$ be an arbitrary sheaf homomorphism on an open set $W \subset B$, $\varepsilon: q^{**}\mathcal{O} \to \mathcal{S}$ a lifting (with $p \circ \varepsilon = \varepsilon^{**}$). Since $\mathcal{S}^*$ is finitely generated, we can find a neighborhood $V(\mathfrak{z}_0) \subset W$ and a sheaf epimorphism $\varepsilon^*: q^*\mathcal{O} \twoheadrightarrow \mathcal{S}^*$ on V for every point $\mathfrak{z}_0 \in W$. Now let $\psi: q^*\mathcal{O} \oplus q^{**}\mathcal{O} \to \mathcal{S}$ be a sheaf morphism on V defined by

$$\psi(\sigma_1, \sigma_2) := j \circ \varepsilon^*(\sigma_1) + \varepsilon(\sigma_2).$$

Since $\mathcal{S}$ is coherent there exists an exact sequence $q\mathcal{O} \xrightarrow{\varphi} q^*\mathcal{O} \oplus q^{**}\mathcal{O} \xrightarrow{\psi} \mathcal{S}$ on a neighborhood $U(\mathfrak{z}_0) \subset V$. Let $\alpha: q\mathcal{O} \to q^{**}\mathcal{O}$ be defined by $\alpha := \mathrm{pr}_2 \circ \varphi$. The theorem will be proved once we show the exactness of the sequence $q\mathcal{O} \xrightarrow{\alpha} q^{**}\mathcal{O} \xrightarrow{\varepsilon^{**}} \mathcal{S}^{**}$. For $\mathfrak{z} \in U$ and $\sigma \in q^{**}\mathcal{O}_{\mathfrak{z}}$ the following statements are equivalent:

1. $\sigma \in \mathrm{Ker}(\varepsilon^{**})$
2. $\varepsilon(\sigma) \in \mathrm{Ker}\, p = \mathrm{Im}\, j$
3. There is a $\sigma_1 \in q^*\mathcal{O}_{\mathfrak{z}}$ with $j \circ \varepsilon^*(\sigma_1) = \varepsilon(\sigma)$
4. $\psi(-\sigma_1, \sigma) = \mathbf{O}$ for a $\sigma_1 \in q^*\mathcal{O}_{\mathfrak{z}}$
5. There is a $\sigma_2 \in q\mathcal{O}_{\mathfrak{z}}$ with $\varphi(\sigma_2) = (-\sigma_1, \sigma)$
6. $\sigma = \mathrm{pr}_2 \circ \varphi(\sigma_2) = \alpha(\sigma_2) \in \mathrm{Im}\, \alpha$. □

Theorem 4.5. *Let $\mathrm{O} \to \mathcal{S}^* \xrightarrow{j} \mathcal{S} \xrightarrow{p} \mathcal{S}^{**}$ be an exact sequence of analytic sheaves over B. If $\mathcal{S}$, $\mathcal{S}^{**}$ are coherent, then $\mathcal{S}^*$ is also coherent.*

Proof. We may regard $\mathcal{S}^*$ as an analytic subsheaf of $\mathcal{S}$, so it suffices to show that $\mathcal{S}^*$ is finitely generated. Let $\mathfrak{z}_0 \in B$ be chosen arbitrarily. Since $\mathcal{S}$ and $\mathcal{S}^{**}$ are coherent there is a neighborhood $W(\mathfrak{z}_0) \subset B$, and over W, a sheaf epimorphism $\varepsilon: q\mathcal{O} \to \mathcal{S}$ and a sheaf epimorphism $\varphi: q^*\mathcal{O} \to q\mathcal{O}$ such that the sequence

$$q^*\mathcal{O} \xrightarrow{\varphi} q\mathcal{O} \xrightarrow{p \circ \varepsilon} \mathcal{S}^{**}$$

is exact.

Then $\varepsilon \circ \varphi(q^*\mathcal{O}) = \varepsilon(\mathrm{Im}\, \varphi) = \varepsilon(\mathrm{Ker}(p \circ \varepsilon)) = \mathrm{Ker}\, p = \mathrm{Im}\, j$, so $\varepsilon \circ \varphi(q^*\mathcal{O}) \simeq \mathcal{S}^*$. Hence $\mathcal{S}^*$ is finitely generated. □

Theorem 4.6

1. *If $\mathcal{S}$ is a coherent sheaf over B and $\mathcal{S}^* \subset \mathcal{S}$ a coherent analytic subsheaf, then $\mathcal{S}/\mathcal{S}^*$ is also coherent.*
2. *If $\mathcal{S}_1, \ldots, \mathcal{S}_\ell$ are coherent analytic sheaves over B then $\mathcal{S}_1 \oplus \cdots \oplus \mathcal{S}_\ell$ is also coherent.*

PROOF

1. There exists a canonical exact sequence $0 \to \mathcal{S}^* \to \mathcal{S} \to \mathcal{S}/\mathcal{S}^* \to 0$. The result follows by Theorem 4.4.

2. For $\ell = 2$, apply Theorem 4.3 to the exact sequence $0 \to \mathcal{S}_1 \xrightarrow{j_1} \mathcal{S}_1 \oplus \mathcal{S}_2 \xrightarrow{p_2} \mathcal{S}_2 \to 0$. The result follows by induction from the isomorphism $\mathcal{S}_1 \oplus \cdots \oplus \mathcal{S}_\ell \simeq (\mathcal{S}_1 \oplus \cdots \oplus \mathcal{S}_{\ell-1}) \oplus \mathcal{S}_\ell$. □

Theorem 4.7. *Let $\varphi: \mathcal{S}_1 \to \mathcal{S}_2$ be a homomorphism of coherent sheaves over B. Then* Ker φ *and* Im φ *are also coherent.*

PROOF. The sequence $0 \to \operatorname{Ker} \varphi \to \mathcal{S}_1 \to \mathcal{S}_2$ is exact, so Ker φ is coherent. Since $\operatorname{Im} \varphi \simeq \mathcal{S}_1/\operatorname{Ker} \varphi$, the coherence of Im φ follows from Theorem 4.6. □

Theorem 4.8 (Serre's five lemma). *Let $\mathcal{S}' \xrightarrow{j_1} \mathcal{S}'' \xrightarrow{j_2} \mathcal{S} \xrightarrow{p_1} \mathcal{S}^* \xrightarrow{p_2} \mathcal{S}^{**}$ be an exact sequence of sheaves. If $\mathcal{S}'$, $\mathcal{S}''$, $\mathcal{S}^*$, $\mathcal{S}^{**}$ are coherent, so also is $\mathcal{S}$.*

PROOF. The sequence $0 \to \mathcal{S}''/\operatorname{Im} j_1 \to \mathcal{S} \to \operatorname{Ker} p_2 \to 0$ is exact and the sheaves $\mathcal{S}''/\operatorname{Im} j_1$ and Ker p_2 are coherent. Hence the result follows from Theorem 4.3. □

Remark. With Serre's five lemma, we can deduce the other theorems:

For example, if the sequence $0 \to \mathcal{S}^* \to \mathcal{S} \to \mathcal{S}^{**}$ is exact, then so is the sequence $0 \to 0 \to \mathcal{S}^* \to \mathcal{S} \to \mathcal{S}^{**}$. If $\mathcal{S}$ and $\mathcal{S}^{**}$ are coherent, then the coherence of $\mathcal{S}^*$ follows from the five lemma.

EXAMPLE. Let $A \subset B$ be an analytic set, $\mathcal{I}(A)$ its ideal sheaf and $\mathcal{H}(A) = \mathcal{O}/\mathcal{I}(A)$. Since $\mathcal{I}(A)$ is coherent, the sheaf $\mathcal{H}(A)$ is also coherent.

If we choose, for example, $A = \{0\} \subset \mathbb{C}^n$, then $\mathcal{I}(A)_0 = \{f_0 : f(0) = 0\}$ is the maximal ideal in $\mathcal{O}_0$, $\mathcal{I}(A)_\mathfrak{z} = \mathcal{O}_\mathfrak{z}$ for $\mathfrak{z} \neq 0$. Therefore

$$\mathcal{H}(A)_\mathfrak{z} = \begin{cases} \mathbb{C} & \text{for } \mathfrak{z} = 0 \\ 0 & \text{otherwise.} \end{cases}$$

In conclusion we note that the general theorems and constructions of this chapter carry over word for word if one admits as the base space an arbitrary topological space instead of a region of $\mathbb{C}^n$.

We define: Let X be a topological space, $\mathcal{R}$ a sheaf of $\mathbb{C}$-algebras over X. A sheaf of $\mathcal{R}$-modules over X is called *coherent* if it is a finite and relation finite sheaf of $\mathcal{R}$-modules.

In particular $\mathcal{R}$ itself is coherent if for every open set $U \subset X$ and each $\mathcal{R}$-homomorphism $\varphi: q\mathcal{R}|U \to \mathcal{R}|U$ the sheaf Ker φ is finite over $\mathcal{R}$.

Theorem 4.9. *Let $\mathscr{R}$ be a coherent sheaf of $\mathbb{C}$-algebras over X, $\mathscr{I} \subset \mathscr{R}$ a coherent ideal sheaf. Then as a sheaf of $\mathbb{C}$-algebras $\mathscr{R}/\mathscr{I}$ is coherent.*

PROOF. We already know that as a sheaf of $\mathscr{R}$-modules $\mathscr{R}/\mathscr{I}$ is coherent. Now let $\pi: \mathscr{R} \to \mathscr{R}/\mathscr{I}$ the canonical projection, $U \subset X$ open, $\varphi: q(\mathscr{R}/\mathscr{I})|U \to (\mathscr{R}/\mathscr{I})|U$ a given $(\mathscr{R}/\mathscr{I})$-homomorphism.

π induces an $\mathscr{R}$-homomorphism $\pi_q: q\mathscr{R} \to q(\mathscr{R}/\mathscr{I})$ and we set $\psi := \varphi \circ \pi_q: q\mathscr{R}|U \to (\mathscr{R}/\mathscr{I})|U$. ψ is an $\mathscr{R}$-homomorphism and hence for every $\mathfrak{z}_0 \in U$ there is an open neighborhood $V(\mathfrak{z}_0) \subset U$ and sections $s_1, \ldots, s_p$, $s_i \in \Gamma(V, \operatorname{Ker} \psi)$ which generate Ker ψ over V. For the sections

$$\tilde{s}_i := \pi_q(s_i) \in \Gamma(V, q(\mathscr{R}/\mathscr{I}))$$

we have

$$\varphi(\tilde{s}_i) = \varphi \circ \pi_q(s_i) = \psi(s_i) = 0,$$

hence $\tilde{s}_i \in \Gamma(V, \operatorname{Ker} \varphi)$.

It is easily verified that the $\tilde{s}_1, \ldots, \tilde{s}_p$ generate the sheaf Ker φ over V as an $(\mathscr{R}/\mathscr{I})$-module.

Corollary. *If $B \subset \mathbb{C}^n$ is a region, $A \subset B$ an analytic subset, then $\mathscr{H}(A)$ and hence $\mathscr{H}(A)|A$ is a coherent sheaf of $\mathbb{C}$-algebras.*

CHAPTER V

Complex Manifolds

1. Complex Ringed Spaces

Let R be a local $\mathbb{C}$-algebra with maximal ideal $\mathfrak{m}$, (see Def. 5.1 in Chapter III), and let $\pi: R \to R/\mathfrak{m} \simeq \mathbb{C}$ be the canonical projection. If $f \in R$, the value of f is the complex number $[f]: = \pi(f)$.

EXAMPLE. Let f be a holomorphic function on a region B, $\mathfrak{z}_0 \in B$ a point. Then $f_{\mathfrak{z}_0} = (W, f)_{\mathfrak{z}_0}$ is an element of the local $\mathbb{C}$-algebra $\mathcal{O}_{\mathfrak{z}_0}$ and $rf \in \Gamma(B, \mathcal{O})$ with $rf(\mathfrak{z}_0) = f_{\mathfrak{z}_0}$.

We introduce the complex valued function $[rf]$ on B by setting

$$[rf](\mathfrak{z}_0): = [rf(\mathfrak{z}_0)].$$

Then $[rf](\mathfrak{z}_0) = [f_{\mathfrak{z}_0}] = \pi(f_{\mathfrak{z}_0}) = f(\mathfrak{z}_0)$, so that $[rf] = f$. Consequently the inverse of the isomorphism $r: M_W \to \Gamma(W, \mathcal{O})$ is given by $r^{-1}(s): = [s]$.

Def. 1.1. A pair $(X, \mathscr{H})$ is called a *complex ringed space* if:

1. X is a topological space;
2. $\mathscr{H}$ is a sheaf of local $\mathbb{C}$-algebras over X.

If $W \subset X$ is an open set, then the set of all complex valued functions on W will be denoted by $\mathscr{F}(W, \mathbb{C})$.

If $f \in \Gamma(W, \mathscr{H})$, then there is an element $[f] \in \mathscr{F}(W, \mathbb{C})$ defined by $[f](x): = [f(x)] \in \mathscr{H}_x/\mathfrak{m}_x \simeq \mathbb{C}$. The correspondence $\Gamma(W, \mathscr{H}) \to \mathscr{F}(W, \mathbb{C})$ given by $f \mapsto [f]$ is a homomorphism of $\mathbb{C}$-algebras but, in general, is neither surjective nor injective.

Comment. If $(X, \mathscr{H})$ is a complex ringed space and $W \subset X$ open, then naturally $(W, \mathscr{H}|W)$ is also a complex ringed space.

Def. 1.2. Let $(X_1, \mathscr{H}_1)$, $(X_2, \mathscr{H}_2)$ be complex ringed spaces. An *isomorphism* between $(X_1, \mathscr{H}_1)$ *and* $(X_2, \mathscr{H}_2)$ is a pair $\varphi = (\tilde{\varphi}, \varphi_*)$ with the following properties:

1. $\tilde{\varphi}: X_1 \to X_2$ is a topological mapping.
2. $\varphi_*: \mathscr{H}_1 \to \mathscr{H}_2$ is a topological mapping.
3. φ_* is stalk-preserving with respect to $\tilde{\varphi}$; that is, the following diagram is commutative:

$$\begin{array}{ccc} \mathscr{H}_1 & \xrightarrow{\varphi_*} & \mathscr{H}_2 \\ \pi_1 \downarrow & & \downarrow \pi_2 \\ X_1 & \xrightarrow{\tilde{\varphi}} & X_2 \end{array}$$

4. For every $x \in X_1$, $\varphi_*|(\mathscr{H}_1)_x : (\mathscr{H}_1)_x \to (\mathscr{H}_2)_{\tilde{\varphi}(x)}$ is a homomorphism of $\mathbb{C}$-algebras.

The existence of an isomorphism φ between $(X_1, \mathscr{H}_1)$ and $(X_2, \mathscr{H}_2)$ is expressed briefly by

$$(X_1, \mathscr{H}_1) \simeq (X_2, \mathscr{H}_2).$$

Theorem 1.1. *Let* $\varphi = (\tilde{\varphi}, \varphi_*) : (X_1, \mathscr{H}_1) \to (X_2, \mathscr{H}_2)$ *be an isomorphism of complex ringed spaces. Then for every open set* $V \subset X_2$ *there is a* $\mathbb{C}$*-algebra isomorphism*

$$\hat{\varphi} : \Gamma(V, \mathscr{H}_2) \to \Gamma(\tilde{\varphi}^{-1}(V), \mathscr{H}_1)$$

defined by $\hat{\varphi}(s) := \varphi_*^{-1} \circ s \circ \tilde{\varphi}$.

PROOF. $\hat{\varphi}(s) : \tilde{\varphi}^{-1}(V) \to \mathscr{H}_1$ is clearly continuous, and

$$\begin{aligned} \pi_1 \circ (\varphi_*^{-1} \circ s \circ \tilde{\varphi}) &= (\pi_1 \circ \varphi_*^{-1}) \circ (s \circ \tilde{\varphi}) = (\tilde{\varphi}^{-1} \circ \pi_2) \circ (s \circ \tilde{\varphi}) \\ &= \tilde{\varphi}^{-1} \circ (\pi_2 \circ s) \circ \tilde{\varphi} = \mathrm{id}_{\tilde{\varphi}^{-1}(V)}. \end{aligned}$$

It is clear that $\hat{\varphi}$ is a homomorphism of $\mathbb{C}$-algebras. An inverse is given by $\hat{\varphi}^{-1}(t) = \varphi_* \circ t \circ \tilde{\varphi}^{-1}$. □

Lemma 1. *If* R *is a local* $\mathbb{C}$*-algebra with the maximal ideal* $\mathfrak{m}$, *then* $a \in \mathfrak{m}$ *if and only if* $a - c \cdot 1 \notin \mathfrak{m}$ *for all* $c \in \mathbb{C} - \{0\}$.

PROOF

1. Let $a \in \mathfrak{m}$, $a - c \cdot 1 \in \mathfrak{m}$. Then also $c \cdot 1 = a - (a - c \cdot 1) \in \mathfrak{m}$. That is, c cannot lie in $\mathbb{C} - \{0\}$.

2. For all $c \in \mathbb{C} - \{0\}$ let $a - c \cdot 1 \notin \mathfrak{m}$. We set $c := \pi(a)$. Then $\pi(a - c \cdot 1) = 0$. Therefore $a - c \cdot 1 \in \mathfrak{m}$, hence $c = 0$ and $a \in \mathfrak{m}$. □

Lemma 2. *Let $\rho:(R_1, \mathfrak{m}_1) \to (R_2, \mathfrak{m}_2)$ be a $\mathbb{C}$-algebra homomorphism between local $\mathbb{C}$-algebras. Then $\rho(\mathfrak{m}_1) \subset \mathfrak{m}_2$. If in particular ρ is an isomorphism, then $\rho(\mathfrak{m}_1) = \mathfrak{m}_2$.*

PROOF

1. If $\sigma \in \mathfrak{m}_1$, then $\sigma - c \cdot 1 \notin \mathfrak{m}_1$ for all $c \in \mathbb{C} - \{0\}$. Therefore for every $c \in \mathbb{C} - \{0\}$ there is a σ_c with $\sigma_c \cdot (\sigma - c \cdot 1) = 1$, and then

$$\rho(\sigma_c) \cdot (\rho(\sigma) - c \cdot 1) = 1.$$

Hence $\rho(\sigma) - c \cdot 1 \notin \mathfrak{m}_2$, so $\rho(\sigma)$ lies in $\mathfrak{m}_2$.

2. If ρ is an isomorphism, then $\rho^{-1}(\mathfrak{m}_2) \subset \mathfrak{m}_1$. Therefore $\mathfrak{m}_2 = \rho\rho^{-1}(\mathfrak{m}_2) \subset \rho(\mathfrak{m}_1) \subset \mathfrak{m}_2$, and $\rho(\mathfrak{m}_1) = \mathfrak{m}_2$. □

Theorem 1.2. *Let $\varphi = (\tilde{\varphi}, \varphi_*):(X_1, \mathscr{H}_1) \to (X_2, \mathscr{H}_2)$ be an isomorphism of complex ringed spaces. For an open set $V \subset X_2$ let $\varphi^*:\mathscr{F}(V, \mathbb{C}) \to \mathscr{F}(\tilde{\varphi}^{-1}(V), \mathbb{C})$ be defined by $\varphi^*(f) := f \circ \tilde{\varphi}$. Then for every $s \in \Gamma(V, \mathscr{H}_2)$, $[\hat{\varphi}(s)] = \varphi^*([s])$ (therefore $[\varphi_*^{-1} \circ s \circ \tilde{\varphi}] = [s] \circ \tilde{\varphi}$).*

PROOF. Let $V \subset X_2$ be open, $y \in V$, $x := \tilde{\varphi}^{-1}(y)$ and $s \in \Gamma(V, \mathscr{H}_2)$. Then $s(y) = ([s](y)) \cdot 1 + \sigma^*$ with $\sigma^* \in (\mathfrak{m}_2)_y$, therefore

$$\varphi_*^{-1}(s(y)) = ([s](y)) \cdot 1 + \varphi_*^{-1}(\sigma^*),$$

with $\varphi_*^{-1}(\sigma^*) \in (\mathfrak{m}_1)_x$. Hence

$$[\varphi_*^{-1} \circ s \circ \tilde{\varphi}](x) = [\varphi_*^{-1}(s(y))] = [s](y) = [s] \circ \tilde{\varphi}(x)$$

follows. □

We thus obtain the following commutative diagram:

$$\begin{array}{ccc}
t \in \Gamma(\tilde{\varphi}^{-1}(V), \mathscr{H}_1) & \xleftarrow{\hat{\varphi}} & \Gamma(V, \mathscr{H}_2) \ni s \\
\downarrow & & \downarrow \\
[t] \in \mathscr{F}(\tilde{\varphi}^{-1}(V), \mathbb{C}) & \xleftarrow{\varphi^*} & \mathscr{F}(V, \mathbb{C}) \ni [s]
\end{array}$$

Since $\hat{\varphi}$ and φ^* are isomorphisms, $t \mapsto [t]$ is injective if and only if $s \mapsto [s]$ is injective.

Def. 1.3. A (*reduced*) *complex space* is a complex ringed space $(X, \mathscr{H})$ with the following properties:

1. X is a Hausdorff space.
2. For every point $x_0 \in X$ there is an open neighborhood $U(x_0) \subset X$ and an analytic set A such that $(U, \mathscr{H}|U) \simeq (A, \mathscr{H}(A))$.

(A lies in an open set $B \subset \mathbb{C}^n$ and $\mathscr{H}(A) := (\mathscr{O}/\mathscr{I}(A))|A$, where $\mathscr{I}(A)$ is the ideal sheaf of A. $\mathscr{H}(A)$ is a coherent sheaf of local $\mathbb{C}$-algebras and hence $\mathscr{H}$ is also coherent.)

A reduced complex space therefore looks locally like an analytic set. If this analytic set has no singularities, then we call the complex space a complex manifold:

Def. 1.4. A *complex manifold* is a complex ringed space $(X, \mathscr{H})$ with the following properties:

1. X is a Hausdorff space
2. For every point $x_0 \in X$ there exists an open neighborhood $U(x_0) \subset X$ and a region $B \subset \mathbb{C}^n$ such that $(U, \mathscr{H}|U) \simeq (B, \mathcal{O})$.

Theorem 1.3. *Let $(X, \mathscr{H})$ be a complex manifold, $W \subset X$ an open subset. Then the mapping $\Gamma(W, \mathscr{H}) \to \mathscr{F}(W, \mathbb{C})$ given by $f \mapsto [f]$ is injective, and for every $f \in \Gamma(W, \mathscr{H})$, $[f]$ is continuous.*

PROOF. There is an open covering $(U_\iota)_{\iota \in I}$ of W and a system $(B_\iota)_{\iota \in I}$ of open sets such that $(U_\iota, \mathscr{H}|U_\iota) \simeq (B_\iota, \mathcal{O}|B_\iota)$. For $f \in \Gamma(W, \mathscr{H})$

$$[f|U_\iota] = [f]|U_\iota.$$

Hence it suffices to prove the proposition for the sets U_ι. It follows immediately from Theorem 1.2 and the equation $r^{-1}(s) = [s]$ that the mapping $\Gamma(U_\iota, \mathscr{H}) \to \mathscr{F}(U_\iota, \mathbb{C})$ is injective. If $f \in \Gamma(U_\iota, \mathscr{H})$, then $\varphi_* \circ f \circ \tilde{\varphi}^{-1} = \hat{\varphi}^{-1}(f)$ lies in $\Gamma(B_\iota, \mathcal{O})$; $[\hat{\varphi}^{-1}(f)]$ is therefore continuous. Hence

$$[f] = [\hat{\varphi}\hat{\varphi}^{-1}(f)] = [\hat{\varphi}^{-1}(f)] \circ \tilde{\varphi}$$

is also continuous. □

Def. 1.5. Let $(X, \mathscr{H})$ be a complex manifold, $W \subset X$ open. A *holomorphic function over W* is an element of the set $[\Gamma(W, \mathscr{H})] = \{[f] \in \mathscr{F}(W, \mathbb{C}): f \in \Gamma(W, \mathscr{H})\}$.

Remarks

1. The mapping $f \mapsto [f]$ defines an isomorphism from $\Gamma(W, \mathscr{H})$ onto the set of holomorphic functions over W.
2. Every holomorphic function is continuous.
3. If $U \subset X$ is open, $B \subset \mathbb{C}^n$ a region and $\varphi:(U, \mathscr{H}) \to (B, \mathcal{O})$ an isomorphism, then for every open subset $V \subset U$ a function $f \in \mathscr{F}(V, \mathbb{C})$ is holomorphic if and only if $f \circ \tilde{\varphi}^{-1}$ is holomorphic.
4. If $U \subset X$ is open, $B \subset \mathbb{C}^n$ a region and $\varphi:(U, \mathscr{H}) \to (B, \mathcal{O})$ an isomorphism, then the pair $(U, \tilde{\varphi})$ is called a complex coordinate system for X. If $(U_1, \tilde{\varphi}_1), (U_2, \tilde{\varphi}_2)$ are two complex coordinate systems with $U_1 \cap U_2 \neq \varnothing$, then $\tilde{\varphi}_{12} := \tilde{\varphi}_1 \circ \tilde{\varphi}_2^{-1}: \tilde{\varphi}_2(U_1 \cap U_2) \to \tilde{\varphi}_1(U_1 \cap U_2)$ is a homeomorphism of open sets in $\mathbb{C}^n$.

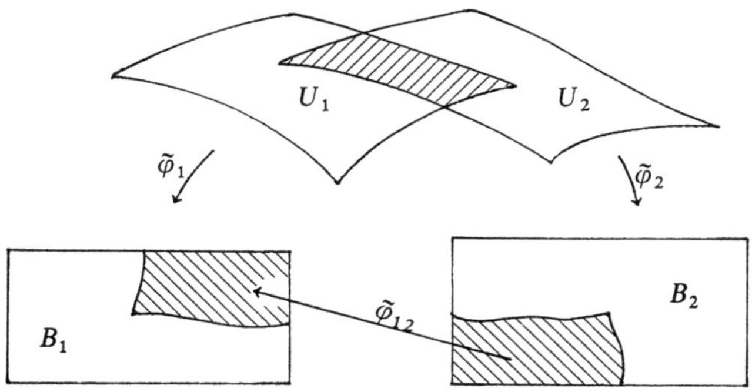

Figure 21. Compatibility of complex coordinate systems.

If f is holomorphic on $\tilde{\varphi}_1(U_1 \cap U_2)$, then $f \circ \tilde{\varphi}_1$ is holomorphic on $U_1 \cap U_2$ (by 3). Therefore, $f \circ \tilde{\varphi}_{12} = (f \circ \tilde{\varphi}_1) \circ \tilde{\varphi}_2^{-1}$ is holomorphic on $\tilde{\varphi}_2(U_1 \cap U_2)$. Then $\tilde{\varphi}_{12}$ is a holomorphic mapping (see Theorem 7.1, Chapter I).

To conclude this section we shall show conversely that we can also define a complex manifold with the help of a suitable system of complex coordinates.

Let X be a Hausdorff space, $(U_\iota)_{\iota \in I}$ an open covering of X. For every U_ι let there be given a homeomorphism $\tilde{\varphi}_\iota$ from U_ι onto a region $B_\iota \subset \mathbb{C}^n$ such that every coordinate transformation

$$\tilde{\varphi}_{\iota_1} \circ \tilde{\varphi}_{\iota_2}^{-1}: \tilde{\varphi}_{\iota_2}(U_{\iota_1} \cap U_{\iota_2}) \to \tilde{\varphi}_{\iota_1}(U_{\iota_1} \cap U_{\iota_2})$$

is holomorphic. One calls the system $\{(U_\iota, \tilde{\varphi}_\iota): \iota \in I\}$ a *complex atlas for* X. Now let $W \subset X$ be open, $f \in \mathscr{F}(W, \mathbb{C})$ and $x_0 \in W$. f is called holomorphic at x_0 if there exists an $\iota_0 \in I$ and a neighborhood $U(x_0) \subset W \cap U_{\iota_0}$ such that $f \circ \tilde{\varphi}_{\iota_0}^{-1}$ is holomorphic in $\tilde{\varphi}_{\iota_0}(U) \subset B_{\iota_0}$. f is called holomorphic on W if f is holomorphic at every point $x \in W$.

Because of the compatibility condition for the coordinate systems $f \circ \tilde{\varphi}_\iota^{-1}$ is holomorphic at $\tilde{\varphi}_\iota(x_0)$ for every ι with $x_0 \in U_\iota$, whenever f is holomorphic at x_0.

Let M_W be the set of holomorphic functions on W, and $r_V^W: M_W \to M_V$ the usual restriction mapping. Then $\{M_W, r_V^W\}$ is a pre-sheaf. The corresponding sheaf $\mathscr{H}$ is called the *sheaf of germs of holomorphic functions over* X.

If $x_0 \in U_\iota \cap W$ and $f \in M_W$, then

$$(W, f)_{x_0} = (W \cap U_\iota, f|W \cap U_\iota)_{x_0}.$$

The system $M_{W \cap U_\iota}$ together with the corresponding restriction mappings form a pre-sheaf for the sheaf $\mathscr{H}|U_\iota$. An isomorphism between the pre-sheaf of $\mathscr{H}|U_\iota$ and the pre-sheaf of $\mathscr{O}|B_\iota$ is defined by $f \mapsto f \circ \tilde{\varphi}_\iota^{-1}$, and this isomorphism induces an isomorphism $(\varphi_\iota)_*: \mathscr{H}|U_\iota \to \mathscr{O}|B_\iota$. $\mathscr{H}$ is thus a sheaf of local $\mathbb{C}$-algebras, $\varphi_\iota := (\tilde{\varphi}_\iota, (\varphi_\iota)_*): (U_\iota, \mathscr{H}) \to (B_\iota, \mathscr{O})$ is an isomorphism of complex ringed spaces, and $(X, \mathscr{H})$ is a complex manifold.

2. Function Theory on Complex Manifolds

Let $(X, \mathscr{H})$ be a complex manifold, $x_0 \in X$. Then there is a neighborhood $U(x_0) \subset X$ and a homeomorphism φ of U onto a region $B \subset \mathbb{C}^n$. The natural number n is independent of the particular choice of φ and one defines $\dim_{x_0}(X) := n$.

Henceforth we always assume that $\dim_x(X) = n =$ constant on all of X. Then $(X, \mathscr{H})$ is called an *n-dimensional complex manifold.*

Theorem 2.1. *Let $(X, \mathscr{H})$ be an n-dimensional complex manifold, $W \subset X$ open. Then $(W, \mathscr{H}|W)$ is also an n-dimensional complex manifold.*

PROOF. It is clear that W is a Hausdorff space and $\mathscr{H}|W$ a sheaf of local $\mathbb{C}$-algebras. For every point $x_0 \in W$ there is a neighborhood $U(x_0) \subset X$ and an isomorphism $\varphi:(U, \mathscr{H}|U) \to (B, \mathcal{O})$. Then $W \cap U$ is a neighborhood of x_0 in W and $(W \cap U, \mathscr{H}|W \cap U) \simeq (\tilde{\varphi}(W \cap U), \mathcal{O})$. □

Def. 2.1. A complex manifold $(X, \mathscr{H})$ is *connected* if the underlying topological space is connected (so there is no decomposition $X = X_1 \cup X_2$ into two disjoint non-empty open subsets).

Theorem 2.2 (Identity theorem). *Let $(X, \mathscr{H})$ be a connected complex manifold, f_1, f_2 holomorphic functions on X and $V \subset X$ a non-empty open subset with $f_1|V = f_2|V$. Then $f_1 = f_2$.*

PROOF. Let $W_1 := \{x \in X : rf_1(x) = rf_2(x)\}$, $W_2 := X - W_1$. W_1 is not empty since V is contained in W_1, and W_1 is open since the set where two sections coincide is always open. Let $x_0 \in W_2$ be an arbitrary point. Then in X there exists an open neighborhood $U(x_0)$ and an isomorphism $(U, \mathscr{H}) \simeq (B, \mathcal{O})$ where B denotes a domain in $\mathbb{C}^n$. If x_0 is not an interior point of W_2, then $U \cap W_1 \neq \varnothing$ is an open neighborhood and $rf_1|U \cap W_1 = rf_2|U \cap W_1$. By the identity theorem in $\mathbb{C}^n$ it now follows that rf_1 and rf_2 coincide on U, and so in particular, that x_0 lies in W_1. That is a contradiction. Every point of W_2 is an interior point of W_2, so W_2 is open. Since X is connected, it follows that $W_1 = X$ and therefore $f_1 = f_2$. □

Theorem 2.3 (Maximum principle). *Let $(X, \mathscr{H})$ be a connected complex manifold, f holomorphic on X, $x_0 \in X$ a point. If $|f|$ has a local maximum at x_0, then f is constant.*

PROOF. There is a neighborhood $U(x_0) \subset X$ and an isomorphism φ: $(U, \mathscr{H}) \xrightarrow{\sim} (B, \mathcal{O})$. Without loss of generality we may assume that $\tilde{\varphi}(x_0) = 0$ and B is a polycylinder about the origin. For $\mathfrak{z} \in B$ and $\mathfrak{z} \neq 0$ let $E_{\mathfrak{z}} := \{t\mathfrak{z} : t \in \mathbb{C}\}$. Then $E_{\mathfrak{z}} \cap B$ is a circular disk in the complex t-plane, and $|(f \circ \tilde{\varphi}^{-1}|E_{\mathfrak{z}} \cap B)|$ has a local maximum at the origin. By the maximum principle of one dimensional complex analysis this means that $f \circ \tilde{\varphi}^{-1}|E_{\mathfrak{z}} \cap B$ is constant. In particular $f(\tilde{\varphi}^{-1}(\mathfrak{z})) = f(x_0)$. Since $\mathfrak{z} \in B$ was chosen arbitrarily it follows that $f|U$ is constant, and by the previously proved identity theorem that holds only if f is constant. □

Theorem 2.4. *Let $(X, \mathscr{H})$ be a connected compact complex manifold. Then every function holomorphic on X is constant.*

PROOF. If f is holomorphic on X, then $|f|$ is continuous on X and therefore attains a maximum on the compact manifold X. By the maximum principle, f is constant. □

EXAMPLE. If we give the Riemann sphere $X := \mathbb{C} \cup \{\infty\}$ the usual topology, one obtains a Hausdorff space.

$\varphi: X - \{\infty\} \to \mathbb{C}$ with $\varphi(x) := x$ and $\psi := X - \{0\} \to \mathbb{C}$ with $\psi(x) := 1/x$ are topological mappings, and the transformations

$$\varphi \circ \psi^{-1}: \mathbb{C} - \{0\} \to \mathbb{C} - \{0\} \quad \text{and} \quad \psi \circ \varphi^{-1}: \mathbb{C} - \{0\} \to \mathbb{C} - \{0\}$$

are holomorphic.

Hence $\{(X - \{\infty\}, \varphi), (X - \{0\}, \psi)\}$ is a covering of X by compatible complex coordinates which induces a sheaf $\mathscr{H}$ on X. X is a one-dimensional complex manifold.

1. X is compact:

Let
$$E_1 := \{z \in X - \{\infty\} : |z| \leqslant 1\},$$
$$E_2 := \{z \in X : |z| \geqslant 1\}.$$

Then E_1 is compact and $(\psi|E_2): E_2 \to E_1$ is a homeomorphism. Therefore E_2 is also compact. The proposition follows because $X = E_1 \cup E_2$.

2. X is connected, since the sets E_1, E_2 are connected and $E_1 \cap E_2 \neq \emptyset$.

By Theorem 2.4 it follows that every function holomorphic on the whole Riemann sphere is constant.

Def. 2.2. An *abstract Riemann surface* is a connected one-dimensional complex manifold.

The Riemann sphere is an abstract Riemann surface. In the next section we shall examine the so-called "concrete Riemann surfaces."

Note. When no confusion can arise, we shall denote a complex manifold simply by X.

Def. 2.3. Let X_1, X_2 be complex manifolds. A continuous mapping $\varphi: X_1 \to X_2$ is called *holomorphic* if for every open set $U \subset X_2$, $g \circ \varphi$ is holomorphic over $\varphi^{-1}(U)$ whenever g is holomorphic over U. If φ is topological and φ and φ^{-1} are holomorphic, then φ is called *biholomorphic.*

Remarks

1. $\mathrm{id}_X: X \to X$ is always biholomorphic.
2. If $\varphi: X_1 \to X_2$ and $\psi: X_2 \to X_3$ are holomorphic mappings, then $\psi \circ \varphi: X_1 \to X_3$ is also holomorphic.
3. Let $f: X \to \mathbb{C}$ be a continuous mapping. f is holomorphic (in the sense of Def. 2.3) if and only if f is a holomorphic function.

PROOF

a. If f is a holomorphic mapping, then $f = \mathrm{id}_{\mathbb{C}} \circ f$ is a holomorphic function over $f^{-1}(\mathbb{C}) = X$.

b. Let f be a holomorphic function, $U \subset \mathbb{C}$ open and g holomorphic over U. Then for every point $x_0 \in f^{-1}(U) \subset X$ there exists a neighborhood $V(x_0) \subset X$ and an isomorphism $\varphi: (V, \mathscr{H}) \to (B, \mathcal{O})$. Since by definition $f \circ \tilde{\varphi}^{-1}$ is holomorphic over B, $(g \circ f) \circ \tilde{\varphi}^{-1} = g \circ (f \circ \tilde{\varphi}^{-1})$ is holomorphic over B, and that means that $g \circ f$ is holomorphic at x_0. □

Theorem 2.5. *A mapping $\psi: X_1 \to X_2$ is biholomorphic if and only if there exists an isomorphism $\varphi: (X_1, \mathscr{H}_1) \to (X_2, \mathscr{H}_2)$ with $\tilde{\varphi} = \psi$.*

PROOF

1. If $\psi: X_1 \to X_2$ is a biholomorphic mapping, then ψ and ψ^{-1} carry holomorphic functions into holomorphic functions and hence induce an isomorphism between the canonical pre-sheaves. Naturally $\tilde{\varphi} = \psi$ for the corresponding isomorphism $\varphi: (X_1, \mathscr{H}_1) \to (X_2, \mathscr{H}_2)$.

2. If $\varphi = (\tilde{\varphi}, \varphi_*): (X_1, \mathscr{H}_1) \to (X_2, \mathscr{H}_2)$ is an isomorphism, then $\tilde{\varphi}$ is a topological mapping. If $U \subset X_2$ is open and g holomorphic over U, then there is an $s \in \Gamma(U, \mathscr{H}_2)$ with $g = [s]$, and $g \circ \tilde{\varphi} = [s] \circ \tilde{\varphi} = [\varphi_*^{-1} \circ s \circ \tilde{\varphi}]$ with $\varphi_*^{-1} \circ s \circ \tilde{\varphi} \in \Gamma(\tilde{\varphi}^{-1}(U), \mathscr{H}_2)$. Therefore $g \circ \tilde{\varphi}$ is holomorphic over $\tilde{\varphi}^{-1}(U)$. Hence $\tilde{\varphi}$ is holomorphic. One shows similarly that $\tilde{\varphi}^{-1}$ is holomorphic. □

Def. 2.4. Let X be a complex manifold. A subset $A \subset X$ is called *analytic* if for every $x_0 \in X$ there is an open neighborhood $U(x_0) \subset X$ and holomorphic functions $f_1, \ldots, f_\ell$ over U such that $U \cap A = \{x \in U: f_1(x) = \cdots = f_\ell(x) = 0\}$.

Theorem 2.6. *Let X be a connected complex manifold and $M \neq X$ an analytic subset of X. Then $\mathring{M} = \varnothing$.*

PROOF. If $\mathring{M} \neq \varnothing$, there exists a point $x_0 \in \partial\mathring{M}$ such that for every open neighborhood $U(x_0) \subset X$ the set $U \cap \mathring{M}$ is open and non-empty. We could choose a connected U such that there exist holomorphic functions $f_1, \ldots, f_d$ on U with $U \cap M = \{x \in U: f_1(x) = \cdots = f_d(x) = 0\}$. Then $f_i | U \cap \mathring{M} = 0$ for $i = 1, \ldots, d$. By the identity theorem it follows that $f_i | U = 0$ for $i = 1, \ldots, d$ and therefore $U \subset \mathring{M}$, a contradiction. □

Def. 2.5. A complex manifold X is called *holomorphically separable* if for every $x_0 \in X$ there are holomorphic functions $f_1, \ldots, f_\ell$ on X such that x_0 lies isolated in the set $\{x \in X : f_1(x) = \cdots = f_\ell(x) = 0\}$.

Remark. One can show that always $\ell \geqslant \dim X$.

EXAMPLE. Let (X, ψ) be a domain over $\mathbb{C}^n$. Then for every $x_0 \in X$ there are open neighborhoods $U(x_0) \subset X$ and $V(\psi(x_0)) \subset \mathbb{C}^n$ such that $\psi | U : U \to V$ is topological. (U, ψ) is therefore a complex coordinate system, and since the identity is always the coordinate transformation, X becomes a complex manifold. The mapping $\psi : X \to \mathbb{C}^n$ is holomorphic (in the sense of Def. 2.3).

A continuous mapping $\psi : X \to Y$ between topological spaces is called *discrete* if for every $y \in Y$, $\psi^{-1}(y)$ is empty or a discrete set in X.

ψ is a discrete mapping.

PROOF. Let $x_0 \in X$, $\mathfrak{z}_0 := \psi(x_0)$ and $x_1 \in \psi^{-1}(\mathfrak{z}_0)$. Then there is a neighborhood $U(x_1) \subset X$ and a neighborhood $V(\mathfrak{z}_0) \subset \mathbb{C}^n$ such that $\psi | U : U \to V$ is topological. But that can only be if $\psi^{-1}(\mathfrak{z}_0) \cap U = \{x_1\}$. Therefore the fiber $\psi^{-1}(\mathfrak{z}_0)$ is a discrete set. □

Hence it follows that every domain over $\mathbb{C}^n$ is holomorphically separable:

PROOF. Let $g_i(z_1, \ldots, z_n) := z_i - z_i^0$ and $f_i := g_i \circ \psi : X \to \mathbb{C}$ for $i = 1, \ldots, n$. $f_1, \ldots, f_n$ are holomorphic functions on X, and x_0 lies isolated in $\{x \in X : f_1(x) = \cdots = f_n(x) = 0\} = \psi^{-1}(\mathfrak{z}_0)$. □

In particular, every domain $G \subset \mathbb{C}^n$ is holomorphically separable. One can generalize the above results in the following manner:

Theorem 2.7. *An n-dimensional complex manifold X is holomorphically separable if and only if there exists a holomorphic discrete mapping $\psi : X \to \mathbb{C}^n$.*

(One direction is clear, for the other see: H. Grauert: "Charakterisierung der holomorph-vollständigen Räume", *Math. Ann.*, *129*: 233–259, 1955.)

Def. 2.6. Let X be a complex manifold.

1. If $K \subset X$ is an arbitrary subset, then

$$\hat{K} := \{x \in X : |f(x)| \leqslant \sup |f(K)|$$

for every holomorphic function f on $X\}$ is called the *holomorphically convex hull of* K.

2. X is called *holomorphically convex if* $\hat{K} \subset X$ is compact whenever $K \subset X$ is compact.

Def. 2.7. X is called a *Stein manifold* if

1. X is holomorphically separable.
2. X is holomorphically convex.

Theorem 2.8. *For a domain* (X, ψ) *over* $\mathbb{C}^n$ *the following properties are equivalent*:

1. *X is a Stein manifold.*
2. *X is holomorphically convex.*
3. *X is a domain of holomorphy.*

The non-trivial equivalence of (2) and (3) was proved in 1953 by Oka. Theorem 2.8 leads us to regard the Stein manifolds as generalizations of domains of holomorphy.

EXAMPLE. If X is a compact complex manifold and $\dim X > 0$, then X is holomorphically convex but not a Stein manifold.

PROOF. If $K \subset X$ is compact, then $\hat{K} \subset X$ is always closed. If X is compact it follows that $\hat{K}$ is also compact. Therefore X is holomorphically convex.

Since X is compact there exists a decomposition of X into finitely many connected components, which are also all compact: $X = X_1 \cup \cdots \cup X_l$. If $x_0 \in X_i$, f is holomorphic on X and $f(x_0) = 0$, then by Theorem 2.4 f vanishes identically on X_i. Therefore each set of the form

$$\{x \in X: f_1(x) = \cdots = f_m(x) = 0\}$$

contains, in addition to the point x_0, the open subset $X_i \subset X$; x_0 is therefore not an isolated point and X cannot be holomorphically separable. □

3. Examples of Complex Manifolds

Concrete Riemann Surfaces

Def. 3.1. A *(concrete) Riemann surface over* $\mathbb{C}$ is a pair (X, φ) with the following properties:

1. X is a Hausdorff space.
2. $\varphi: X \to \mathbb{C}$ is a continuous mapping.
3. For every $x_0 \in X$ there is an open neighborhood $U(x_0) \subset X$, a connected open set $V \subset \mathbb{C}$ and a topological mapping $\psi: V \to U$ such that
   a. $\varphi \circ \psi: V \to \mathbb{C}$ is holomorphic.
   b. $(\varphi \circ \psi)'$ does not vanish on any open subset of V.

One also calls the mapping ψ a *local uniformization of the Riemann surface* (X, φ).

Theorem 3.1. *Let (X, φ) be a Riemann surface over $\mathbb{C}$. Then X has a canonical one-dimensional complex manifold structure, and $\varphi: X \to \mathbb{C}$ is a holomorphic mapping.*

PROOF

1. Let $x_0 \in X$, $z_0 := \varphi(x_0) \in \mathbb{C}$. Then there is a neighborhood $U(x_0) \subset X$ and a connected neighborhood $V(z_0) \subset \mathbb{C}$ as well as a topological mapping $\psi: V \to U$ with the local uniformization property. (U, ψ^{-1}) is therefore a complex coordinate system for X at x_0.

Now two such coordinate systems $(U_1, \psi_1^{-1}), (U_2, \psi_2^{-1})$ may be given. Then $\psi := \psi_1^{-1} \circ \psi_2 : \psi_2^{-1}(U_1 \cap U_2) \to \psi_1^{-1}(U_1 \cap U_2)$ is a topological mapping. If we set $f_\lambda(t) := \varphi \circ \psi_\lambda(t)$ for $t \in V_\lambda$, $\lambda = 1, 2$, then f_λ is a holomorphic function on V_λ whose derivative does not vanish on any open subset of V_λ. Let $t_0 \in \psi_1^{-1}(U_1 \cap U_2)$ be chosen so that $f_1'(t_0) \neq 0$. Then there is a neighborhood $U(t_0) \subset \psi_1^{-1}(U_1 \cap U_2)$ and an open set $W \subset \mathbb{C}$ such that $f_1|U: U \to W$ is biholomorphic. Let

$$g_1 := (f_1|U)^{-1} = (\varphi \circ \psi_1|U)^{-1}: W \to U.$$

The mapping

$$\psi_1|U: U \to V := \psi_1(U) \subset U_1 \cap U_2$$

is topological, and so is

$$\varphi|V = g_1^{-1} \circ (\psi_1|U)^{-1} = ((\psi_1|U) \circ g_1)^{-1}: V \to W.$$

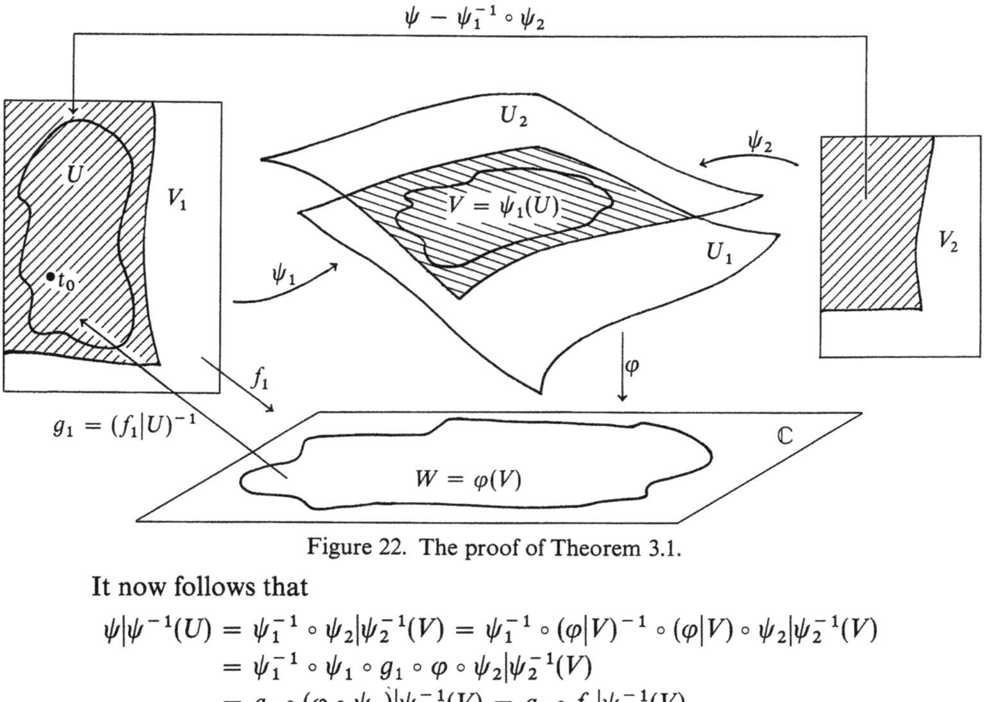

Figure 22. The proof of Theorem 3.1.

It now follows that

$$\begin{aligned}\psi|\psi^{-1}(U) &= \psi_1^{-1} \circ \psi_2|\psi_2^{-1}(V) = \psi_1^{-1} \circ (\varphi|V)^{-1} \circ (\varphi|V) \circ \psi_2|\psi_2^{-1}(V)\\ &= \psi_1^{-1} \circ \psi_1 \circ g_1 \circ \varphi \circ \psi_2|\psi_2^{-1}(V)\\ &= g_1 \circ (\varphi \circ \psi_2)|\psi_2^{-1}(V) = g_1 \circ f_2|\psi_2^{-1}(V),\end{aligned}$$

which is a holomorphic function. By the identity theorem $D: = \{t \in V_1: f'_1(t) = 0\}$ is a discrete set, and so is

$$D': = \psi^{-1}(\psi_1^{-1}(U_1 \cap U_2) \cap D).$$

If

$$s_0 \in \varphi_2^{-1}(U_1 \cap U_2) - D' = \psi^{-1}(\psi_1^{-1}(U_1 \cap U_2) - D),$$

then

$$t_0: = \psi(s_0) \in \psi_1^{-1}(U_1 \cap U_2) - D,$$

therefore

$$f'_1(t_0) \neq 0.$$

As we just showed, there is a neighborhood $U(t_0)$ such that $\psi|\psi^{-1}(U)$ is holomorphic. In particular ψ is holomorphic at $s_0 = \psi^{-1}(t_0)$.

A continuous mapping which is holomorphic outside a discrete set must, however, be everywhere holomorphic by the Riemann extension theorem.

Hence the coordinate transformations are holomorphic, so X is canonically a complex manifold.

2. Let $B \subset \mathbb{C}$ be open, g holomorphic on B. Then $W: = \varphi^{-1}(B)$ is open in X and $g \circ (\varphi|W): W \to \mathbb{C}$ is continuous. If (U, ψ^{-1}) is a coordinate system, then $f: = \varphi \circ \psi: \psi^{-1}(U) \to \mathbb{C}$ is holomorphic, and so is $(g \circ \varphi) \circ \psi = g \circ f$, which means that $g \circ \varphi$ is holomorphic on W. □

Now let (X, φ) be a Riemann surface, $\psi: V \to U$ a local uniformization and $f: = \varphi \circ \psi$. By assumption the set $D: = \{t \in V: f'(t) = 0\}$ is discrete in V. Let $t_0 \in V$ and $x_0: = \psi(t_0) \in U$.

Case 1. $f'(t_0) \neq 0$.

Then there are neighborhoods $V_1(t_0) \subset V$ and $W(f(t_0)) \subset \mathbb{C}$ such that $f|V_1: V_1 \to W$ is biholomorphic. $U_1: = \psi(V_1)$ is open in U, and $\psi_1: = \psi \circ (f|V_1)^{-1}: W \to U_1$ is topological. Moreover, $\varphi \circ \psi_1 = \text{id}_W$. Therefore there exists a local uniformization $\psi_1: W \to U_1$ with $x_0 \in U_1$ and $\varphi \circ \psi_1 = \text{id}_W$. Thus, via φ, U_1 is a sheet over W.

In this situation we say that X is *unbranched at* x_0. Clearly X is unbranched everywhere outside a discrete set.

Case 2. $f'(t_0) = 0$.

Let $z_0: = f(t_0)$. Then at $t_0, f - z_0$ has a zero of order $k \geqslant 2$, that is, on V

$$f(t) = z_0 + (t - t_0)^k \cdot h(t),$$

where h is a holomorphic function with $h(t_0) \neq 0$.

There exists a neighborhood $V_1(t_0) \subset V$ and a holomorphic function g on V_1 with $g^k = h$. In particular $g(t_0) \neq 0$. Let $\tau: V_1 \to \mathbb{C}$ be defined by $\tau(t): = (t - t_0) \cdot g(t)$. Then $\tau'(t_0) = g(t_0) \neq 0$, so there is a neighborhood $V_2(t_0) \subset V_1$ and an open set $W_2 \subset \mathbb{C}$ such that $\tau|V_2: V_2 \to W_2$ is biholomorphic. Clearly

$$\psi \circ (\tau|V_2)^{-1}: W_2 \to U_2: = \psi(V_2) \subset X$$

is topological and

$$\varphi \circ (\psi \circ (\tau|V_2)^{-1}) = f \circ (\tau|V_2)^{-1}$$

holomorphic, and $(f \circ (\tau|V_2)^{-1})'$ vanishes at most on a discrete set. Moreover

$$f \circ (\tau|V_2)^{-1}(s) = f \circ (\tau|V_2)^{-1}(\tau(t)) = f(t) = z_0 + ((t - t_0) \cdot g(t))^k$$
$$= z_0 + \tau(t)^k = z_0 + s^k.$$

Therefore there exists a local uniformization $\psi_2: W_2 \to U_2$ with $x_0 \in U_2$ and $\varphi \circ \psi_2(t) = \varphi(x_0) + t^k$.

Since the coordinate transformation has a non-vanishing derivative, the order k of a zero is not affected by a change of chart, that is, k depends only on the point x_0. We say that x_0 is a *branch point of order k*. ψ_2, where $\varphi \circ \psi_2(t) = \varphi(x_0) + t^k$, is called the *distinguished uniformization*.

X then locally represents a branched k-fold covering over $\varphi(x_0)$, in the sense that there lies exactly one point of X over $\varphi(x_0)$, while over every point $z \neq \varphi(x_0)$ in some neighborhood of $\varphi(x_0)$ there lie exactly k points of X.

EXAMPLE. Let $X: = \{(w, z) \in \mathbb{C}^2 : w^2 = z^3\}$. With the topology induced by $\mathbb{C}^2$, X becomes a Hausdorff space, and the mapping $\varphi: X \to \mathbb{C}$ with $\varphi(w, z): = z$ is continuous.

In order to show that X is a Riemann surface over $\mathbb{C}$, we must specify the local uniformization. Let $\psi: \mathbb{C} \to X$ be defined by $\psi(t): = (t^3, t^2)$.

a. ψ is injective. If $\psi(t_1) = \psi(t_2)$, then $t_1^2 = t_2^2$ and $t_1^3 = t_2^3$. If $\psi(t) = 0$ then $t = 0$. If $t_1 \neq 0$, then also $t_2 \neq 0$, so we can divide and then $t_1 = t_2$.

b. ψ is surjective. If $0 \neq (w, z) \in X$, then $z \neq 0$ so there exist two complex numbers t_1, t_2 such that $\{t_1, t_2\} = \{t : t^2 = z\}$. Then $t_1 = -t_2$ so that $w^2 = z^3 = (t_1^3)^2 = (t_2^3)^2$; therefore $w \in \{t_1^3, t_2^3\}$, so either $\psi(t_1) = (w, z)$ or $\psi(t_2) = (w, z)$.

c. $\psi: \mathbb{C} \to X$ is topological. The continuity of ψ is clear. Because of the continuity of the roots, ψ^{-1} is continuous.

Hence there is a global uniformization for X, given by ψ. ($\varphi \circ \psi(t) = t^2$ is holomorphic, and has a derivative which does not vanish identically anywhere.)

Let $\tau: X \to \mathbb{C}$ be defined by $\tau: = \psi^{-1}$. Since $\tau \circ \psi = \mathrm{id}_{\mathbb{C}}$, τ is a holomorphic function on X. τ cannot be holomorphically extended into $\mathbb{C}^2$. Otherwise, suppose, $g(w, z) = \sum_{\nu, \mu} a_{\nu\mu} w^\nu z^\mu$ is a holomorphic function on $\mathbb{C}^2$ (for example, in a neighborhood of $0 \in \mathbb{C}^2$). Then

$$(g|X)(w, z) = g(\psi(t)) = \sum_{\nu, \mu} a_{\nu\mu} t^{3\nu + 2\mu}$$
$$= a_{00} + a_{01}t^2 + a_{10}t^3 + a_{02}t^4 + a_{11}t^5 + a_{20}t^6 + \cdots.$$

If $\hat{\tau}$ were a holomorphic extension of τ, then we would have $t = \tau \circ \psi(t) = (\hat{\tau}|X)(\psi(t))$. But that cannot be.

Complex Submanifolds

Let X be a complex manifold, (U, φ) a coordinate system on X. If $x_0 \in U$ and f holomorphic on a neighborhood $V(x_0) \subset U$, then we define the partial derivatives of f at x_0 with respect to φ by

$$(D_\nu f)_\varphi(x_0) := \frac{\partial(f \circ \varphi^{-1})}{\partial z_\nu}(\varphi(x_0)).$$

Now suppose we have another coordinate system (U', φ') with $V \subset U \cap U'$. Then the functional matrix

$$\mathfrak{M}_{(\varphi' \circ \varphi^{-1})}(\varphi(x_0)) = \left((a_{\nu\mu})\begin{matrix} \nu = 1, \ldots, n \\ \mu = 1, \ldots, n\end{matrix}\right)$$

has a non-vanishing determinant, and:

$$(D_\nu f)_\varphi(x_0) = \sum_{\mu=1}^{n} a_{\nu\mu} \cdot (D_\mu f)_{\varphi'}(x_0).$$

Therefore, if $f_1, \ldots, f_d$ are holomorphic functions on V, the natural number

$$\operatorname{rk}_{x_0}(f_1, \ldots, f_d) := \operatorname{rk}\left(((D_\nu f_\mu)_\varphi(x_0))\begin{matrix} \mu = 1, \ldots, d \\ \nu = 1, \ldots, n\end{matrix}\right)$$

is independent of φ.

Def. 3.2. Let X be an n-dimensional complex manifold, $A \subset X$ analytic. A is called *free of singularities of the codimension d* if for every point $x_0 \in A$ there exists a neighborhood $U(x_0) \subset X$ and holomorphic functions $f_1, \ldots, f_d$ on U such that:

1. $A \cap U = \{x \in U : f_1(x) = \cdots = f_d(x) = 0\}$
2. $rk_x(f_1, \ldots, f_d) = d$ for all $x \in U$.

Theorem 3.2. *An analytic set $A \subset X$ is free of singularities of codimension d if and only if for every point $x_0 \in A$ there exists a neighborhood $U(x_0) \subset X$ and an isomorphism $\varphi = (\tilde{\varphi}, \varphi_*) : (U, \mathscr{H}) \to (B, \mathscr{O})$ such that $\tilde{\varphi}(U \cap A) = \{(w_1, \ldots, w_n) \in B : w_1 = \cdots = w_d = 0\}$.*

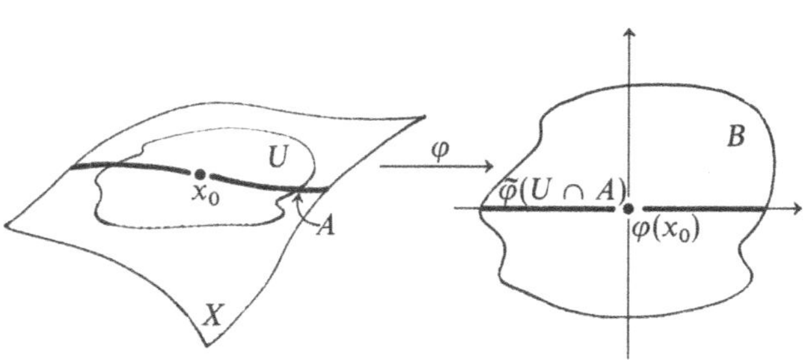

Figure 2 . Illustration for Theorem 3.2.

PROOF. Let $x_0 \in A$ be given, $U(x_0) \subset X$ a neighborhood and $\varphi:(U, \mathscr{H}) \to (\hat{B}, \mathcal{O})$ an isomorphism. $A \cap U$ is singularity free of codimension d if and only if $\tilde{\varphi}(A \cap U)$ is a regular analytic set $\hat{B}$ of the complex dimension $n - d$. That, however, is equivalent to the existence, for every $\mathfrak{z} \in \tilde{\varphi}(A \cap U)$ of a neighborhood $V(\mathfrak{z}) \subset \hat{B}$ and an isomorphism $\psi:(V, \mathcal{O}) \to (B, \mathcal{O})$ such that

$$\tilde{\psi}(\tilde{\varphi}(A \cap U) \cap V) = \{\mathfrak{w} \in B : w_1 = \cdots = w_d = 0\}$$

(see Theorem 6.15 in Chapter III).

$$\varphi_1 := \psi \circ \varphi:(\tilde{\varphi}^{-1}(V(\tilde{\varphi}(x_0))), \mathscr{H}) \to (B, \mathcal{O})$$

is such a function. □

Theorem 3.3. *Let $A \subset X$ be an analytic set, free of singularities of codimension d. Then X induces a canonical $(n - d)$-dimensional manifold structure on A, and the natural imbedding $j_A : A \hookrightarrow X$ is holomorphic.*

PROOF. A, with the relative topology induced by X, is clearly a Hausdorff space. A function f defined on an open set $W \subset A$ will be called holomorphic if for every $x \in W$ there is an open neighborhood $U(x) \subset X$ and a holomorphic function $\hat{f}$ on U such that $U \cap A \subset W$ and $\hat{f}|U \cap A = f|U \cap A$. If X is an open set in $\mathbb{C}^n$ and A is a part of a $(n - d)$-dimensional plane, then this new notion of holomorphy on A agrees with the earlier notion. The set of holomorphic functions defines a pre-sheaf on A.

If $x_0 \in A$, then there exists a neighborhood $U(x_0) \subset X$ and an isomorphism $\varphi:(U, \mathscr{H}) \to (B, \mathcal{O})$ such that

$$\tilde{\varphi}(U \cap A) = \{\mathfrak{w} \in B : w_1 = \cdots = w_d = 0\} = B \cap (\{0\} \times \mathbb{C}^{n-d})$$

The pre-sheaf of holomorphic functions on $U \cap A$ is mapped isomorphically by φ onto the pre-sheaf of locally holomorphically continuable functions on $B \cap (\{0\} \times \mathbb{C}^{n-d}) = B'$, and the latter coincides with the pre-sheaf of holomorphic functions on the region $B' \subset \mathbb{C}^{n-d}$. For the sheaf $\mathscr{H}'$ of germs of holomorphic functions on A it is then true that $(U \cap A, \mathscr{H}') \simeq (B', \mathcal{O})$. Hence A is a complex manifold.

If $U \subset X$ is open and f holomorphic on U, then by definition $f \circ j_A = f|A \cap U$ is also holomorphic, and hence $j_A : A \hookrightarrow X$ is a holomorphic mapping. □

Remark. An analytic set $A \subset X$, free of singularities, is also called a *complex submanifold of* X. The example $X = \{(w, z) \in \mathbb{C}^2 : w^2 = z^3\}$ considered in part (a) is not a submanifold of $\mathbb{C}^2$.

Cartesian Products

Theorem 3.4. *Let $X_1, \ldots, X_\ell$ be complex manifolds, $n_i := \dim X_i$ for $i = 1, \ldots, \ell$ and $n := n_1 + \cdots + n_\ell$. Then there is an n-dimensional manifold structure on $X := X_1 \times \cdots \times X_\ell$, such that all projections $p_i : X \to X_i$ are holomorphic.*

PROOF. With the sets $W = W_1 \times \cdots \times W_\ell$, $W_i \subset X_i$ open as the basis for the topology of X, X becomes a Hausdorff space.

If $x_0 = (x_1, \ldots, x_\ell) \in X$, then there exist neighborhoods $U_i(x_i) \subset X_i$ and isomorphisms $\varphi_i:(U_i, \mathscr{H}_i) \to (B_i, \mathcal{O}_i)$. Let

$$U: = U_1 \times \cdots \times U_\ell, \qquad \tilde{\varphi}_1 \times \cdots \times \tilde{\varphi}_\ell: U \to B: = B_1 \times \cdots \times B_\ell$$

be defined by

$$(\tilde{\varphi}_1 \times \cdots \times \tilde{\varphi}_\ell)(x_1', \ldots, x_\ell'): = (\tilde{\varphi}_1(x_1'), \ldots, \tilde{\varphi}_\ell(x_\ell')).$$

Then $(U, (\tilde{\varphi}_1 \times \cdots \times \tilde{\varphi}_\ell))$ is a complex coordinate system at x_0. If $(V, (\tilde{\psi}_1 \times \cdots \times \tilde{\psi}_\ell))$ is another coordinate system, then the transformation

$$(\tilde{\varphi}_1 \times \cdots \times \tilde{\varphi}_\ell) \circ (\tilde{\psi}_1 \times \cdots \times \tilde{\psi}_\ell)^{-1} = (\tilde{\varphi}_1 \circ \tilde{\psi}_1^{-1} \times \cdots \times \tilde{\varphi}_\ell \circ \tilde{\psi}_\ell^{-1})$$

is holomorphic. Therefore X is an n-dimensional complex manifold. Suppose $W \subset X_1$ is open, and g is holomorphic on W. Then

$$V: = p_1^{-1}(W) = W \times X_2 \times \cdots \times X_\ell.$$

Let $x_0 \in V$ and

$$(U_1 \times \cdots \times U_\ell, \tilde{\varphi}_1 \times \cdots \times \tilde{\varphi}_\ell)$$

be a coordinate system for X at x_0. Then

$$\begin{aligned}(g \circ p_1) \circ (\tilde{\varphi}_1 \times \cdots \times \tilde{\varphi}_\ell)^{-1}(z_1, \ldots, z_\ell) &= g \circ \tilde{\varphi}_1^{-1}(z_1) \\ &= (g \circ \tilde{\varphi}_1^{-1}) \circ \mathrm{pr}_1(z_1, \ldots, z_n),\end{aligned}$$

and

$$(g \circ \tilde{\varphi}_1^{-1}) \circ \mathrm{pr}_1 : B_1 \times \cdots \times B_\ell \to \mathbb{C}$$

is holomorphic. Therefore $g \circ p_1 : X \to \mathbb{C}$ is also holomorphic; that is, p_1 is a holomorphic mapping. The proof for $p_2, \ldots, p_\ell$ is similar. □

Theorem 3.5. *Let X be an n-dimensional complex manifold. Then the diagonal $D: = \{(x, x): x \in X\} \subset X \times X$ is an analytic subset free of singularities of codimension n.*

PROOF

1. Since X is a Hausdorff space, the diagonal $D \subset X \times X$ is closed. Therefore D is analytic at each point $(x, y) \in X \times X - D$.

2. Let $(x_0, x_0) \in D$. Then there is a neighborhood $U(x_0) \subset X$ and an isomorphism $\varphi:(U, \mathscr{H}) \simeq (B, \mathcal{O})$ and then $\hat{U}: = U \times U$ is a neighborhood of (x_0, x_0) in $X \times X$, which is biholomorphically equivalent to $B \times B$. Therefore there exist coordinates $z_1, \ldots, z_n, w_1, \ldots, w_n$ (with $z_\nu: = \mathrm{pr}_\nu \circ \tilde{\varphi}$, $w_\nu: = \mathrm{pr}_{n+\nu} \circ \tilde{\varphi}$) in $\hat{U}$ such that $D \cap \hat{U} = \{(x, x) \in X \times X : (z_i - w_i)(x, x) = 0 \text{ for } i = 1, \ldots, n\}$. Moreover

$$\begin{aligned}\operatorname{rk}_{(x,x)}(z_1 - w_1, \ldots, z_n - w_n) &= \operatorname{rk}\left(((D_\nu(z_i - w_i))_{\tilde{\varphi}}(x, x)) \begin{array}{l} i = 1, \ldots, n \\ \nu = 1, \ldots, 2n \end{array}\right) \\ &= \operatorname{rk}\left(\begin{array}{cc|cc} 1 & 0 & -1 & 0 \\ & \ddots & & \ddots \\ 0 & 1 & 0 & -1 \end{array}\right) = n\end{aligned}$$

which was to be proved. □

Theorem 3.6. *Let X be a complex manifold, $D \subset X \times X$ the diagonal. Then the diagonal mapping $d: X \to D$ by $d(x) := (x, x)$ is biholomorphic.*

PROOF. d is bijective, and the inverse mapping $d^{-1} = p_1|D$ is holomorphic. It remains to be shown that d is holomorphic. Let $W \subset D$ be open, g holomorphic on W, $(x_0, x_0) \in W$. Then there exists a neighborhood $U(x_0) \subset X$ and a holomorphic function $\hat{g}$ on $U \times U$ such that $(U \times U) \cap D \subset W$ and $\hat{g}|(U \times U) \cap D = g|(U \times U) \cap D$. Without loss of generality we may assume that there is an isomorphism $\varphi: (U, \mathscr{H}) \to (B, \mathscr{O})$. The mapping $d^*: B \to B \times B$ with $d^*(\mathfrak{z}) := (\mathfrak{z}, \mathfrak{z})$ is holomorphic and

$$(g \circ d) \circ \tilde{\varphi}^{-1}(\mathfrak{z}) = \hat{g} \circ d \circ \tilde{\varphi}^{-1}(\mathfrak{z}) = \hat{g} \circ (\tilde{\varphi} \times \tilde{\varphi})^{-1} \circ d^*(\mathfrak{z}).$$

Therefore $(g \circ d) \circ \tilde{\varphi}^{-1}$ and hence $g \circ d$ is holomorphic. □

Complex Projective Spaces

We define a relation on $\mathbb{C}^{n+1} - \{0\}$ by setting $\mathfrak{z}_1 \sim \mathfrak{z}_2$ if and only if there exists a $t \in \mathbb{C} - \{0\}$ with $\mathfrak{z}_2 = t \cdot \mathfrak{z}_1$.

It is clear that "$\sim$" is an equivalence relation, and we denote the equivalence class of $\mathfrak{z}_0$ by $G(\mathfrak{z}_0) = \{\mathfrak{z} = t\mathfrak{z}_0 : t \in \mathbb{C} - \{0\}\}$. $G(\mathfrak{z}_0)$ is simply the complex line through 0 and $\mathfrak{z}_0$ with the origin removed.

Def. 3.3. The set $\mathbb{P}^n := \{G(\mathfrak{z}) : \mathfrak{z} \in \mathbb{C}^{n+1} - \{0\}\}$ is called the *n-dimensional complex projective space* and the mapping $\pi: \mathbb{C}^{n+1} - \{0\} \to \mathbb{P}^n$ with $\pi(\mathfrak{z}) := G(\mathfrak{z})$ is called the *natural projection.*

π is a surjective mapping, and we give $\mathbb{P}^n$ the finest topology in which π is continuous. A set $U \subset \mathbb{P}^n$ is therefore open if and only if $\pi^{-1}(U) \subset \mathbb{C}^{n+1} - \{0\}$ is open.

Let $W_i := \{\mathfrak{z} = (z_1, \ldots, z_{n+1}) \in \mathbb{C}^{n+1} : z_i = 1\}$, for $i = 1, \ldots, n + 1$. Then W_i is an affine hyperplane in $\mathbb{C}^{n+1} - \{0\}$, and in particular, it is an n-dimensional complex submanifold. Let

$$W_i^* := \{\mathfrak{z} = (z_1, \ldots, z_{n+1}) \in \mathbb{C}^{n+1} : z_i \neq 0\}.$$

A holomorphic mapping $\alpha_i: W_i^* \to \mathbb{C}^n$ is defined by

$$\alpha_i(z_1, \ldots, z_{n+1}) := \frac{1}{z_i}(z_1, \ldots, \hat{z}_i, \ldots, z_{n+1}).$$

Then $\alpha_i|W_i : W_i \to \mathbb{C}^n$ is biholomorphic with

$$(\alpha_i|W_i)^{-1}(z_1, \ldots, z_n) = (z_1, \ldots, z_{i-1}, 1, z_i, \ldots, z_n).$$

If $W \subset W_i$ is open, then for $\mathfrak{z} \in W_i^*$, $\mathfrak{z} \in \alpha_i^{-1}(\alpha_i(W))$ if and only if there is a $\mathfrak{z}' \in W$ with $\alpha_i(\mathfrak{z}) = \alpha_i(\mathfrak{z}')$, and therefore with $1/(z_i)\mathfrak{z} = \mathfrak{z}'$. That happens if and only if $\mathfrak{z} \in \pi^{-1}\pi(W)$. Therefore $\pi^{-1}\pi(W) = \alpha_i^{-1}\alpha_i(W)$ is open, which means that:

1. The system of sets $U_i := \pi(W_i)$ forms an open covering of $\mathbb{P}^n$.
2. $\pi|W_i : W_i \to U_i$ is an open mapping.

If $\pi(\mathfrak{z}) = \pi(\mathfrak{z}')$ for $\mathfrak{z}, \mathfrak{z}' \in W_i$, then there is a $t \in \mathbb{C}$ with $\mathfrak{z}' = t \cdot \mathfrak{z}$; therefore $1 = z_i' = tz_i = t$, so $\mathfrak{z} = \mathfrak{z}'$. Hence $\pi|W_i : W_i \to U_i$ is injective, so with the preceding considerations, it is topological. Hence for each i, $\varphi_i := \alpha_i \circ (\pi|W_i)^{-1} : U_i \to \mathbb{C}^n$ is a complex coordinate system for $\mathbb{P}^n$. Moreover

$$\begin{aligned}(\pi|W_i)^{-1} \circ \pi(z_1, \ldots, z_{n+1}) &= (\pi|W_i)^{-1} \circ \pi\left(\frac{1}{z_i}(z_1, \ldots, z_{n+1})\right) \\ &= \frac{1}{z_i}(z_1, \ldots, z_{n+1}) \\ &= (\alpha_i|W_i)^{-1}\left(\frac{1}{z_i}(z_1, \ldots, \hat{z}_i, \ldots, z_{n+1})\right) \\ &= (\alpha_i|W_i)^{-1} \circ \alpha_i(z_1, \ldots, z_{n+1}),\end{aligned}$$

and hence the coordinate transformations

$$\varphi_j \circ \varphi_i^{-1} : \varphi_i(U_i \cap U_j) \to \varphi_j(U_i \cap U_j)$$

have the form

$$\begin{aligned}\varphi_j \circ \varphi_i^{-1}(z_1, \ldots, z_n) &= \alpha_j \circ (\pi|W_j)^{-1} \circ \pi \circ (\alpha_i|W_i)^{-1}(z_1, \ldots, z_n) \\ &= \alpha_j(z_1, \ldots, z_{i-1}, 1, z_i, \ldots, z_n) \\ &= \frac{1}{z_j}(z_1, \ldots, \hat{z}_j, \ldots, z_{i-1}, 1, z_i, \ldots, z_n),\end{aligned}$$

which is a holomorphic mapping.

We still must show that $\mathbb{P}^n$ is a Hausdorff space. Let $x_1, x_2 \in \mathbb{P}^n$, $x_1 \neq x_2$.

1. If both points lie in the same coordinate neighborhood U_i, it is trivial to find disjoint neighborhoods.

2. Suppose x_1, x_2 are not elements of the same coordinate neighborhood. Then for arbitrary points $\mathfrak{z}_i \in \pi^{-1}(x_i)$ we have

$$z_j^{(1)} \cdot z_j^{(2)} = 0, \qquad j = 1, \ldots, n + 1.$$

Without loss of generality, then, we may assume that

$$\begin{aligned}\mathfrak{z}_1 &= (1, z_2^{(1)}, \ldots, z_s^{(1)}, 0, \ldots, 0), \quad \text{with} \quad z_j^{(1)} \neq 0 \text{ for } \quad j = 2, \ldots, s. \\ \mathfrak{z}_2 &= (0, \ldots, 0, z_{s+1}^{(2)}, \ldots, z_n^{(2)}, 1).\end{aligned}$$

Let

$$V_1 := \pi(\{(1, w_2, \dots, w_{n+1}) \in \mathbb{C}^{n+1} : |w_{n+1}| < 1\}),$$
$$V_2 := \pi(\{(w_1, \dots, w_n, 1) \in \mathbb{C}^{n+1} : |w_1| < 1\}).$$

V_1 is an open neighborhood of x_1, V_2 is an open neighborhood of x_2, and $V_1 \cap V_2 = \varnothing$.

Theorem 3.7. *The n-dimensional complex projective space is an n-dimensional complex manifold, and the natural projection $\pi : \mathbb{C}^{n+1} - \{0\} \to \mathbb{P}^n$ is holomorphic.*

PROOF. In order to complete the proof we have to demonstrate the holomorphy of π. Let $W \subset X$ be open, g holomorphic in W. Without loss of generality we may assume that $W \subset U_1$. Then $g \circ \varphi_1^{-1} = g \circ \pi \circ (\alpha_1 | W_1)^{-1} : \mathbb{C}^n \to \mathbb{C}$ is holomorphic, and so is $g \circ \pi = (g \circ \varphi_1^{-1}) \circ (\alpha_1 | W_1)$. □

Theorem 3.8. *$\mathbb{P}^n$ is compact.*

PROOF. Let $S := \{\mathfrak{z} \in \mathbb{C}^{n+1} : \|\mathfrak{z}\| = 1\} = S^{2n+1}$. For $\mathfrak{z} \in \mathbb{C}^{n+1} - \{0\}$, $\hat{\mathfrak{z}} := (1/\|\mathfrak{z}\|) \cdot \mathfrak{z}$ lies in S and $\pi(\mathfrak{z}) = \pi(\hat{\mathfrak{z}})$. Therefore $\pi | S : S \to \mathbb{P}^n$ is a surjective continuous mapping. Since S is compact and $\mathbb{P}^n$ is separated, it follows that $\mathbb{P}^n$ is also compact. □

The 1-dimensional complex projective space $\mathbb{P}^1$ is covered by two coordinate neighborhoods U_1, U_2. Here $U_1 = \pi(\{\mathfrak{z} = (1, z_2) : z_2 \in \mathbb{C}\})$, and $U_2 - U_1 = \pi(\{\mathfrak{z} = (0, z_2) : z_2 \in \mathbb{C} - \{0\}\}) = \{G(0, 1)\}$ consists of a single point. Hence $\mathbb{P}^1 = U_1 \cup \{G(0, 1)\}$.

Theorem 3.9. *Let $X = \mathbb{C} \cup \{\infty\}$ be the Riemann sphere. A biholomorphic mapping $\varphi : X \to \mathbb{P}^1$ is defined by $\varphi(\infty) := G(0, 1)$ and $\varphi(z) := \varphi_1^{-1}(z) = \pi(1, z)$.*

PROOF. It is clear that φ is bijective. On X one has two coordinate systems $\psi_1 : X - \{\infty\} \to \mathbb{C}$, and $\psi_2 : X - \{0\} \to \mathbb{C}$. Let $X_1 := X - \{\infty\}$, $X_2 := X - \{0\}$. Then

$$(\varphi | X_\lambda) \circ \psi_\lambda^{-1}(z) = \begin{cases} \pi(1, z) & \text{for} \quad \lambda = 1, \\ \pi(z, 1) & \phantom{\text{for}} \quad \lambda = 2. \end{cases}$$

Therefore $\varphi | X_\lambda : X_\lambda \to U_\lambda$ is biholomorphic for $\lambda = 1, 2$, and so φ is biholomorphic. □

The n-dimensional Complex Torus

Let $\mathfrak{c}_1, \dots, \mathfrak{c}_{2n} \in \mathbb{C}^n$ be linearly independent as real vectors. Then

$$\Gamma := \left\{\mathfrak{z} = \sum_{\lambda=1}^{2n} k_\lambda \mathfrak{c}_\lambda : k_\lambda \in \mathbb{Z} \text{ for } \lambda = 1, \dots, 2n\right\}$$

is a subgroup of the additive group of $\mathbb{C}^n$ (a translation group). Two points of $\mathbb{C}^n$ will be called *equivalent* if there is a translation of Γ carrying one into another; that is,

$$\mathfrak{z} \sim \mathfrak{z}' \Leftrightarrow \mathfrak{z} - \mathfrak{z}' \in \Gamma.$$

This is in fact an equivalence relation, and we give the set T^n of all equivalence classes the finest topology in which the canonical projection $\pi_T : \mathbb{C}^n \to T^n$ is continuous. We call the topological space T^n an n-dimensional complex torus. Any two n-dimensional tori are homeomorphic. For $\mathfrak{z}_0 \in \Gamma$ and $U \subset \mathbb{C}^n$ let $U + \mathfrak{z}_0 := \{\mathfrak{z} + \mathfrak{z}_0 : \mathfrak{z} \in U\}$. If U is open, then $U + \mathfrak{z}_0$ is open for every $\mathfrak{z}_0 \in \Gamma$, and so is $\pi_T^{-1}\pi_T(U) = \{\mathfrak{z} \in \mathbb{C}^n : \mathfrak{z} - \mathfrak{z}' \in \Gamma \text{ for a } \mathfrak{z}' \in U\} = \bigcup_{\mathfrak{z}_0 \in \Gamma} (U + \mathfrak{z}_0)$. Thus π_T is an open mapping. Let $\mathfrak{z}_0 \in \mathbb{C}^n$ be an arbitrary point. Then the set

$$F_{\mathfrak{z}_0} := \left\{\mathfrak{z} = \mathfrak{z}_0 + \sum_{\nu=1}^{2n} r_\nu \mathfrak{c}_\nu : r_\nu \in \mathbb{R} \text{ and } -\frac{1}{2} < r_\nu < \frac{1}{2} \quad \nu = 1, \ldots, 2n\right\}$$

is open in $\mathbb{C}^n$.

For two points $\mathfrak{z}, \mathfrak{z}' \in F_{\mathfrak{z}_0}$, $\mathfrak{z} - \mathfrak{z}' = \sum_{\nu=1}^{2n} (r_\nu - r'_\nu)\mathfrak{c}_\nu$ with $|r_\nu - r'_\nu| < 1$ for $\nu = 1, \ldots, 2n$. Therefore $\mathfrak{z}$ and $\mathfrak{z}'$ can only be equivalent if they are equal, that is

$$\pi_T | F_{\mathfrak{z}_0} : F_{\mathfrak{z}_0} \to U_{\mathfrak{z}_0} := \pi_T(F_{\mathfrak{z}_0}) \subset T^n$$

is injective. Hence

$$\varphi_{\mathfrak{z}_0} := (\pi_T | F_{\mathfrak{z}_0})^{-1} : U_{\mathfrak{z}_0} \to F_{\mathfrak{z}_0}$$

is a complex coordinate system for the torus, and the set of all $U_{\mathfrak{z}_0}$ covers the entire torus.

Theorem 3.10 *The n-dimensional complex torus T^n is a compact n-dimensional complex manifold and the canonical projection $\pi_T : \mathbb{C}^n \to T^n$ is holomorphic. [Since one can show that the complex structure on T^n depends on the vectors $\mathfrak{c}_1, \ldots, \mathfrak{c}_{2n}$, we also write: $T^n = T^n(\mathfrak{c}_1, \ldots, \mathfrak{c}_{2n})$.]*

PROOF

1. Any two complex charts for T^n are holomorphically compatible.

$$\varphi_{\mathfrak{z}_1} \circ \varphi_{\mathfrak{z}_2}^{-1} = \varphi_{\mathfrak{z}_1} \circ \pi_T : \varphi_{\mathfrak{z}_2}(U_{\mathfrak{z}_1} \cap U_{\mathfrak{z}_2}) \to \varphi_{\mathfrak{z}_1}(U_{\mathfrak{z}_1} \cap U_{\mathfrak{z}_2})$$

is a topological mapping, where

$$\varphi_{\mathfrak{z}_1} \circ \varphi_{\mathfrak{z}_2}^{-1}(\mathfrak{z}) = \mathfrak{z} + \sum_{\nu=1}^{2n} k_\nu(\mathfrak{z})\mathfrak{c}_\nu$$

and the functions k_ν are integer valued. Since $\{\mathfrak{c}_1, \ldots, \mathfrak{c}_{2n}\}$ is a (real) basis of $\mathbb{C}^n$, the k_ν must be continuous, and therefore locally constant. But then $\varphi_{\mathfrak{z}_1} \circ \varphi_{\mathfrak{z}_2}^{-1}$ is holomorphic.

2. T^n is a Hausdorff space: Let $x_1 = \pi_T(\mathfrak{z}_1) \neq \pi_T(\mathfrak{z}_2) = x_2$. Then we can write:

$$\mathfrak{z}_1 - \mathfrak{z}_2 = \sum_{\nu=1}^{2n} k_\nu \mathfrak{c}_\nu + \sum_{\nu=1}^{2n} r_\nu \mathfrak{c}_\nu,$$

with $k_\nu \in \mathbb{Z}$ and $0 \leqslant r_\nu < 1$ for $\nu = 1, \ldots, 2n$. Moreover not all r_ν can vanish simultaneously. Suppose $r_1 \neq 0$ and let $\varepsilon > 0$ be chosen so that $2\varepsilon < r_1 < 1 - 2\varepsilon$.

$$U: = \left\{\mathfrak{z} = \sum_{\nu=1}^{2n} \tilde{r}_\nu \mathfrak{c}_\nu : |\tilde{r}_\nu| < \varepsilon\right\}$$

is open, and hence

$$U_1(\mathfrak{z}_1): = U + \mathfrak{z}_1 \text{ and } U_2(\mathfrak{z}_2): = U + \mathfrak{z}_2$$

are open neighborhoods. Suppose $\pi_T(U_1) \cap \pi_T(U_2) \neq \varnothing$, so there are points $\mathfrak{z}' \in U_1$, $\mathfrak{z}'' \in U_2$ with $\mathfrak{z}' \sim \mathfrak{z}''$. But then we have

$$\mathfrak{z}' = \mathfrak{z}_1 + \sum_{\nu=1}^{2n} r'_\nu \mathfrak{c}_\nu \quad \text{and} \quad \mathfrak{z}'' = \mathfrak{z}_2 + \sum_{\nu=1}^{2n} r''_\nu \mathfrak{c}_\nu \quad \text{with} \quad |r'_\nu| < \varepsilon \quad \text{and} \quad |r''_\nu| < \varepsilon,$$

therefore

$$\mathfrak{z}' - \mathfrak{z}'' = (\mathfrak{z}_1 - \mathfrak{z}_2) + \sum_{\nu=1}^{2n} (r'_\nu - r''_\nu)\mathfrak{c}_\nu = \sum_{\nu=1}^{2n} k_\nu \mathfrak{c}_\nu + \sum_{\nu=1}^{2n} (r_\nu + (r'_\nu - r''_\nu))\mathfrak{c}_\nu.$$

Since

$$1 > r_1 + 2\varepsilon > |r_1 + (r'_1 - r''_1)| > r_1 - 2\varepsilon > 0,$$

$r_1 + (r'_1 - r''_1)$ cannot be an integer. That is a contradiction, so $\pi_T(U_1)$ and $\pi_T(U_2)$ are disjoint.

3. If $\mathfrak{z} \in \mathbb{C}^n$, then $\mathfrak{z}$ is equivalent to a point

$$\mathfrak{z}' \in F: = \left\{\mathfrak{z} = \sum_{\nu=1}^{2n} r_\nu \mathfrak{c}_\nu : -\frac{1}{2} \leqslant r_\nu \leqslant \frac{1}{2}\right\}.$$

F is compact, π_T is continuous, T^n is a Hausdorff space, and: $\pi_T(F) = T^n$. Hence it follows that T^n is compact.

4. $\pi_T : \mathbb{C}^n \to T^n$ is holomorphic. If $W \subset T^n$ is open, if g is holomorphic in W and if $\mathfrak{z}_0 \in V: = \pi_T^{-1}(W)$, then $g \circ \pi_T | V \cap F_{\mathfrak{z}_0} = g \circ \varphi_{\mathfrak{z}_0}^{-1} | V \cap F_{\mathfrak{z}_0}$ is holomorphic. □

Hopf Manifolds

Let $\rho > 1$ be a real number, $\Gamma_H: = \{\rho^k : k \in \mathbb{Z}\}$. Γ_H is a subgroup of the multiplicative group of the positive real numbers. Two elements $\mathfrak{z}_1, \mathfrak{z}_2 \in \mathbb{C}^n - \{0\}$ are considered equivalent if there is a $\rho^k \in \Gamma_H$ with $\mathfrak{z}_2 = \rho^k \mathfrak{z}_1$. The set H of all equivalence classes will be given the finest topology in which the canonical projection $\pi_H : \mathbb{C}^n - \{0\} \to H$ is continuous. We obtain complex coordinate systems for H in the following manner.

Let

$$F_r := \{\mathfrak{z} \in \mathbb{C}^n - \{0\} : r < \|\mathfrak{z}\| < \rho r\}$$

for arbitrary real numbers $r > 0$. Then $\bigcup_{r \in \mathbb{R}_+} F_r = \mathbb{C}^n - \{0\}$, and we can show that

$$\pi_H | F_r : F_r \to U_r := \pi(F_r) \subset H$$

is topological. (U_r, φ_r) is therefore a complex chart. In a manner similar to that of the preceding examples we can prove:

Theorem 3.11. *H is a compact n-dimensional complex manifold (the so-called Hopf manifold), and $\pi_H : \mathbb{C}^n - \{0\} \to H$ is holomorphic.*

If, for $\mathfrak{z}_1, \mathfrak{z}_2 \in \mathbb{C}^n - \{0\}$, $\pi_H(\mathfrak{z}_1) = \pi_H(\mathfrak{z}_2)$, then there is a $k \in \mathbb{Z}$ with $\mathfrak{z}_2 = \rho^k \mathfrak{z}_1$. But then $G(\mathfrak{z}_2) = G(\mathfrak{z}_1)$. Therefore there is a mapping $h : H \to \mathbb{P}^n$ defined by $h(\pi_H(\mathfrak{z})) := G(\mathfrak{z})$. We obtain the following commutative diagram.

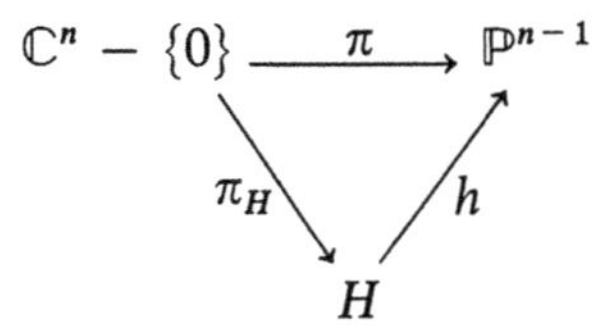

Since π_H is locally biholomorphic, it follows that h is holomorphic. □

Meromorphic Functions and Projective-Algebraic Manifolds

Let X be an arbitrary complex manifold.

Theorem 3.12. *Let $U \subset X$ be open, $x_0 \in U$. Let g, h be holomorphic functions on U with $g(x_0) = h(x_0) = 0$. If the germs g_{x_0}, h_{x_0} are relatively prime, then for every complex number c there exists a point x arbitrarily close to x_0 with $h(x) \neq 0$ and $g(x)/h(x) = c$.*

PROOF. Without loss of generality we can assume that U is a polycylinder in $\mathbb{C}^n$ and $x_0 = 0$. By the Weierstrass preparation theorem one can further assume that g_{x_0}, h_{x_0} are elements of $\mathcal{O}'_{x_0}[z_1]$. If we denote the quotient field of $\mathcal{O}'_{x_0}$ by Q'_{x_0}, then it follows from Theorem 4.2 of Chapter III that g_{x_0}, h_{x_0} are already relatively prime in $Q'_{x_0}[z_1]$. By Theorem 6.6 of Chapter III there exists a greatest common divisor of g_{x_0}, h_{x_0} which can be written as a linear combination of g_{x_0}, h_{x_0} with coefficients in $\mathcal{O}'_{x_0}[z_1]$, and that greatest common divisor clearly must be a unit in $Q'_{x_0}[z_1]$. Thus there exists a neighborhood $V(0) \subset U$ and there are holomorphic functions g_1, h_1 on V as well as a nowhere vanishing function d independent of z_1 such that on V

$$d = g_1 g + h_1 h.$$

Now suppose that the theorem is false for $c = 0$. Then there is a neighborhood $W(0) \subset V$ such that for every $\mathfrak{z} \in W$, $g(\mathfrak{z}) = 0$ implies $h(\mathfrak{z}) = 0$. Since the zeroes of a polynomial depend continuously on the coefficients and since the polynomial $g(z_1, 0)$ has a zero at $z_1 = 0$, for suitably small $\mathfrak{z}' \in W \cap (\{0\} \times \mathbb{C}^{n-1})$ there is always a z_1 with $(z_1, \mathfrak{z}') \in W$ and $g(z_1, \mathfrak{z}') = 0$. But then $h(z_1, \mathfrak{z}') = 0$ also and consequently $d(\mathfrak{z}') = 0$. Therefore d vanishes identically near $x_0 = 0$, which is a contradiction. The assertion is thus proved for $c = 0$, and if we replace g by $g - c \cdot h$ and g/h by $(g - c \cdot h)/h$, we obtain the theorem for arbitrary c. □

Def. 3.4. A *meromorphic function on* X is a pair (A, f) with the following properties:

1. A is a subset of X.
2. f is a holomorphic function on $X - A$.
3. For every point $x_0 \in A$ there is a neighborhood $U(x_0) \subset X$ and holomorphic functions g, h on U such that:
   a. $A \cap U = \{x \in U | h(x) = 0\}$
   b. The germs $g_{x_0}, h_{x_0} \in \mathcal{O}_{x_0}$ are relatively prime.
   c. $f(x) = g(x)/h(x)$ for every $x \in U - A$.

Remark. If (A, f) is a meromorphic function on X it follows immediately from the definition that A is either empty or a 1-codimensional analytic set. We call A *the set of poles of the meromorphic function* (A, f).

Theorem 3.13. *Let $Y \subset X$ be an open dense subset, f a holomorphic function on Y. For every point $x_0 \in X - Y$ let there be a neighborhood $U(x_0) \subset X$ and holomorphic functions h, g on U such that g_{x_0} and h_{x_0} are relatively prime and for every $x \in Y$, $g(x) = f(x) \cdot h(x)$.*

Finally, let A be the set of all points $x_0 \in X - Y$ such that given a real number $r > 0$ and a neighborhood $V(x_0) \subset X$, there is an $x \in V \cap Y$ with $|f(x)| > r$.

Then there exists a uniquely determined holomorphic extension $\hat{f}$ of f to $X - A$ such that $(A, \hat{f})$ is a meromorphic function.

PROOF. Let $x_0 \in X - Y$. By assumption there exists a neighborhood $U(x_0) \subset X$ and holomorphic functions g, h on U which are relatively prime at x_0, such that $g(x) = f(x) \cdot h(x)$ for $x \in U \cap Y$.

If $h(x_0) \neq 0$, then g/h is bounded in a neighborhood of x_0. Therefore x_0 does not lie in A.

If $h(x_0) = 0$ and $g(x_0) \neq 0$, then $f = g/h$ assumes arbitrarily large values near x_0. Furthermore, if $h(x_0) = g(x_0) = 0$, f is also not bounded near x_0, by Theorem 3.12. Thus x_0 lies in A.

Hence $A \cap U = \{x \in U : h(x) = 0\}$.

(g/h) is a continuation of f on $U - A$. We can carry out this construction at every point of $X - Y$. Y is dense in X, so by the identity theorem the local continuation is already uniquely determined by f, and so we obtain a global holomorphic continuation $\hat{f}$ of f to $X - A$. It follows directly from the construction that $(A, \hat{f})$ is a meromorphic function. □

Theorem 3.13 allows us to define the sum and product of meromorphic functions:

If (A, f), (A', f') are meromorphic functions on X, then $Y := X - (A \cup A')$ is open and dense in X, and at every point of $A \cup A'$ we can write $f + f'$ and $f \cdot f'$ as the reduced fraction of two holomorphic functions. There are analytic sets $A_1, A_2 \subset X$ and meromorphic functions (A_1, f_1), (A_2, f_2) on X with $A_1, A_2 \subset A \cup A'$ and $f_1|Y = f + f'$, $f_2|Y = f \cdot f'$.

One sets

$$(A, f) + (A', f') := (A_1, f_1)$$
$$(A, f) \cdot (A', f') := (A_2, f_2).$$

If X is connected, then the meromorphic functions on X form a field. We can think of any holomorphic function f on X as a meromorphic function $(\varnothing, f)$.

Examples

1. Let $X = \mathbb{C} \cup \{\infty\}$, the Riemann sphere with the canonical coordinates $\psi_1 : X_1 \to \mathbb{C}$, $\psi_2 : X_2 \to \mathbb{C}$ (see Theorem 3.9). Let p and q be two relatively prime polynomials in $\mathbb{C}[z]$, and let $N_q := \{x \in X_1 : q(x) = 0\}$. Then $Y := X_1 - N_q$ is a dense open subset of X and $f(x) := p(x)/q(x)$ defines a holomorphic function f on Y. Let

$$P := \begin{cases} \{\infty\} & \text{if } \deg(q) < \deg(p) \\ \varnothing & \text{if } \deg(q) \geqslant \deg(p), \end{cases} \qquad A := N_q \cup P.$$

We want to show that there is a holomorphic function $\hat{f}$ on $X - A$ with $\hat{f}|Y = f$, such that $(A, \hat{f})$ is meromorphic. It suffices to show that there exists a neighborhood $U(\infty) \subset X$ and holomorphic functions g, h on U with

$$\frac{g}{h}\Big|U \cap Y = f|U \cap Y.$$

For then, since $A = \{x \in X : f \text{ is bounded in no neighborhood of } x\}$, the existence of an $\hat{f}$ with the desired properties follows from Theorem 3.13.

Now let $U := \{x \in X_2 : p(x) \neq 0,\ q(x) \neq 0\}$, set $g := 1/q$ and $h := 1/p$ on $U - \{\infty\}$, and $g(\infty) = h(\infty) := 0$. Then g, h are continuous functions on U. $\tilde{p} := g \circ \psi_2^{-1}$ and $\tilde{q} := h \circ \psi_2^{-1}$ are continuous on $\psi_2(U)$ and holomorphic except at the origin. By the Riemann extension theorem $\tilde{p}$, $\tilde{q}$ are actually holomorphic on the whole set $\psi_2(U)$, hence also g, h on U. Moreover, g/h and f concide on $U \cap Y$. This finishes the proof. □

The meromorphic function $(A, \hat{f})$ is written in short as p/q, since both the values and the poles of f are uniquely determined by p and q.

2. Let $X = \mathbb{C}^2$, $A := \{(z_1, z_2) \in \mathbb{C}^2 : z_2 = 0\}$. Then $f(z_1, z_2) := z_1/z_2$ is a holomorphic function on $X - A$ and (A, f) is a meromorphic function on X.

For $\mathfrak{z}_0 = (z_1^{(0)}, 0) \in A$ set $\mathfrak{z}_n := (z_1^{(0)} + (1/n), 1/n^2)$. Then the sequence of points $\mathfrak{z}_n$ (outside A) tends towards $\mathfrak{z}_0$; the values $f(\mathfrak{z}_n) = z_1^{(0)} \cdot n^2 + n$ are unbounded (for arbitrary $z_1^{(0)}$). At the point $\mathfrak{z}_0 = 0$, the case of an "indeterminate point" (which cannot occur in the one-dimensional case) arises, as numerator and denominator vanish simultaneously. The function assumes every possible value in an arbitrary neighborhood of the indeterminate point.

3. Let $\mathfrak{z}_0 \in \mathbb{C}^n - \{0\}$ be a fixed vector, $G = G(\mathfrak{z}_0)$. If $\mathfrak{z} \in G$, then $\rho^k\mathfrak{z} \in G$ also. Hence one can also divide G by Γ_H. $T = G/\Gamma_H$ is a 1-dimensional complex torus and at the same time a submanifold of H. If f is a meromorphic function on H, then $f \circ \pi_H$ is meromorphic on $\mathbb{C}^n - \{0\}$. For $n \geqslant 2$ there is a continuity theorem for meromorphic functions, which in the present case says that $f \circ \pi_H$ can be continued to a meromorphic function $\hat{f}$ on $\mathbb{C}^n$. Naturally $\hat{f}|G$ is then also meromorphic. If $\mathfrak{z}_0$ were a pole of $\hat{f}|G$, then all the points $\rho^k\mathfrak{z}_0 \in G$ would also be poles of $\hat{f}|G$, and these points cluster about the origin. Since this cannot be, we must either have $\hat{f}|G \equiv \infty$ or $\hat{f}|G$ holomorphic. If $\hat{f}|G$ is holomorphic, then $f|T$ is also holomorphic and therefore constant (since T is compact). The submanifolds $T \subset H$ are precisely the fibers of the holomorphic mapping $h: H \to \mathbb{P}^{n-1}$. Hence we can show that there exists a meromorphic function g on $\mathbb{P}^{n-1}$ with $g \circ h = f$. In other words, on the n-dimensional manifold H there are no "more" meromorphic functions than on the $(n-1)$-dimensional manifold $\mathbb{P}^{n-1}$.

Def. 3.5. An n-dimensional compact complex manifold is called *projective-algebraic* if there exists an $N \in \mathbb{N}$ and an analytic subset $A \subset \mathbb{P}^N$ which is free of singularities and of codimension $N - n$ such that $X \simeq A$.

By a theorem of Chow every projective-algebraic manifold is already "algebraic" in the sense that it can be described by polynomial equations. Furthermore:

Theorem 3.14. *Let X be a projective-algebraic manifold. Then for arbitrary points $x_1, x_2 \in X$ with $x_1 \neq x_2$ there is always a meromorphic function f on X which is holomorphic at x_1 and x_2 with $f(x_1) \neq f(x_2)$.*

This means that there are "many" meromorphic functions on projective-algebraic manifolds. The Hopf manifold is not projective-algebraic. One can also interpret this topologically.

For a topological space X let $H_i(X, \mathbb{R})$ be the i-th homology group of X with coefficients in $\mathbb{R}$. If X is a $2n$-dimensional compact real manifold, then

$$B_i(X) := \dim_{\mathbb{R}} H_i(X, \mathbb{R}) \begin{cases} < \infty & \text{for} \quad i = 0, \ldots, 2n \\ = 0 & \text{for} \quad i > 2n \end{cases}$$

We call $B_i(X)$ the ith Betti number and associate with X the *Betti polynomial* $P(X) := \sum_{i=0}^{2n} B_i(X)t^i$. For cartesian products there is the formula

$$P(X \times Y) = P(X) \cdot P(Y).$$

Theorem 3.15. *If X is a projective-algebraic manifold, then the Betti numbers satisfy*

$$B_{2i+1}(X) \in 2\mathbb{Z},$$
$$B_{2i}(X) \neq 0.$$

This theorem is proved within the framework of the theory of "Kähler manifolds." It constitutes a necessary condition which is not fulfilled for a Hopf manifold. We can easily convince ourselves that H is homeomorphic to $S^{2n-1} \times S^1$. But for spheres S^k, $P(S^k) = 1 + t^k$. Hence it follows that

$$P(H) = P(S^{2n-1}) \cdot P(S^1) = (1 + t^{2n-1}) \cdot (1 + t) = 1 + t + t^{2n-1} + t^{2n}.$$

For $n \geqslant 2$ therefore $B_0(H) = 1$, $B_1(H) = 1$, and $B_2(H) = 0$. On the other hand, the n-dimensional complex torus satisfies this necessary condition for projective-algebraic manifolds:

$$T^n \simeq \underbrace{S^1 \times \cdots \times S^1}_{2n\text{-times}} \qquad \text{(topological)}$$

therefore

$$P(T^n) = (1 + t)^{2n} = \sum_{i=0}^{2n} \binom{2n}{i} t^i.$$

Hence $B_i(T^n) = \binom{2n}{i}$, and the condition is satisfied. Nevertheless not every torus is projective-algebraic. This property depends very critically on the vectors $\mathfrak{c}_1, \ldots, \mathfrak{c}_{2n}$ which define the torus. It can be shown that the so-called "period relations" (which involve only the vectors $\mathfrak{c}_1, \ldots, \mathfrak{c}_{2n}$) furnish a sufficient condition.

4. Closures of $\mathbb{C}^n$

Def. 4.1. Let X and Y be connected n-dimensional complex manifolds. If Y is compact and $X \subset Y$ is open, then we call Y a *closure of* X.

Example. The coordinate neighborhood $U_1 \subset \mathbb{P}^n$ is isomorphic to $\mathbb{C}^n$. Hence $\mathbb{P}^n$ is a closure of $\mathbb{C}^n$. (We use the notation of Section 3, above.)

A holomorphic function f is defined on $U_1 = \mathbb{C}^n$ by

$$f(\pi(1, z_2, \ldots, z_{n+1})) := \sum_{|\nu|=0}^{p} a_{\nu_2 \ldots \nu_{n+1}} z_2^{\nu_2} \cdots z_{n+1}^{\nu_{n+1}}$$

(with $a_{\nu_2 \ldots \nu_{n+1}} \in \mathbb{C}$ and $|\nu| := \nu_2 + \cdots + \nu_{n+1}$). If

$$x = \pi(1, z_2, \ldots, z_{n+1}) = \pi(w_1, \ldots, 1, \ldots, w_{n+1}) \in U_1 \cap U_i,$$

then

$$z_2 = \frac{w_2}{w_1}, \ldots, z_i = \frac{1}{w_1}, \ldots, z_{n+1} = \frac{w_{n+1}}{w_1}.$$

Hence there is a meromorphic function f_i with $f_i|U_i \cap U_1 = f|U_i \cap U_1$ given by

$$f_i(\pi(w_1, \ldots, 1, \ldots, w_{n+1})): = \sum_{|\nu|=0}^{p} a_{\nu_2 \cdots \nu_{n+1}} \left(\frac{w_2}{w_1}\right)^{\nu_2} \cdots \left(\frac{1}{w_1}\right)^{\nu_i} \cdots \left(\frac{w_{n+1}}{w_1}\right)^{\nu_{n+1}}.$$

$\hat{f}$ with $\hat{f}|U_i: = f_i$ is then a meromorphic function on $\mathbb{P}^n$ with $\hat{f}|\mathbb{C}^n = f$.

Def. 4.2. Let Y be a closure of $\mathbb{C}^n$. Y is called a *regular closure of* $\mathbb{C}^n$ if every polynomial defined on $\mathbb{C}^n$ extends to a meromorphic function on Y.

Clearly $\mathbb{P}^n$ is a regular closure of $\mathbb{C}^n$.

Theorem 4.1. *If Y is a regular closure of $\mathbb{C}^n$, then $Y - \mathbb{C}^n$ is an analytic set of codimension* 1.

PROOF. Let $z_1, \ldots, z_n$ be the coordinates of $\mathbb{C}^n$. By hypothesis they can be continued to meromorphic functions $f_1, \ldots, f_n$ on Y.

The set P_i of poles of f_i is an analytic set of codimension 1, and so is $P: = \bigcup_{i=1}^{n} P_i$. Hence it suffices to show that $Y - \mathbb{C}^n = P$.

Let $\mathfrak{z}_0 \in \partial\mathbb{C}^n \subset Y$. Then there is a sequence $(\mathfrak{z}_i)$ in $\mathbb{C}^n$ with $\lim_{i\to\infty} \mathfrak{z}_i = \mathfrak{z}_0$. This means that $(z_k^{(i)})$ is unbounded for at least one $k \in \{1, \ldots, n\}$. We can find a subsequence $(z_k^{(\nu_i)})$ with $\lim_{i\to\infty} |z_k^{(\nu_i)}| = \infty$. Hence, for $i \to \infty$, $f_k(\mathfrak{z}_{\nu_i})$ tends to infinity, so $\mathfrak{z}_0$ is a pole of f_k. Thus $\mathfrak{z}_0$ lies in P, and since $\mathfrak{z}_0 \in \partial\mathbb{C}^n$ was chosen arbitrarily, $\partial\mathbb{C}^n \subset P$.

An analytic set of codimension 1 cannot separate a manifold; that is, $Y - P$ is connected. Hence for every point $\mathfrak{z}_0 \in (Y - \mathbb{C}^n) - P$ there is a path $\varphi:[0, 1] \to Y - P$ with $\varphi(0) = 0$ and $\varphi(1) = \mathfrak{z}_0$. Since such a path always meets the boundary $\partial\mathbb{C}^n$, we must have $(Y - \mathbb{C}^n) - P = \varnothing$. □

Remark. For $n \geqslant 2$ Bieberbach has constructed an injective holomorphic mapping $\beta:\mathbb{C}^n \to \mathbb{C}^n$ whose functional determinant equals 1 everywhere and whose image $U: = \beta(\mathbb{C}^n)$ has the property that there exist interior points in $\mathbb{C}^n - U$. We can regard U as an open subset of $\mathbb{P}^n$. Then $\mathbb{P}^n$ is a closure of $\mathbb{C}^n \simeq U$, but this closure is not regular since $\mathbb{C}^n - U$ contains interior points. A 1-codimensional analytic set cannot have interior points (Theorem 2.6)!

As a further example we consider the Osgood closure of $\mathbb{C}^n$.

Let

$$\bar{\mathbb{C}}^n: = \underbrace{\mathbb{P}^1 \times \cdots \times \mathbb{P}^1}_{n\text{-times}}.$$

Each factor $\mathbb{P}^1$ is isomorphic to the Riemann sphere, which has the canonical coordinates ψ_1, ψ_2. We obtain coordinates on $\bar{\mathbb{C}}^n$ by letting $U_{\nu_1 \cdots \nu_n} := U_{\nu_1} \times \cdots \times U_{\nu_n}$ and

$$\psi_{\nu_1 \cdots \nu_n} := \psi_{\nu_1} \times \cdots \times \psi_{\nu_n} : U_{\nu_1 \cdots \nu_n} \to \mathbb{C}^n.$$

(with $\nu_\lambda \in \{1, 2\}$). $\bar{\mathbb{C}}^n$ is compact, and $\mathbb{C}^n \simeq U_{1 \cdots 1} \subset \bar{\mathbb{C}}^n$ is an open subset. Therefore $\bar{\mathbb{C}}^n$ is a closure of $\mathbb{C}^n$ and we can see directly that this closure is regular.

If Y is a closure of $\mathbb{C}^n$, we call the elements of $Y - \mathbb{C}^n$ *infinitely distant points*. In special cases one can describe the infinitely distant points more exactly.

1. Let $Y = \mathbb{P}^n$ be the usual projective closure of $\mathbb{C}^n$. Then

$$\mathbb{C}^n \simeq U_1 = \pi(\{(1, z_2, \ldots, z_{n+1}) \in \mathbb{C}^{n+1}\}) = \pi(\{(z_1, \ldots, z_{n+1}) \in \mathbb{C}^{n+1} : z_1 \neq 0\}),$$

Therefore

$$\mathbb{P}^n - \mathbb{C}^n = \pi(\{(0, z_2, \ldots, z_{n+1}) \in \mathbb{C}^{n+1} - \{0\}\}),$$

and this set is isomorphic to $\mathbb{P}^{n-1}$.

2. Let $Y = \bar{\mathbb{C}}^n$ be the Osgood closure of $\mathbb{C}^n$. Then $\mathbb{C}^n \simeq U_{1 \cdots 1} = U_1 \times \cdots \times U_1$ and

$$\begin{aligned} \bar{\mathbb{C}}^n - \mathbb{C}^n &= \{(x_1, \ldots, x_n) \in \mathbb{P}^1 \times \cdots \times \mathbb{P}^1 \text{ there is an } i \text{ with } x_i \notin U_1\} \\ &= \{(x_1, \ldots, x_n) \in \mathbb{P}^1 \times \cdots \times \mathbb{P}^1 : \text{there is an } i \text{ with } x_i = \infty\} \\ &= (\{\infty\} \times \mathbb{P}^1 \times \cdots \times \mathbb{P}^1) \cup \cdots \cup (\mathbb{P}^1 \times \cdots \times \mathbb{P}^1 \times \{\infty\}). \end{aligned}$$

In the first case the set of infinitely distant points is free of singularities of codimension 1; and in the second case it is the finite union of analytic subsets of codimension 1 which are free of singularities and each of which is isomorphic to $\mathbb{C}^{n-1}$. For $n = 1$, $\bar{\mathbb{C}} = \mathbb{P}^1$. One can prove that only this closure exists in the 1-dimensional case. For $n = 2$, $\bar{\mathbb{C}}^2 - \mathbb{C}^2 = (\{\infty\} \times \mathbb{P}^1) \cup (\mathbb{P}^1 \times \{\infty\})$ with $(\{\infty\} \times \mathbb{P}^1) \cap (\mathbb{P}^1 \times \{\infty\}) = \{(\infty, \infty)\}$. The analytic set $\bar{\mathbb{C}}^2 - \mathbb{C}^2$ has a singularity at the point (∞, ∞) as can easily be demonstrated.

Def. 4.3. Let X and Y be connected n-dimensional complex manifolds, let $M \subset X$ and $N \subset Y$ be closed proper subsets, and let $\pi: X - M \to Y - N$ be a biholomorphic mapping. Then (X, M, π, N, Y) is called a *modification*.

For example $(\mathbb{P}^n, \mathbb{P}^{n-1}, \mathrm{id}_{\mathbb{C}^n}, \bar{\mathbb{C}}^n - \mathbb{C}^n, \bar{\mathbb{C}}^n)$ is a modification. We can therefore use modifications to describe transformations between distinct closures of $\mathbb{C}^n$.

Def. 4.4. Let $\varphi: X \to Y$ be a holomorphic mapping between connected complex manifolds, $\dim X = n$ and $\dim Y = m$. Then

$$E(\varphi) := \{x \in X : \dim_x(\varphi^{-1}(\varphi(x))) > n - m\}$$

is called the *set of degeneracy of* φ.

If $\dim X = \dim Y$, then as can be shown, $E(\varphi) = \{x \in X : x \text{ is not an isolated point of } \varphi^{-1}(\varphi(x))\}$.

Both the following theorems were proved by Remmert:

Theorem 4.2. *If $\varphi: X \to Y$ is a holomorphic mapping between connected complex manifolds, then $E(\varphi)$ is an analytic subset of X.*

Theorem 4.3 (Projection theorem). *If $\varphi: X \to Y$ is a proper holomorphic mapping between complex manifolds and $M \subset X$ is an analytic subset, then $\varphi(M) \subset Y$ is also analytic.*

Def. 4.5. A modification (X, M, π, N, Y) is called *proper* if π can be continued to a proper holomorphic mapping $\hat{\pi}: X \to Y$ such that $M = E(\hat{\pi})$.

Theorem 4.4. *Let (X, M, π, N, Y) be a proper modification, $\hat{\pi}: X \to Y$ a continuation of π in the sense of Def. 4.5. Then M and N are analytic sets, and $\hat{\pi}(M) = N$.*

PROOF. By Theorem 4.2, $M = E(\hat{\pi})$ is analytic, and by Theorem 4.3, $N^*: = \hat{\pi}(M)$ is analytic. It remains to show that $N = N^*$:

1. Suppose there is a $y_0 \in N^* - N$. We set $x_0: = \pi^{-1}(y_0) \in X - M$ and choose an $x_0^* \in M$ with $\hat{\pi}(x_0^*) = y_0$. Then we can find open neighborhoods $U(x_0)$, $V(x_0^*)$ and $W(y_0)$ such that:

a. $U \cap V = \varnothing$
b. $W \subset Y - N$
c. $\pi(U) = W$
d. $\hat{\pi}(V) \subset W$.

But from this it follows that $V - M \subset X - M$ is open and non-empty, and $\pi(V - M) = \hat{\pi}(V - M)$ lies in W. Therefore

$$V - M = \pi^{-1}\pi(V - M) \subset \pi^{-1}(W) = U.$$

That is a contradiction; and so $N^* \subset N$.

2. $Y - N$ is open and non-empty, so for every point $y_0 \in \partial(Y - N)$ there is a sequence (y_i) in $Y - N$ with $\lim_{i \to \infty} y_i = y_0$. The set $K: = \{y_0, y_1, y_2, \ldots\}$ is compact, and since $\hat{\pi}$ is proper, $K^*: = \hat{\pi}^{-1}(K)$ is also compact. In particular, K^* contains the uniquely determined points $x_i \in X - M$ with $\pi(x_i) = y_i$. We can find a subsequence (x_{ν_i}) of (x_i) which converges to a point $x_0 \in K^*$. Since $\hat{\pi}$ is continuous, we must have $\hat{\pi}(x_0) = y_0$; therefore $x_0 \in M$ and $y_0 \in \hat{\pi}(M) = N^*$. Hence we have shown that $\partial(Y - N)$ lies in N^*.

Suppose there is a point $y_0 \in N - N^*$. Then since N^* is analytic, we can connect y_0 with a point $y_0^* \in Y - N$ by a path running entirely in $Y - N^*$. Each such path, however, intersects $\partial(Y - N)$, and hence N^*. That is a contradiction; so $N \subset N^*$, and thus $N = N^*$. □

The most important special case is the Hopf σ-process:

Theorem 4.5. *Let $G \subset \mathbb{C}^n$ be a domain with $0 \in G$, $\pi:\mathbb{C}^n - \{0\} \to \mathbb{P}^{n-1}$ the natural projection. Then $X := \{(\mathfrak{z}, x) \in (G - \{0\}) \times \mathbb{P}^{n-1}: x = \pi(\mathfrak{z})\} \cup (\{0\} \times \mathbb{P}^{n-1})$ is an analytic set of codimension $n - 1$ in $G \times \mathbb{P}^{n-1}$ which is free of singularities, therefore an n-dimensional complex manifold.*

PROOF. Let $\varphi_i: U_i \to \mathbb{C}^{n-1}$ be the canonical coordinate system of $\mathbb{P}^{n-1}$. If $\mathfrak{z} = (z_1, \ldots, z_n) \in G - \{0\}$ and $x = \pi(\mathfrak{z}) \in U_1$, then

$$x = \pi\left(1, \frac{z_2}{z_1}, \ldots, \frac{z_n}{z_1}\right).$$

Therefore $w_\lambda(x) = z_{\lambda+1}/z_1$ for $\lambda = 1, \ldots, n - 1$, where we denote the coordinates on U_1 by w_λ. Hence it follows that

$$\begin{aligned} &X \cap (G \times U_1) \\ &= \left\{(\mathfrak{z}, x) \in G \times U_1 : z_1 \neq 0, w_\lambda(x) = \frac{z_{\lambda+1}}{z_1} \text{ for } \lambda = 1, \ldots, n - 1\right\} \cup (\{0\} \times U_1) \\ &= \{(z_1, \ldots, z_n; x) \in G \times U_1 : z_1 \cdot w_1(x) - z_2 \\ &= \cdots = z_1 \cdot w_{n-1}(x) - z_n = 0\}. \end{aligned}$$

There is an analogous representation for $U_2, \ldots, U_n$. Therefore X is an analytic set in $G \times \mathbb{P}^{n-1}$.

Since clearly $rk_{(\mathfrak{z}, x)}(z_1 \cdot w_1 - z_2, \ldots, z_1 \cdot w_{n-1} - z_n) = n - 1$ on all of U_1 and an analogous statement can be made for $U_2, \ldots, U_n$, X is free of singularities of codimension $n - 1$. □

Theorem 4.6. *Let $X \subset G \times \mathbb{P}^{n-1}$ be the analytic set described in Theorem 4.5, $\varphi: X \to G$ the holomorphic mapping induced by the product projection $\mathrm{pr}_1: G \times \mathbb{P}^{n-1} \to G$, $\psi := \varphi|(X - (\{0\} \times \mathbb{P}^{n-1}))$. Then $(X, \{0\} \times \mathbb{P}^{n-1}, \psi, \{0\}, G)$ is a proper modification. It is called the "σ-process."*

PROOF

1. $\psi': G - \{0\} \to X - (\{0\} \times \mathbb{P}^{n-1})$ with $\psi'(\mathfrak{z}) := (\mathfrak{z}, \pi(\mathfrak{z}))$ is clearly holomorphic, and:

$$\begin{aligned} \psi \circ \psi'(\mathfrak{z}) &= \mathrm{pr}_1(\mathfrak{z}, \pi(\mathfrak{z})) = \mathfrak{z}, \\ \psi' \circ \psi(\mathfrak{z}, x) &= \psi' \circ \psi(\mathfrak{z}, \pi(\mathfrak{z})) = \psi'(\mathfrak{z}) = (\mathfrak{z}, \pi(\mathfrak{z})). \end{aligned}$$

Therefore $\psi' = \psi^{-1}$, and $\psi: X - (\{0\} \times \mathbb{P}^{n-1}) \to G - \{0\}$ is biholomorphic.

2. φ is a holomorphic continuation of ψ, and $\varphi^{-1}(\mathfrak{z}) = \{(\mathfrak{z}, x) \in G \times \mathbb{P}^{n-1}: (\mathfrak{z}, x) \in X\} = \{(\mathfrak{z}, \pi(\mathfrak{z}))\}$ for $\mathfrak{z} \in G - \{0\}$, $\varphi^{-1}(0) = \{0\} \times \mathbb{P}^{n-1}$. Therefore $E(\varphi) = \{0\} \times \mathbb{P}^{n-1} \simeq \mathbb{P}^{n-1}$.

3. If $K \subset G$ is compact, then $K \times \mathbb{P}^{n-1}$ is compact, and so is $\varphi^{-1}(K) = (K \times \mathbb{P}^{n-1}) \cap X$. Therefore φ is proper. □

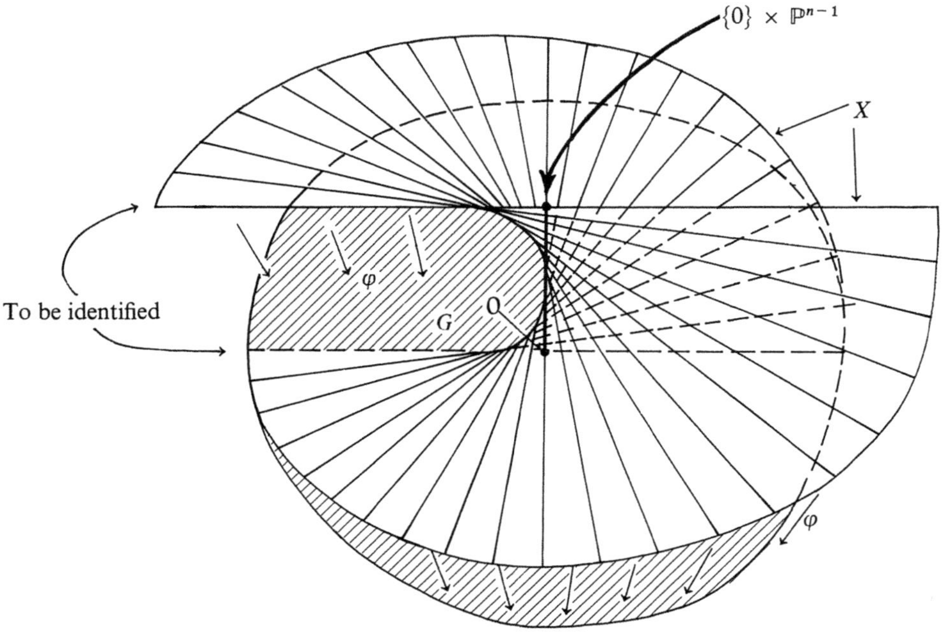

Figure 24. The Hopf σ-process.

Remark. Clearly we can regard $\mathbb{P}^{n-1}$ as the set of all directions in $\mathbb{C}^n$. By the σ-process these directions are separated in the following sense:

If one approaches the origin in $G - \{0\}$ from the direction $x_0 \in \mathbb{P}^{n-1}$, say along a path w, then one approaches the point $(0, x_0)$ along the directly lifted path $\psi^{-1} \circ w$ in $X - \mathbb{P}^{n-1}$.

It can be shown that the σ-process is invariant under biholomorphic mappings. Hence it can also be performed on complex manifolds.

CHAPTER VI

Cohomology Theory

1. Flabby Cohomology

In this chapter we apply, with the help of cohomology groups, the methods and results of sheaf theory to complex manifolds.

X will always be an n-dimensional complex manifold and R a commutative ring with 1. If $\mathscr{S}$ is a sheaf of R-modules over X and $U \subset X$ is open, then we let $\hat{\Gamma}(U, \mathscr{S})$ denote the set of all functions $s: U \to \mathscr{S}$ with $\pi \circ s = \mathrm{id}_U$ (where $\pi: \mathscr{S} \to X$ is the sheaf projection), and we call these not necessarily continuous functions *generalized sections*. Clearly $\Gamma(U, \mathscr{S})$ is an R-submodule of $\hat{\Gamma}(U, \mathscr{S})$.

If $\varphi: \mathscr{S}_1 \to \mathscr{S}_2$ is a homomorphism of R-module sheaves, then $\varphi_*: \hat{\Gamma}(U, \mathscr{S}_1) \to \hat{\Gamma}(U, \mathscr{S}_2)$ with $\varphi_*(s) := \varphi \circ s$ is an R-module homomorphism.

Theorem 1.1. *($\hat{\Gamma}_U: \mathscr{S} \rightsquigarrow \hat{\Gamma}(U, \mathscr{S}), \varphi \rightsquigarrow \varphi_*$) is an exact covariant functor from the category of R-module sheaves over X to the category of R-modules. Therefore:*

1. *if $\mathscr{S}$ is an R-module sheaf, then $\hat{\Gamma}(U, \mathscr{S})$ is an R-module;*
2. *if $\varphi: \mathscr{S}_1 \to \mathscr{S}_2$ is a homomorphism of R-module sheaves, then φ_*: $\hat{\Gamma}(U, \mathscr{S}_1) \to \hat{\Gamma}(U, \mathscr{S}_2)$ is a homomorphism of R-modules;*
3. a. $(\mathrm{id}_{\mathscr{S}})_* = \mathrm{id}_{\hat{\Gamma}(U, \mathscr{S})}$;
   b. $(\varphi \circ \psi)_* = \varphi_* \circ \psi_*$;
4. *if $\mathscr{S}_1 \xrightarrow{\psi} \mathscr{S}_2 \xrightarrow{\varphi} \mathscr{S}_3$ is an exact sequence of R-module sheaves, then $\hat{\Gamma}(U, \mathscr{S}_1) \xrightarrow{\psi_*} \hat{\Gamma}(U, \mathscr{S}_2) \xrightarrow{\varphi_*} \hat{\Gamma}(U, \mathscr{S}_3)$ is an exact sequence of R-modules.*

The proof is completely trivial.

For $U \subset X$ open, we set $M_U := \hat{\Gamma}(U, \mathscr{S})$; if $U, V \subset X$ are open with $V \subset U$, then we define $r_V^U: M_U \to M_V$ by $r_V^U(s) := s|V$. Then $\{M_U, r_V^U\}$ is a pre-sheaf and we denote the corresponding sheaf by $W(\mathscr{S})$.

Theorem 1.2.

1. *The canonical mapping $r: M_U \to \Gamma(U, W(\mathscr{S}))$ is an R-module isomorphism.*
2. *The canonical injection $i_U: \Gamma(U, \mathscr{S}) \hookrightarrow \hat{\Gamma}(U, \mathscr{S})$ induces an injective sheaf homomorphism $\varepsilon: S \to W(\mathscr{S})$ with $\varepsilon_*|\Gamma(U, \mathscr{S}) = r \circ i_U$.*

PROOF. (1) is proved exactly as is Theorem 2.3 in Chapter IV. To prove (2):

Clearly $i_U(s)|V = i_V(s|V)$ for $s \in \Gamma(U, \mathscr{S})$. If we identify the sheaf induced by $\{\Gamma(U, \mathscr{S}), r_V^U\}$ with the sheaf $\mathscr{S}$, then it follows from Theorem 2.1 of Chapter IV that there exists exactly one sheaf morphism $\varepsilon: \mathscr{S} \to W(\mathscr{S})$ with $\varepsilon_*(s) = ri_U(s)$ for $s \in \Gamma(U, \mathscr{S})$. If $\sigma \in \mathscr{S}_x$ and $\varepsilon(\sigma) = \mathbf{O}_x$, then there exists a neighborhood $U(x) \subset X$ and an $s \in \Gamma(U, \mathscr{S})$ with $s(x) = \sigma$. Therefore $\mathbf{O}_x = \varepsilon(\sigma) = \varepsilon \circ s(x) = \varepsilon_*(s)(x) = ri_U(s)(x)$, with $ri_U(s) \in \Gamma(U, W(\mathscr{S}))$. Then there exists a neighborhood $V(x) \subset U$ with $ri_U(s)|V = \mathbf{O}$; therefore $i_U(s)|V = \mathbf{O}$ by (1), and then clearly $s|V = \mathbf{O}$. Hence $\sigma = s(x) = \mathbf{O}_x$. □

Let $\varphi: \mathscr{S}_1 \to \mathscr{S}_2$ be a sheaf homomorphism. Then for open sets $U, V \subset X$ with $U \subset V$ and $s \in \hat{\Gamma}(U, \mathscr{S}_1)$ we have $\varphi_*(s)|V = \varphi_*(s|V)$. By Theorem 2.1 of Chapter IV φ induces exactly one sheaf homomorphism $W\varphi: W(\mathscr{S}_1) \to W(\mathscr{S}_2)$ with $(W\varphi)_*(rs) = r(\varphi_*(s))$.

Let $s \in \Gamma(U, \mathscr{S}_1)$. If $\varepsilon_\lambda: \mathscr{S}_\lambda \hookrightarrow W(\mathscr{S}_\lambda)$ are the canonical injections (for $\lambda = 1, 2$), then

$$(W\varphi) \circ \varepsilon_1 \circ s = (W\varphi)_*(ri_U^{(1)}(s)) = r(\varphi_*(i_U^{(1)}(s))) = r(i_U^{(2)}(\varphi_* s)) = \varepsilon_2 \circ \varphi \circ s.$$

Hence it follows that $(W\varphi) \circ \varepsilon_1 = \varepsilon_2 \circ \varphi$.

Def. 1.1. Let $\mathscr{S}$ be a sheaf of R-modules over X. $\mathscr{S}$ is called *flabby* if for every open set

$$r_U^X: \Gamma(X, \mathscr{S}) \to \Gamma(U, \mathscr{S}) \qquad \text{is surjective.}$$

Theorem 1.3. *If $\mathscr{S}$ is a sheaf of R-modules over X, then $W(\mathscr{S})$ is a flabby sheaf.*

PROOF. We can identify $\Gamma(U, W(\mathscr{S}))$ with $\hat{\Gamma}(U, \mathscr{S})$. If $s \in \hat{\Gamma}(U, \mathscr{S})$ then we define $s^* \in \hat{\Gamma}(X, \mathscr{S})$ by

$$s^*(x) := \begin{cases} s(x) & \text{for } x \in U; \\ 0 & \text{for } x \in X - U. \end{cases}$$

Clearly $r_U^X s^* = s$. □

Theorem 1.4. *$(W: \mathscr{S} \rightsquigarrow W(\mathscr{S}), \varphi \rightsquigarrow W\varphi)$ is an exact covariant functor from the category of R-module sheaves over X to itself.*

PROOF

1. Let $\psi:\mathscr{S}_1 \to \mathscr{S}, \varphi:\mathscr{S} \to \mathscr{S}_2$ be sheaf homomorphisms and $s \in \Gamma(U, \mathscr{S}_1)$. Then $W(\varphi \circ \psi) \circ rs = r((\varphi \circ \psi)_* s) = r(\varphi_*(\psi_* s)) = W\varphi \circ (r(\psi_* s)) = W\varphi \circ W\psi \circ rs$.

2. $W(\mathrm{id}_{\mathscr{S}}) \circ rs = r((\mathrm{id}_{\mathscr{S}})_* s) = rs$, for $s \in \Gamma(U, \mathscr{S})$.

3. Let $\mathscr{S}_1 \xrightarrow{\psi} \mathscr{S} \xrightarrow{\varphi} \mathscr{S}_2$ be exact.
   a. Then $W\varphi \circ W\psi = W(\varphi \circ \psi) = W(0) = 0$.
   b. Let $\sigma \in W(\mathscr{S})_x$ and $W\varphi(\sigma) = \mathbf{O}_x$. Then there exists a neighborhood $U(x) \subset X$ and an $s \in \hat{\Gamma}(U, \mathscr{S})$ with $rs(x) = \sigma$, therefore $W\varphi \circ rs(x) = \mathbf{O}_x$.

Hence there exists a neighborhood $V(x) \subset U$ with $\mathbf{O} = W\varphi \circ rs|V = r(\varphi \circ s)|V$; therefore $(\varphi \circ s)|V = \mathbf{O}$. We can construct an $s_1 \in \hat{\Gamma}(V, \mathscr{S}_1)$ with $\psi \circ s_1 = s|V$ pointwise. Then $W\psi \circ rs_1 = r(\psi \circ s_1) = rs|V$, and therefore $W\psi(rs_1(x)) = \sigma$. □

Def. 1.2. Let $\mathscr{S}$ be a sheaf of R-modules. A *resolution of* $\mathscr{S}$ is an exact sequence of sheaves of R-modules:

$$0 \to \mathscr{S} \to \mathscr{S}_0 \to \mathscr{S}_1 \to \mathscr{S}_2 \to \cdots.$$

If the sheaves $\mathscr{S}_0, \mathscr{S}_1, \mathscr{S}_2, \ldots$ are all flabby, then we speak of a flabby resolution.

We now show how to assign a canonical flabby resolution to any sheaf $\mathscr{S}$.

1. The sequence $0 \to \mathscr{S} \xrightarrow{\varepsilon} W(\mathscr{S})$ is exact. Let $W_0(\mathscr{S}) := W(\mathscr{S})$.
2. Suppose we have constructed an exact sequence $0 \to \mathscr{S} \to W_0(\mathscr{S}) \xrightarrow{d_0} \cdots \xrightarrow{d_{\ell-1}} W_\ell(\mathscr{S})$, with flabby sheaves $W_0(\mathscr{S}), W_1(\mathscr{S}), \ldots, W_\ell(\mathscr{S})$.

   Then there is an exact sequence

$$W_\ell(\mathscr{S}) \xrightarrow{q} W_\ell(\mathscr{S})/\mathrm{Im}(d_{\ell-1}) \xrightarrow{j} W(W_\ell(\mathscr{S})/\mathrm{Im}(d_{\ell-1}))$$

Let $W_{\ell+1}(\mathscr{S}) := W(W_\ell(\mathscr{S})/\mathrm{Im}(d_{\ell-1}))$,

$$d_\ell := j \circ q.$$

Clearly $\mathrm{Ker}(d_\ell) = \mathrm{Ker}(q) = \mathrm{Im}(d_{\ell-1})$; that is, the extended sequence $0 \to \mathscr{S} \to W_0(\mathscr{S}) \to \cdots \to W_\ell(\mathscr{S}) \to W_{\ell+1}(\mathscr{S})$ remains exact. Thus we construct an exact sequence $W_0(\mathscr{S}) \to W_1(\mathscr{S}) \to W_2(\mathscr{S}) \to \cdots$ by induction. We write $\mathfrak{W}(\mathscr{S})$ as an abbreviation. The exact sequence $0 \to \mathscr{S} \xrightarrow{\varepsilon} \mathfrak{W}(\mathscr{S})$ is called the *canonical flabby resolution of* $\mathscr{S}$.

Theorem 1.5. *Let $\varphi:\mathscr{S}_1 \to \mathscr{S}_2$ be a homomorphism of sheaves of R-modules over X. Then there are canonical homomorphisms $W_i\varphi: W_i(\mathscr{S}_1) \to W_i(\mathscr{S}_2)$ with $(W_{i+1}\varphi) \circ d_i = d_i \circ (W_i\varphi)$ for $i \in \mathbb{N}_0$ and $(W_0\varphi) \circ \varepsilon = \varepsilon \circ \varphi$.*

PROOF. We proceed by induction. Let $W_0\varphi := W\varphi$. If $W_0\varphi, W_1\varphi, \ldots, W_\ell\varphi$ have been constructed, then we have the following commutative diagram.

$$\begin{array}{ccccccc}
W_{\ell-1}(\mathscr{S}_1) & \xrightarrow{d_{\ell-1}} & W_\ell(\mathscr{S}_1) & \overset{q_1}{\twoheadrightarrow} & W_\ell(\mathscr{S}_1)/\mathrm{Im}(d_{\ell-1}) & \overset{j_1}{\hookrightarrow} & W_{\ell+1}(\mathscr{S}_1) \\
\downarrow W_{\ell-1}\varphi & & \downarrow W_\ell\varphi & & & & \\
W_{\ell-1}(\mathscr{S}_2) & \xrightarrow{d_{\ell-1}} & W_\ell(\mathscr{S}_2) & \overset{q_2}{\twoheadrightarrow} & W_\ell(\mathscr{S}_2)/\mathrm{Im}(d_{\ell-1}) & \overset{j_2}{\hookrightarrow} & W_{\ell+1}(\mathscr{S}_2)
\end{array}$$

It can be completed by a homomorphism

$$\psi: W_\ell(\mathscr{S}_1)/\mathrm{Im}(d_{\ell-1}) \to W_\ell(\mathscr{S}_2)/\mathrm{Im}(d_{\ell-1}) \qquad \text{with} \qquad \psi \circ q_1 = q_2 \circ W_\ell\varphi.$$

(If $q_1(\sigma) = \mathbf{O}$, then there exists a σ^* with $d_{\ell-1}(\sigma^*) = \sigma$; therefore $W_\ell\varphi(\sigma) = d_{\ell-1} \circ (W_{\ell-1}\varphi)(\sigma^*)$, so $q_2 \circ W_\ell\varphi(\sigma) = \mathbf{O}$.) We define $W_{\ell+1}\varphi := W\psi$. All diagrams remain commutative. □

The system of homomorphisms $W_i\varphi$ is denoted by $\mathfrak{W}(\varphi)$. We can regard $\mathfrak{W}(\varphi): \mathfrak{W}(\mathscr{S}_1) \to \mathfrak{W}(\mathscr{S}_2)$ as a "homomorphism between flabby resolutions."

Clearly $\mathfrak{W}(\mathrm{id}_{\mathscr{S}}) = \mathrm{id}_{\mathfrak{W}(\mathscr{S})}$, $\mathfrak{W}(\psi \circ \varphi) = \mathfrak{W}(\psi) \circ \mathfrak{W}(\varphi)$.

Therefore $(\mathfrak{W}: \mathscr{S} \rightsquigarrow \mathfrak{W}(\varphi))$ is a covariant functor. We need the next two lemmas in order to show that $\mathfrak{W}$ is also an exact functor.

Lemma 1. *Let the following diagram of sheaves of R-modules be commutative, have exact rows and columns, and moreover let the mapping φ_0 be surjective:*

$$\begin{array}{ccccc}
 & & \mathscr{S}_1 & \overset{\varphi_0}{\twoheadrightarrow} & \mathscr{S}_2 \\
 & & \downarrow \psi_1 & & \downarrow \psi_2 \\
\mathscr{S}_3 & \xrightarrow{\varphi_1} & \mathscr{S}_4 & \xrightarrow{\varphi_2} & \mathscr{S}_5 \\
\downarrow \psi_3 & & \downarrow \psi_4 & & \downarrow \psi_5 \\
\mathscr{S}_6 & \xrightarrow{\varphi_3} & \mathscr{S}_7 & \xrightarrow{\varphi_4} & \mathscr{S}_8 \\
 & & \downarrow \psi_6 & & \\
 & & \mathscr{S}_9 & &
\end{array}$$

If $\sigma \in \mathscr{S}_6$ and $\psi_6 \circ \varphi_3(\sigma) = \mathbf{O}$, then there exists a $\hat{\sigma} \in \mathscr{S}_3$ with

$$\varphi_3(\sigma - \psi_3(\hat{\sigma})) = \mathbf{O}.$$

PROOF. Let $\sigma_1 := \varphi_3(\sigma) \in \mathscr{S}_7$.

1. Because $\psi_6(\sigma_1) = \mathbf{O}$ there exists a $\sigma_2 \in \mathscr{S}_4$ with $\psi_4(\sigma_2) = \sigma_1$.
2. $\psi_5(\varphi_2(\sigma_2)) = \varphi_4(\psi_4(\sigma_2)) = \varphi_4(\varphi_3(\sigma)) = \mathbf{O}$; therefore there exists a $\sigma_3 \in \mathscr{S}_2$ with $\psi_2(\sigma_3) = \varphi_2(\sigma_2)$, and there is a $\sigma_4 \in \mathscr{S}_1$ with $\varphi_0(\sigma_4) = \sigma_3$.
3. $\varphi_2 \circ \psi_1(\sigma_4) = \psi_2 \circ \varphi_0(\sigma_4) = \varphi_2(\sigma_2)$; therefore $\varphi_2(\sigma_2 - \psi_1(\sigma_4)) = \mathbf{O}$. Hence there is a $\sigma_5 \in \mathscr{S}_3$ with $\varphi_1(\sigma_5) = \sigma_2 - \psi_1(\sigma_4)$.
4. Let $\hat{\sigma} := \sigma_5$.

Then

$$\varphi_3(\sigma - \psi_3(\hat{\sigma})) = \varphi_3(\sigma) - \psi_4 \circ \varphi_1(\sigma_5) = \varphi_3(\sigma) - \psi_4(\sigma_2) = \varphi_3(\sigma) - \sigma_1 = \mathbf{O}. \ \square$$

Lemma 2. *In the sequence*

$$\mathscr{S}_1 \overset{\varphi_1}{\twoheadrightarrow} \mathscr{S}_2 \overset{\varphi_2}{\rightarrow} \mathscr{S}_3 \overset{\varphi_3}{\rightarrow} \mathscr{S}_4 \overset{\varphi_4}{\hookrightarrow} \mathscr{S}_5$$

let φ_1 *be surjective,* φ_4 *injective and* $\operatorname{Ker} \varphi_3 = \operatorname{Im} \varphi_2$. *Then*

$$\mathscr{S}_1 \xrightarrow{\varphi_2 \circ \varphi_1} \mathscr{S}_3 \xrightarrow{\varphi_4 \circ \varphi_3} \mathscr{S}_5$$

is exact.

PROOF. $\operatorname{Im}(\varphi_2 \circ \varphi_1) = \operatorname{Im} \varphi_2 = \operatorname{Ker} \varphi_3 = \operatorname{Ker}(\varphi_4 \circ \varphi_3)$. □

Theorem 1.6. $\mathfrak{W}$ *is an exact functor.*

PROOF

1. Let $0 \to \mathscr{S}' \to \mathscr{S} \to \mathscr{S}'' \to 0$ be exact. We show by induction that

$$0 \to W_\ell(\mathscr{S}') \to W_\ell(\mathscr{S}) \to W_\ell(\mathscr{S}'') \to 0$$

is exact. For $\ell = 0$ this has already been proved in Theorem 1.4. Therefore let $\ell \geqslant 1$. We consider the case $\ell = 1$; the general case is handled entirely analogously.

The following diagram is commutative:

$$\begin{array}{ccccccccc}
 & & 0 & \longrightarrow & 0 & \longrightarrow & 0 & \longrightarrow & 0 \\
 & & \downarrow & & \downarrow & & \downarrow & & \downarrow \\
0 & \longrightarrow & \mathscr{S}' & \overset{\varphi_1'}{\hookrightarrow} & \mathscr{S} & \overset{\varphi_1''}{\twoheadrightarrow} & \mathscr{S}'' & \longrightarrow & 0 \\
 & & \downarrow{\scriptstyle \psi_1'} & & \downarrow{\scriptstyle \psi_1} & & \downarrow{\scriptstyle \psi_1''} & & \downarrow \\
0 & \longrightarrow & W_0(\mathscr{S}') & \overset{\varphi_2'}{\hookrightarrow} & W_0(\mathscr{S}) & \overset{\varphi_2''}{\twoheadrightarrow} & W_0(\mathscr{S}'') & \longrightarrow & 0 \\
 & & \downarrow{\scriptstyle \psi_2'} & & \downarrow{\scriptstyle \psi_2} & & \downarrow{\scriptstyle \psi_2''} & & \downarrow \\
0 & \longrightarrow & \mathscr{Q}' & \overset{\varphi_3'}{\longrightarrow} & \mathscr{Q} & \overset{\varphi_3''}{\longrightarrow} & \mathscr{Q}'' & \longrightarrow & 0
\end{array}$$

(with $\mathscr{Q} := W_0(\mathscr{S})/\mathscr{S}$, $\mathscr{Q}'$ and $\mathscr{Q}''$ similarly). All columns and the three top rows are exact.

a. Since φ_2'' and ψ_2'' are surjective φ_3'' is also surjective.
b. Since ψ_2' is surjective and $\varphi_2'' \circ \varphi_2' = \mathbf{0}$, also $\varphi_3'' \circ \varphi_3' = \mathbf{0}$.
c. Let $\sigma \in \mathscr{Q}$ with $\varphi_3''(\sigma) = \mathbf{0}$. Then there exists a $\sigma^* \in W_0(\mathscr{S})$ with $\psi_2(\sigma^*) = \sigma$; therefore $\psi_2'' \circ \varphi_2''(\sigma^*) = \mathbf{0}$.

By Lemma 1 there is a $\hat{\sigma} \in \mathscr{S}$ with $\varphi_2''(\sigma^* - \psi_1(\hat{\sigma})) = \mathbf{0}$. Therefore there exists a $\sigma' \in W_0(\mathscr{S}')$ with $\varphi_2'(\sigma') = \sigma^* - \psi_1(\hat{\sigma})$. It follows that $\psi_2'(\sigma') \in \mathscr{Q}'$ and

$$\varphi_3' \circ \psi_2'(\sigma') = \psi_2(\sigma^* - \psi_1(\hat{\sigma})) = \sigma.$$

d. Let $\sigma' \in \mathscr{Q}'$ with $\varphi_3'(\sigma') = \mathbf{0}$. Then there is a $\sigma^* \in W_0(\mathscr{S}')$ with $\psi_2'(\sigma^*) = \sigma'$. Hence $\psi_2 \circ \varphi_2'(\sigma^*) = \mathbf{0}$.

By Lemma 1 there is a $\hat{\sigma} \in \mathcal{S}'$ with $\varphi'_2(\sigma^* - \psi'_1(\hat{\sigma})) = \mathbf{O}$. Since φ'_2 is injective it follows that $\sigma^* - \psi'_1(\hat{\sigma}) = \mathbf{O}$, hence

$$\mathbf{O} = \psi'_2(\sigma^* - \psi'_1(\hat{\sigma})) = \psi'_2(\sigma^*) = \sigma'.$$

Thus, the last row of the diagram is exact, and by Theorem 1.4 we now have the exactness of the sequence $0 \to W_1(\mathcal{S}') \to W_1(\mathcal{S}) \to W_1(\mathcal{S}'') \to 0$.

2. Now let $\mathcal{S}' \xrightarrow{\varphi} \mathcal{S} \xrightarrow{\psi} \mathcal{S}''$ be exact.
Then we obtain the following exact sequences:

$$0 \to \operatorname{Ker} \varphi \to \mathcal{S}' \to \mathcal{Q}' \to 0 \text{ (with } \mathcal{Q}' := \mathcal{S}'/\operatorname{Ker} \varphi)$$
$$0 \to \mathcal{Q}' \to \mathcal{S} \to \operatorname{Im} \psi \to 0$$
$$0 \to \operatorname{Im} \psi \to \mathcal{S}'' \to \mathcal{Q}'' \to 0 \text{ (with } \mathcal{Q}'' := \mathcal{S}''/\operatorname{Im} \psi).$$

Applying (1), we obtain an exact sequence of the form

$$W_\ell(\mathcal{S}') \twoheadrightarrow W_\ell(\mathcal{Q}') \hookrightarrow W_\ell(\mathcal{S}) \twoheadrightarrow W_\ell(\operatorname{Im} \psi) \hookrightarrow W_\ell(\mathcal{S}''),$$

where the first mapping is surjective, the last mapping is injective and the sequence in the center is exact. By Lemma 2 it follows that $W_\ell(\mathcal{S}') \to W_\ell(\mathcal{S}) \to W_\ell(\mathcal{S}'')$ is exact. But that means that $\mathfrak{W}$ is exact. □

Def. 1.3. A *cochain complex over* R is a sequence of R-module homomorphisms

$$M^\bullet : M^0 \xrightarrow{d^0} M^1 \xrightarrow{d^1} M^2 \xrightarrow{d^2} M^3 \to \cdots$$

with $d^i \circ d^{i-1} = 0$ for $i \in \mathbb{N}$.

$Z^n(M^\bullet) := \operatorname{Ker} d^n$ is called the *n-th group of the cocycles*,
$B^n(M^\bullet) := \operatorname{Im} d^{n-1}$ is called the *n-th group of the coboundaries*.

We set $B^0(M^\bullet) := 0$. Then clearly $B^n(M^\bullet) \subset Z^n(M^\bullet)$, and $H^n(M^\bullet) := Z^n(M^\bullet)/B^n(M^\bullet)$ is called the *n-th cohomology group of the complex* $M^\bullet$.

Remark. Clearly $M^\bullet$ is exact at (the location) $n > 0$ if and only if $H^n(M^\bullet) = 0$. In this sense, one says that the cohomology groups measure the deviation of the complex $M^\bullet$ from exactness.

Def. 1.4. An *augmented cochain complex* is a triple $(E, \varepsilon, M^\bullet)$ with the following properties:

1. E is an R-module.
2. $M^\bullet$ is a cochain complex.
3. $\varepsilon : E \to M^0$ is an R-module monomorphism with $\operatorname{Im} \varepsilon = \operatorname{Ker} d^0$.

Remark. If $(E, \varepsilon, M^\bullet)$ is an augmented complex, then

$$E \simeq \operatorname{Im} \varepsilon = \operatorname{Ker} d^0 = Z^0(M^\bullet) \simeq H^0(M^\bullet).$$

If $H^\ell(M^\bullet) = 0$ for $\ell \geqslant 1$, we call the complex *acyclic*.

Theorem 1.7. *($\Gamma:\mathcal{S} \rightsquigarrow \Gamma(X, \mathcal{S}), \varphi \rightsquigarrow \varphi_*$) is a left-exact functor; that is, if*

$$0 \to \mathcal{S}' \xrightarrow{\varphi} \mathcal{S} \xrightarrow{\psi} \mathcal{S}'' \to 0$$

is exact, then so is

$$0 \to \Gamma(X, \mathcal{S}') \xrightarrow{\varphi_*} \Gamma(X, \mathcal{S}) \xrightarrow{\psi_*} \Gamma(X, \mathcal{S}'')$$

exact.

PROOF. Since $\hat{\Gamma}$ is exact (see Theorem 1.1), it is clear that φ_* is injective and $\psi_* \circ \varphi_* = 0$. Now let $s \in \Gamma(X, \mathcal{S})$ with $\mathbf{0} = \psi_*(s) = \psi \circ s$. Then there exists a generalized section $s' \in \hat{\Gamma}(X, \mathcal{S}')$ with $\varphi_*(s') = s$. We must still show that s' is continuous. For every point $x \in X$ there is a neighborhood $U(x)$ and an $s^* \in \Gamma(U, \mathcal{S}')$ with $(\varphi \circ s^*)(x) = s(x)$. Therefore there is a neighborhood $V(x) \subset U$ with $\varphi \circ s^*|V = s|V$.

Since φ is injective, it follows from $\varphi \circ s^*|V = \varphi \circ s'|V$ that $s^*|V = s'|V$, so that s' is continuous at x. □

Theorem 1.8. *Let $\mathcal{S}$ be a sheaf of R-modules over X,*

$$W^\bullet(\mathcal{S}): \Gamma(X, W_0(\mathcal{S})) \to \Gamma(X, W_1(\mathcal{S})) \to \Gamma(X, W_2(\mathcal{S})) \to \cdots.$$

Then $(\Gamma(X, \mathcal{S}), \varepsilon_*, W^\bullet(\mathcal{S}))$ is an augmented cochain complex.

PROOF. Clearly $W^\bullet(\mathcal{S})$ is a complex, $\varepsilon_*:\Gamma(X, \mathcal{S}) \to \Gamma(X, W_0(\mathcal{S}))$ an R-module monomorphism, and $(d_0)_* \circ \varepsilon_* = 0$.

Consider the mapping

$$d_0: W_0(\mathcal{S}) \overset{q}{\twoheadrightarrow} W_0(\mathcal{S})/\mathrm{Im}(\varepsilon) \overset{j}{\hookrightarrow} W(W_0(\mathcal{S})/\mathrm{Im}(\varepsilon)) = W_1(\mathcal{S}).$$

Let $s \in \Gamma(X, W_0(\mathcal{S}))$ and $\mathbf{0} = d_0 \circ s = j \circ q \circ s$. Then $q \circ s = \mathbf{0}$; therefore $s(x) \in \mathrm{Im}(\varepsilon)$ for every $x \in X$. Since $\mathrm{Im}(\varepsilon) \simeq \mathcal{S}, \Gamma(X, \mathrm{Im}(\varepsilon)) \simeq \Gamma(X, \mathcal{S})$; so there is an $s^* \in \Gamma(X, \mathcal{S})$ with $\varepsilon_*(s^*) = s$. □

Def. 1.5. Let $\mathcal{S}$ be a sheaf of R-modules over X. Then we define

$$Z^\ell(X, \mathcal{S}) := Z^\ell(W^\bullet(\mathcal{S})), B^\ell(X, \mathcal{S}) := B^\ell(W^\bullet(\mathcal{S})).$$

We call

$$H^\ell(X, \mathcal{S}) := Z^\ell(X, \mathcal{S})/B^\ell(X, \mathcal{S}) = H^\ell(W^\bullet(\mathcal{S}))$$

the *ℓ-th cohomology group of X with values in $\mathcal{S}$.*

Remark. Clearly $H^0(X, \mathcal{S}) \simeq \Gamma(X, \mathcal{S})$.

Theorem 1.9. *If $0 \to \mathcal{S}' \xrightarrow{\varphi} \mathcal{S} \xrightarrow{\psi} \mathcal{S}'' \to 0$ is an exact sequence of sheaves of R-modules, and if $\mathcal{S}'$ is a flabby sheaf, then*

$$0 \to \Gamma(X, \mathcal{S}') \xrightarrow{\varphi_*} \Gamma(X, \mathcal{S}) \xrightarrow{\psi_*} \Gamma(X, \mathcal{S}'') \to 0$$

is exact.

PROOF. We need only show that ψ_* is surjective. Let $s'' \in \Gamma(X, \mathscr{S}'')$ be given.

1. If x_1, x_2 are points in X, then there are neighborhoods $U(x_1), V(x_2) \subset X$ and sections $s \in \Gamma(U, \mathscr{S})$, $s^* \in \Gamma(V, \mathscr{S})$ with $\psi \circ s = s''|U$ and $\psi \circ s^* = s''|V$.

If $U \cap V = \varnothing$, then this defines a section over $U \cup V$, whose image is $s''|U \cup V$. Suppose $U \cap V \neq \varnothing$. The sequence

$$0 \to \Gamma(U \cap V, \mathscr{S}') \to \Gamma(U \cap V, \mathscr{S}) \to \Gamma(U \cap V, \mathscr{S}'')$$

is exact, and since $\psi \circ (s - s^*)|U \cap V = \mathbf{0}$, there is an $s' \in \Gamma(U \cap V, \mathscr{S}')$ with $\varphi \circ s' = (s - s^*)|U \cap V$.

Since $\mathscr{S}'$ is flabby, we can extend s' to an element $\hat{s} \in \Gamma(V, \mathscr{S}')$. Let

$$s_1(x) := \begin{cases} s(x) & \text{for} \quad x \in U \\ (\varphi \circ \hat{s} + s^*)(x) & \text{for} \quad x \in V. \end{cases}$$

Then s_1 lies in $\Gamma(U \cup V, \mathscr{S})$ and $\psi \circ s_1 = s''|U \cup V$. In this case there is also a section over $U \cup V$ whose image is $s''|U \cup V$.

2. We consider the system $\mathfrak{M}$ of all pairs $(\tilde{U}, \tilde{s})$ with the following properties:

a. $\tilde{U} \subset X$ is open with $U \subset \tilde{U}$
b. $\tilde{s} \in \Gamma(\tilde{U}, S)$ with $\tilde{s}|U = s$ and $\psi \circ \tilde{s} = s''|\tilde{U}$.

In $\mathfrak{M}$ we consider all subsystems $(\tilde{U}_\iota, \tilde{s}_\iota)_{\iota \in I}$ with the following properties:

For $(\iota_1, \iota_2) \in I \times I$ either $\tilde{U}_{\iota_1} \subset \tilde{U}_{\iota_2}$ and $\tilde{s}_{\iota_2}|\tilde{U}_{\iota_1} = \tilde{s}_{\iota_2}$, or $\tilde{U}_{\iota_2} \subset \tilde{U}_{\iota_1}$ and $\tilde{s}_{\iota_1}|\tilde{U}_{\iota_2} = \tilde{s}_{\iota_2}$. For each such system the pair $(\tilde{U}, \tilde{s})$ with $\tilde{U} := \bigcup_{\iota \in I} \tilde{U}_\iota$ and $\tilde{s}|\tilde{U}_\iota := \tilde{s}_\iota$ is again an element of $\mathfrak{M}$. Zorn's lemma[1] implies that there exists a "maximal element" (U_0, s_0) in $\mathfrak{M}$. By (1) U_0 cannot be a proper subset of X. This completes the proof. □

As a consequence we have:

Theorem 1.10. *Let $\mathscr{S}$ be a flabby sheaf of R-modules over X and $0 \to \mathscr{S} \to \mathscr{S}_0 \to \mathscr{S}_1 \to \cdots$ a flabby resolution of $\mathscr{S}$. Then the sequence*

$$0 \to \Gamma(X, \mathscr{S}) \to \Gamma(X, \mathscr{S}_0) \to \Gamma(X, \mathscr{S}_1) \to \cdots$$

is exact.

[1] Let X be a non-empty set with a relation $\leqslant$ such that:

1. $x \leqslant x$ for all $x \in X$.
2. if $x \leqslant y$ and $y \leqslant z$, then $x \leqslant z$ for all $x, y, z \in X$.
3. if $x \leqslant y$ and $y \leqslant x$, then $x = y$ for all $x, y \in X$.

A chain in X is a set $K \subset X$ with the property that for any two elements $x, y \in K$ either $x \leqslant y$ or $y \leqslant x$. Zorn's lemma says that if there exists an upper bound for every chain $K \subset X$ (an element $s \in X$ with $x \leqslant s$ for all $x \in K$), then there exists a maximal element in X (an element $x_0 \in X$ such that for $x \in X$ it always follows from $x_0 \leqslant x$ that $x = x_0$).

PROOF. Let $\mathscr{B}_\lambda := \operatorname{Im}(\varphi_\lambda : \mathscr{S}_{\lambda-1} \to \mathscr{S}_\lambda)$ for $\lambda = 0, 1, 2, \ldots$ and $\mathscr{S}_{-1} := \mathscr{S}$.

1. We show by induction that all $\mathscr{B}_\lambda$ are flabby: For $\mathscr{B}_0 \simeq \mathscr{S}$ this is true by assumption; suppose we have proved that $\mathscr{B}_0, \mathscr{B}_1, \ldots, \mathscr{B}_{\ell-1}$ are flabby sheaves.

For $U \subset X$ open, the exactness of the sequence $0 \to \Gamma(U, \mathscr{B}_{\ell-1}) \to \Gamma(U, \mathscr{S}_{\ell-1}) \to \Gamma(U, \mathscr{B}_\ell) \to 0$ follows from the exactness of the sequence $0 \to \mathscr{B}_{\ell-1} \hookrightarrow \mathscr{S}_{\ell-1} \to \mathscr{B}_\ell \to 0$ by Theorem 1.9. Let $s \in \Gamma(U, \mathscr{B}_\ell)$.

Then there is an $s' \in \Gamma(U, \mathscr{S}_{\ell-1})$ with $\varphi_\ell \circ s' = s$. Since $\mathscr{S}_{\ell-1}$ is flabby there is an $s^* \in \Gamma(X, \mathscr{S}_{\ell-1})$ with $s^*|U = s'$. But now $\varphi_\ell \circ s^* \in \Gamma(X, \mathscr{B}_\ell)$ and $\varphi_\ell \circ s^*|U = s$. Therefore, $\mathscr{B}_\ell$ is flabby.

2. The following sequences are exact.

$$0 \to \mathscr{B}_{\ell-1} \to \mathscr{S}_{\ell-1} \to \mathscr{B}_\ell \to 0$$
$$0 \to \mathscr{B}_\ell \to \mathscr{S}_\ell \to \mathscr{B}_{\ell+1} \to 0$$
$$0 \to \mathscr{B}_{\ell+1} \to \mathscr{S}_{\ell+1} \to \mathscr{B}_{\ell+2} \to 0.$$

By Theorem 1.9 the associated sequences of the modules of sections are exact. We can combine these into a sequence which satisfies the conditions of Lemma 2:

$$\Gamma(X, \mathscr{S}_{\ell-1}) \twoheadrightarrow \Gamma(X, \mathscr{B}_\ell) \to \Gamma(X, \mathscr{S}_\ell) \to \Gamma(X, \mathscr{B}_{\ell+1}) \hookrightarrow \Gamma(X, \mathscr{S}_{\ell+1}).$$

Then the sequence $\Gamma(X, \mathscr{S}_{\ell-1}) \to \Gamma(X, \mathscr{S}_\ell) \to \Gamma(X, \mathscr{S}_{\ell+1})$ is exact, as was to be shown. □

Thus we have obtained

Theorem 1.11. *If $\mathscr{S}$ is a flabby sheaf over X, then the complex $W^\bullet(\mathscr{S})$ is acyclic; therefore $H^\ell(X, \mathscr{S}) = 0$ for $\ell \geq 1$.*

EXAMPLE. Let $\mathscr{I}(A)$ be the ideal sheaf of the analytic set $A = \{0\} \in \mathbb{C}^n$. Then $\mathscr{H}(A) = \mathscr{O}/\mathscr{I}(A)$ is a coherent analytic sheaf over $\mathbb{C}^n$, in particular, a sheaf of $\mathbb{C}$-modules. Clearly $\mathscr{H}(A)$ is flabby, and

$$H^0(\mathbb{C}^n, \mathscr{H}(A)) \simeq \mathbb{C}, H^\ell(\mathbb{C}^n, \mathscr{H}(A)) = 0 \qquad \text{for } \ell \geq 1.$$

2. The Čech Cohomology

Let X be a complex manifold, R a commutative ring with 1, and $\mathscr{S}$ a sheaf of R-modules. Moreover, let $\mathfrak{U} = (U_\iota)_{\iota \in I}$ be an open covering of X, with $U_\iota \neq \varnothing$ for every $\iota \in I$. We define

$$U_{\iota_0 \cdots \iota_\ell} := U_{\iota_0} \cap \ldots \cap U_{\iota_\ell},$$
$$I_\ell := \{(\iota_0, \ldots, \iota_\ell) : U_{\iota_0 \cdots \iota_\ell} \neq \varnothing\}.$$

Let $\mathfrak{S}_n$ be the set of permutations of the set $\{0, 1, 2, \ldots, n-1\}$. For $\sigma \in \mathfrak{S}_n$, let

$$\operatorname{sgn}(\sigma) := \begin{cases} +1 & \text{if } \sigma \text{ is the product of an even number of transpositions} \\ -1 & \text{otherwise.} \end{cases}$$

Def. 2.1. An ℓ-dimensional (alternating) *cochain* over $\mathfrak{U}$ with values in $\mathscr{S}$ is a mapping

$$\xi : I_\ell \to \bigcup_{(\iota_0, \ldots, \iota_\ell)} \Gamma(U_{\iota_0 \cdots \iota_\ell}, \mathscr{S})$$

with the following properties:

1. $\xi(\iota_0, \ldots, \iota_\ell) \in \Gamma(U_{\iota_0 \cdots \iota_\ell}, \mathscr{S})$.
2. $\xi(\iota_{\sigma(0)}, \ldots, \iota_{\sigma(\ell)}) = \operatorname{sgn}(\sigma)\xi(\iota_0, \ldots, \iota_\ell)$ for $\sigma \in \mathfrak{S}_{\ell+1}$.

The set of all ℓ-dimensional alternating cochains over $\mathfrak{U}$ with values in $\mathscr{S}$ is denoted by $C^\ell(\mathfrak{U}, \mathscr{S})$. $C^\ell(\mathfrak{U}, \mathscr{S})$ becomes an R-module by setting

$$(\xi_1 + \xi_2)(\iota_0, \ldots, \iota_\ell) := \xi_1(\iota_0, \ldots, \iota_\ell) + \xi_2(\iota_0, \ldots, \iota_\ell)$$

and

$$(r \cdot \xi)(\iota_0, \ldots, \iota_\ell) := r \cdot \xi(\iota_0, \ldots, \iota_\ell).$$

Theorem 2.1. *$\delta^\ell : C^\ell(\mathfrak{U}, \mathscr{S}) \to C^{\ell+1}(\mathfrak{U}, \mathscr{S})$ with*

$$(\delta^\ell\xi)(\iota_0, \ldots, \iota_\ell) := \sum_{\lambda=0}^{\ell+1} (-1)^{\lambda+1}(\xi(\iota_0, \ldots, \hat{\iota}_\lambda, \ldots, \iota_{\ell+1}) | U_{\iota_0 \cdots \iota_{\ell+1}})$$

is an R-module homomorphism with $\delta^\ell \circ \delta^{\ell-1} = 0$.

PROOF

1. First we show that $\delta^\ell\xi$ is alternating. It suffices to consider transpositions.

$$\begin{aligned}
&(\delta^\ell\xi)(\iota_0, \ldots, \iota_\nu, \iota_{\nu+1}, \ldots, \iota_{\ell+1}) \\
&= \sum_{\lambda \neq \nu, \nu+1} (-1)^{\lambda+1}\xi(\iota_0, \ldots, \hat{\iota}_\lambda, \ldots, \iota_{\ell+1}) + (-1)^{\nu+1}\xi(\iota_0, \ldots, \hat{\iota}_\nu, \ldots, \iota_{\ell+1}) \\
&\quad + (-1)^{\nu+2}\xi(\iota_0, \ldots, \hat{\iota}_{\nu+1}, \ldots, \iota_{\ell+1}) \\
&= - \sum_{\lambda \neq \nu, \nu+1} (-1)^{\lambda+1}\xi(\iota_0, \ldots, \hat{\iota}_\lambda, \ldots, \iota_{\nu+1}, \iota_\nu, \ldots, \iota_{\ell+1}) \\
&\quad + (-1)^{\nu+1}\xi(\iota_0, \ldots, \iota_{\nu+1}, \hat{\iota}_\nu, \ldots, \iota_{\ell+1}) \\
&\quad + (-1)^{\nu+2} \cdot \xi(\iota_0, \ldots, \hat{\iota}_{\nu+1}, \iota_\nu, \ldots, \iota_{\ell+1}) \\
&= -\delta\,\xi(\iota_0, \ldots, \iota_{\nu+1}, \iota_\nu, \ldots, \iota_{\ell+1})
\end{aligned}$$

2. It is clear that δ^ℓ is a homomorphism. Moreover,

$$\begin{aligned}
&(\delta^{\ell+1} \circ \delta^\ell\xi)(\iota_0, \ldots, \iota_{\ell+2}) \\
&\qquad = \sum_{\lambda=0}^{\ell+2} (-1)^{\lambda+1}(\delta^\ell\xi)(\iota_0, \ldots, \hat{\iota}_\lambda, \ldots, \iota_{\ell+2}) \\
&\qquad = \sum_{\lambda=0}^{\ell+2} (-1)^{\lambda+1}\Bigg[\sum_{\eta=0}^{\lambda-1} (-1)^{\eta+1}\xi(\iota_0, \ldots, \hat{\iota}_\eta, \ldots, \hat{\iota}_\lambda, \ldots, \iota_{\ell+2}) \\
&\qquad\quad + \sum_{\eta=\lambda+1}^{\ell+2} (-1)^{\eta}\xi(\iota_0, \ldots, \hat{\iota}_\lambda, \ldots, \hat{\iota}_\eta, \ldots, \iota_{\ell+2})\Bigg] \\
&\qquad = \sum_{\eta<\lambda} (-1)^{\lambda+\eta}\xi(\iota_0, \ldots, \hat{\iota}_\eta, \ldots, \hat{\iota}_\lambda, \ldots, \iota_{\ell+2}) \\
&\qquad\quad + \sum_{\lambda<\eta} (-1)^{\lambda+\eta+1}\xi(\iota_0, \ldots, \hat{\iota}_\lambda, \ldots, \hat{\iota}_\eta, \ldots, \iota_{\ell+2}) = 0. \quad \square
\end{aligned}$$

Def. 2.2. $\delta: = \delta^\ell: C^\ell(\mathfrak{U}, \mathscr{S}) \to C^{\ell+1}(\mathfrak{U}, \mathscr{S})$ is called the *coboundary operator.* We denote by $C^\bullet(\mathfrak{U}, \mathscr{S})$ the *Čech complex*

$$C^0(\mathfrak{U}, \mathscr{S}) \xrightarrow{\delta} C^1(\mathfrak{U}, \mathscr{S}) \xrightarrow{\delta} C^2(\mathfrak{U}, \mathscr{S}) \to \cdots.$$

$\varepsilon: \Gamma(X, \mathscr{S}) \to C^0(\mathfrak{U}, \mathscr{S})$ is defined by $(\varepsilon s)(\iota): = s|U_\iota$.

Theorem 2.2. *$(\Gamma(X, \mathscr{S}), \varepsilon, C^\bullet(\mathfrak{U}, \mathscr{S}))$ is an augmented cochain complex.*

PROOF. Clearly ε is an R-module homomorphism. If $\varepsilon s = 0$, then $s|U_\iota = \mathbf{O}$ for every $\iota \in I$; therefore $s = \mathbf{O}$. Hence ε is injective.

Let $\xi \in C^0(\mathfrak{U}, \mathscr{S})$ and $\delta\xi = 0$. Since

$$(\delta\xi)(\iota_0, \iota_1) = (-\xi(\iota_1) + \xi(\iota_0))|U_{\iota_0\iota_1}$$

this is equivalent to $\xi(\iota_0)|U_{\iota_0\iota_1} = \xi(\iota_1)|U_{\iota_0\iota_1}$. Therefore there is a section $s \in \Gamma(X, \mathscr{S})$ with $\varepsilon s = \xi$ defined by $s|U_\iota: = \xi(\iota)$. □

Def. 2.3. The elements of $Z^\ell(\mathfrak{U}, \mathscr{S}): = Z^\ell(C^\bullet(\mathfrak{U}, \mathscr{S}))$, resp. $B^\ell(\mathfrak{U}, \mathscr{S}): = B^\ell(C^\bullet(\mathfrak{U}, \mathscr{S}))$ are called *(alternating) ℓ-dimensional cocycles*, resp. *coboundaries, over $\mathfrak{U}$ with values in $\mathscr{S}$.* $H^\ell(\mathfrak{U}, \mathscr{S}): = Z^\ell(\mathfrak{U}, \mathscr{S})/B^\ell(\mathfrak{U}, \mathscr{S}) = H^\ell(C^\bullet(\mathfrak{U}, \mathscr{S}))$ is the *ℓ-th Čech cohomology group of $\mathfrak{U}$ with values in $\mathscr{S}$.* In particular $H^0(\mathfrak{U}, \mathscr{S}) \simeq \Gamma(X, \mathscr{S})$.

If we choose the covering of $\mathfrak{U}$ too coarse, then all the higher cohomology groups vanish:

Theorem 2.3. *If X itself belongs to the elements of the covering $\mathfrak{U}$, then $H^\ell(\mathfrak{U}, \mathscr{S}) = 0$ for $\ell \geqslant 1$.*

PROOF. If $\mathfrak{U} = (U_\iota)_{\iota \in I}$, then there is a $\rho \in I$ with $X = U_\rho$. Let $\xi \in Z^\ell(\mathfrak{U}, \mathscr{S})$, $\ell \geqslant 1$. There is an element $\eta \in C^{\ell-1}(\mathfrak{U}, \mathscr{S})$ defined by

$$\eta(\iota_0, \ldots, \iota_{\ell-1}): = \xi(\rho, \iota_0, \ldots, \iota_{\ell-1}).$$

Since $\delta\xi = 0$ we have

$$0 = \delta\xi(\rho, \iota_0, \ldots, \iota_\ell) = -\xi(\iota_0, \ldots, \iota_\ell) + \sum_{\lambda=0}^{\ell} (-1)^\lambda \xi(\rho, \iota_0, \ldots, \hat{\iota}_\lambda, \ldots, \iota_\ell);$$

therefore

$$\begin{aligned}\delta(-\eta)(\iota_0, \ldots, \iota_\ell) &= -\sum_{\lambda=0}^{\ell} (-1)^{\lambda+1}\eta(\iota_0, \ldots, \hat{\iota}_\lambda, \ldots, \iota_\ell) \\ &= \sum_{\lambda=0}^{\ell} (-1)^\lambda \xi(\rho, \iota_0, \ldots, \hat{\iota}_\lambda, \ldots, \iota_\ell) = \xi(\iota_0, \ldots, \iota_\ell).\end{aligned}$$

In other words, $\delta(-\eta) = \xi$, so $\xi \in B^\ell(\mathfrak{U}, \mathscr{S})$. □

Theorem 2.4. *Let $\mathfrak{U}$ be an arbitrary covering of X and $\mathscr{S}$ a flabby sheaf. Then $H^\ell(\mathfrak{U}, \mathscr{S}) = 0$ for $\ell \geqslant 1$.*

PROOF. We proceed by induction on ℓ: Let $\xi \in Z^\ell(\mathfrak{U}, \mathscr{S})$, $\ell \geqslant 1$. If $U \subset X$ is open, then we set $U \cap \mathfrak{U} := \{U \cap U_\iota \neq \varnothing : U_\iota \in \mathfrak{U}\}$ and

$$(\xi|U)(\iota_0, \ldots, \iota_\ell) := \xi(\iota_0, \ldots, \iota_\ell)|U \cap U_{\iota_0 \cdots \iota_\ell}$$

With this notation we have $\xi|U \in Z^\ell(U \cap \mathfrak{U}, \mathscr{S})$.

For arbitrary $x_0 \in X$, there is an $\iota_0 \in I$ and an open neighborhood $U(x_0) \subset U_{\iota_0}$. But then $U \in U \cap \mathfrak{U}$, so $H^\ell(U \cap \mathfrak{U}, \mathscr{S}) = 0$ for $\ell \geqslant 1$, and there is an $\eta \in C^{\ell-1}(U \cap \mathfrak{U}, \mathscr{S})$ with $\delta\eta = \xi|U$.

If $V \subset X$ is an open set for which there is an $\eta' \in C^{\ell-1}(V \cap \mathfrak{U}, \mathscr{S})$ with $\delta\eta' = \xi|V$, we set

$$s := (\eta - \eta')|U \cap V \in Z^{\ell-1}(U \cap V \cap \mathfrak{U}, \mathscr{S}).$$

If $\ell = 1$, then s lies in $\Gamma(U \cap V, \mathscr{S})$, and since $\mathscr{S}$ is flabby, we can extend s to an $\hat{s} \in \Gamma(V, \mathscr{S})$. Then set

$$s^*(x) := \begin{cases} \eta(x) & x \in U \\ \eta'(x) + \hat{s}(x) & x \in V. \end{cases}$$

Clearly $s^* \in \Gamma(U \cup V, \mathscr{S})$ and $\delta s^* = \xi|U \cup V$ (because $\delta\hat{s} = 0$).

If $\ell > 1$, then by the induction hypothesis there is a $\gamma \in C^{\ell-2}(U \cap V \cap \mathfrak{U}, \mathscr{S})$ with $\delta\gamma = s$. Since $\mathscr{S}$ is flabby,

$$\gamma(\iota_0, \ldots, \iota_{\ell-2}) \in \Gamma(U \cap V \cap U_{\iota_0 \cdots \iota_{\ell-2}}, \mathscr{S})$$

can be extended to an element

$$\hat{\gamma}(\iota_0, \ldots, \iota_{\ell-2}) \in \Gamma(V \cap U_{\iota_0 \cdots \iota_{\ell-2}}, \mathscr{S}).$$

Let

$$\eta^*(\iota_0, \ldots, \iota_{\ell-1})(x) := \begin{cases} \eta(\iota_0, \ldots, \iota_{\ell-1})(x) & \text{for} \quad x \in U \cap U_{\iota_0 \ldots \iota_{\ell-1}} \\ (\eta' + \delta\hat{\gamma})(\iota_0, \ldots, \iota_{\ell-1})(x) & \text{for} \quad x \in V \cap U_{\iota_0 \ldots \iota_{\ell-1}} \end{cases}$$

Then $\eta^* \in C^{\ell-1}((U \cup V) \cap \mathfrak{U}, \mathscr{S})$ and $\delta\eta^* = \xi|U \cup V$.

By Zorn's lemma there must be a "maximal element" (U_0, s_0) for $\ell = 1$, resp. (U_0, η_0) for $\ell > 1$ with $s_0 \in \Gamma(U_0, \mathscr{S})$ and $\delta s_0 = \xi|U_0$, resp. $\eta_0 \in C^\ell(\mathfrak{U}, \mathscr{S})$ and $\delta\eta_0 = \xi|U_0$. But an element is only maximal if $U_0 = X$; therefore $\xi \in B^\ell(\mathfrak{U}, \mathscr{S})$. □

Remark. Let $\mathfrak{U}$ be a covering of X and $\mathscr{S}$ a sheaf of R-modules, $\xi \in C^1(\mathfrak{U}, \mathscr{S})$. It is worth noting the following criteria:

1. $\xi \in Z^1(\mathfrak{U}, \mathscr{S})$ if and only if

$$\xi(\iota_0, \iota_2) = \xi(\iota_0, \iota_1) + \xi(\iota_1, \iota_2)$$

on $U_{\iota_0\iota_1\iota_2}$.

2. $\xi \in B^1(\mathfrak{U}, \mathscr{S})$ if and only if for all ι there exists an $\rho(\iota) \in \Gamma(U_\iota, \mathscr{S})$ with

$$\xi(\iota_0, \iota_1) = \rho(\iota_0) - \rho(\iota_1)$$

on $U_{\iota_0\iota_1}$.

The first condition is also called the *compatibility condition.*

Def. 2.4. A system $(U_\iota, f_\iota)_{\iota \in I}$ is called a *Cousin I distribution on X* if

1. $\mathfrak{U} := (U_\iota)_{\iota \in I}$ is an open covering of X;
2. f_ι is meromorphic on U_ι;
3. $f_{\iota_0} - f_{\iota_1}$ is holomorphic on $U_{\iota_0 \iota_1}$ for all ι_0, ι_1.

A solution of a Cousin I distribution is a meromorphic function f on X such that $f_\iota - f$ is holomorphic on U_ι.

Theorem 2.5. *Let $(U_\iota, f_\iota)_{\iota \in I}$ be a Cousin I distribution on X, $\mathscr{H}$ the structure sheaf of X, $\mathfrak{U} := (U_\iota)_{\iota \in I}$. Then*

1. $\gamma(\iota_0, \iota_1) := (f_{\iota_0} - f_{\iota_1})|U_{\iota_0 \iota_1}$ *defines an element* $\gamma \in Z^1(\mathfrak{U}, \mathscr{H})$.
2. $(U_\iota, f_\iota)_{\iota \in I}$ *is solvable if and only if γ lies in* $B^1(\mathfrak{U}, \mathscr{H})$.

PROOF

1. Clearly

$$\gamma(\iota_0, \iota_1) + \gamma(\iota_1, \iota_2) = (f_{\iota_0} - f_{\iota_1}) + (f_{\iota_1} - f_{\iota_2}) = f_{\iota_0} - f_{\iota_2} = \gamma(\iota_0, \iota_2)$$

on $U_{\iota_0 \iota_1 \iota_2}$.

2. a. Let $(U_\iota, f_\iota)_{\iota \in I}$ be solvable. Then there is a meromorphic function f on X such that $(f_\iota - f)|U_\iota$ is holomorphic. Let

$$\rho(\iota) := (f_\iota - f)|U_\iota \in \Gamma(U_\iota, \mathscr{H}).$$

ρ lies in $C^0(\mathfrak{U}, \mathscr{H})$ and

$$\rho(\iota_0) - \rho(\iota_1) = (f_{\iota_0} - f) - (f_{\iota_1} - f) = f_{\iota_0} - f_{\iota_1} = \gamma(\iota_0, \iota_1)$$

on $U_{\iota_0 \iota_1}$.

b. If γ lies in $B^1(\mathfrak{U}, \mathscr{H})$, then for every $\iota \in I$ there is a $\rho(\iota) \in \Gamma(U_\iota, \mathscr{H})$ such that $\rho(\iota_0) - \rho(\iota_1) = \gamma(\iota_0, \iota_1)$ on $U_{\iota_0 \iota_1}$. Then

$$f_{\iota_0} - f_{\iota_1} = \gamma(\iota_0, \iota_1) = \rho(\iota_0) - \rho(\iota_1),$$

so $f_{\iota_0} - \rho(\iota_0) = f_{\iota_1} - \rho(\iota_1)$ on $U_{\iota_0 \iota_1}$. Then there is a meromorphic function f on X defined by $f|U_\iota := f_\iota - \rho(\iota)$ with

$$(f_\iota - f)|U_\iota = \rho(\iota) \in \Gamma(U_\iota, \mathscr{H}).$$ □

Corollary. *If $H^1(\mathfrak{U}, \mathscr{H}) = 0$, then every Cousin I distribution belonging to the covering $\mathfrak{U}$ is solvable.*

EXAMPLE. Let $X = \mathbb{C}$. A *Mittag–Leffler distribution on* $\mathbb{C}$ is a discrete point sequence (z_ν) together with principal parts f_ν which define a meromorphic function in $\mathbb{C}$.

Now let $U_0 := \mathbb{C} - \{z_\nu : \nu \in \mathbb{N}\}$, $f_0 := 0$ and U_ν be an open neighborhood of z_ν which contains no point z_μ with $\mu \neq \nu$. Then $f_\nu|(U_\nu - \{z_\nu\})$ is holomorphic.

Hence $\mathfrak{U} = (U_\nu)_{\nu\in\mathbb{N}}$ is an open covering of $\mathbb{C}$ and $(f_\nu - f_\mu)|U_{\nu\mu}$ is always holomorphic. Therefore $(U_\nu, f_\nu)_{\nu\in\mathbb{N}}$ is a Cousin I distribution. Each solution of this Cousin I distribution is a solution of the Mittag–Leffler problem.

3. Double Complexes

Def. 3.1. A *double complex* is a system (C_{ij}) of R-modules (with $i, j \in \mathbb{N}_0$) and R-module homomorphisms, $d': C_{ij} \to C_{i+1,j}$ and $d'': C_{ij} \to C_{i,j+1}$, such that

1. $d'd' = 0$
2. $d''d'' = 0$
3. $d'd'' = -d''d'$

(thus

$$d := d' + d'': \bigoplus_{i+j=n} C_{ij} \to \bigoplus_{i+j=n+1} C_{ij} \quad \text{with} \quad d \circ d = 0).$$

A double complex is therefore an (anticommutative) diagram of the following form:

$$\begin{array}{ccccccc}
C_{00} & \xrightarrow{d''} & C_{01} & \xrightarrow{d''} & C_{02} & \xrightarrow{d''} & \cdots \\
\downarrow{\scriptstyle d'} & & \downarrow{\scriptstyle d'} & & \downarrow{\scriptstyle d'} & & \\
C_{10} & \xrightarrow{d''} & C_{11} & \xrightarrow{d''} & C_{12} & \xrightarrow{d''} & \cdots \\
\downarrow{\scriptstyle d'} & & \downarrow{\scriptstyle d'} & & \downarrow{\scriptstyle d'} & & \\
C_{20} & \xrightarrow{d''} & C_{21} & \xrightarrow{d''} & C_{22} & \xrightarrow{d''} & \cdots \\
\downarrow{\scriptstyle d'} & & \downarrow{\scriptstyle d'} & & \downarrow{\scriptstyle d'} & &
\end{array}$$

Def. 3.2.

$$Z_{ij} := \{\xi \in C_{ij} \ \text{with} \ d'\xi = 0 \ \text{and} \ d''\xi = 0\}$$

$$B_{0j} := d''(\{\xi \in C_{0,j-1} \ \text{with} \ d'\xi = 0\}) \quad \text{for} \quad j \geqslant 1,$$
$$B_{i0} := d'(\{\xi \in C_{i-1,0} \ \text{with} \ d''\xi = 0\}) \quad \text{for} \quad i \geqslant 1,$$
$$B_{00} := 0 \ \text{and} \ B_{ij} := d'd''C_{i-1,j-1} \quad \text{for} \quad i, j \geqslant 1.$$

We call the elements of Z_{ij} *cycles of bidegree* (i, j); the elements of B_{ij} are called *boundaries of bidegree* (i, j).

Clearly B_{ij} is an R-submodule of Z_{ij} for all i, j and we define the homology group of the double complex of bidegree (i, j) by $H_{ij} := Z_{ij}/B_{ij}$. Let the canonical projection be denoted by $q_{ij}: Z_{ij} \to H_{ij}$.

Theorem 3.1. *Let $(M, \varepsilon_1, A^\bullet)$, $(M, \varepsilon_2, B^\bullet)$ be two augmented cochain complexes. Let there be given a double complex $(C_{\nu\mu}, d', d'')$ and homomorphisms $d'_j: A^j \to C_{0j}$ and $d''_i: B^i \to C_{i0}$ such that*

(1) $d''_0 \circ \varepsilon_2 = d'_0 \circ \varepsilon_1, d'' \circ d'_j = d'_{j+1} \circ d$ and $d' \circ d''_i = d''_{i+1} \circ d,$

where d denotes the operation in $A^\bullet$ and $B^\bullet$;

(2) *$(A^j, d'_j, C_{\bullet j})$ and $(B^i, d''_i, C_{i\bullet})$ are augmented cochain complexes. Then $H^j(A^\bullet) \simeq H_{0j}$ and $H^i(B^\bullet) \simeq H_{i0}$.*

PROOF

1. $Z_{0j} = \{\xi \in C_{0j}: d'\xi = 0 \text{ and } d''\xi = 0\} = \{\xi \in C_{0j}$: There is an $\eta \in A^j$ with $d'_j\eta = \xi$, $d''\xi = 0\} = \{\xi \in C_{0j}$: There is an $\eta \in A^j$ with $d'_j\eta = \xi$ and $d'_{j+1}(d\eta) = 0\} = \{\xi \in C_{0j}$: There is an $\eta \in A^j$ with $d'_j\eta = \xi$ and $d\eta = 0\} = d'_j(Z^j(A^\bullet))$.

2. $B_{0j} = \{d''\xi: \xi \in C_{0,j-1}$ with $d'\xi = 0\} = \{d''\xi$: There is an $\eta \in A^{j-1}$ with $d'_{j-1}\eta = \xi\} = d'_j(B^j(A^\bullet))$ for $j \geqslant 1$ and $B_{00} = 0 = d'_0(B^0(A^\bullet))$.

3. Since d'_j is always injective, it follows that

$$H_{0j} = Z_{0j}/B_{0j} \simeq Z^j(A^\bullet)/B^j(A^\bullet) = H^j(A^\bullet).$$

One shows that $H_{i0} \simeq H^i(B^\bullet)$ analogously. □

EXAMPLE. Let X be a complex manifold, $\mathscr{S}$ a sheaf of R-modules over X, $\mathfrak{U}$ an open covering of X. If $\mathfrak{W}(\mathscr{S}): \mathscr{S}_0 \to \mathscr{S}_1 \to \mathscr{S}_2 \to \cdots$ is the canonical flabby resolution of $\mathscr{S}$, then let

$$W^\bullet(\mathscr{S}): \Gamma(X, \mathscr{S}_0) \xrightarrow{d} \Gamma(X, \mathscr{S}_1) \xrightarrow{d} \Gamma(X, \mathscr{S}_2) \xrightarrow{d} \cdots.$$

$(\Gamma(X, \mathscr{S}), \varepsilon_*, W^\bullet(\mathscr{S}))$ is an augmented cochain complex.

If one sets

$$C^\bullet(\mathfrak{U}, \mathscr{S}): C^0(\mathfrak{U}, \mathscr{S}) \xrightarrow{\delta} C^1(\mathfrak{U}, \mathscr{S}) \xrightarrow{\delta} C^2(\mathfrak{U}, \mathscr{S}) \to \cdots,$$

then $(\Gamma(X, \mathscr{S}), \varepsilon, C^\bullet(\mathfrak{U}, \mathscr{S}))$ is also an augmented cochain complex.

Now let

$$C_{ij} := C^i(\mathfrak{U}, \mathscr{S}_j), d' := \delta_{(j)} = \delta: C^i(\mathfrak{U}, \mathscr{S}_j) \to C^{i+1}(\mathfrak{U}, \mathscr{S}_j),$$

$d'' := (-1)^i d_*: C^i(\mathfrak{U}, \mathscr{S}_j) \to C^i(\mathfrak{U}, \mathscr{S}_{j+1})$ (with $d_*\xi(\iota_0, \ldots, \iota_i) := d_*(\xi(\iota_0, \ldots, \iota_i))$). Clearly $d'd' = 0$ and $d''d'' = 0$. Moreover it is true that

$$(d_*\delta\xi)(\iota_0, \ldots, \iota_{i+1}) = d_*\left(\sum_{\lambda=0}^{i+1} (-1)^{\lambda+1} \xi(\iota_0, \ldots, \hat{\iota}_\lambda, \ldots, \iota_{i+1})\right)$$
$$= \sum_{\lambda=0}^{i+1} (-1)^{\lambda+1} d_*\xi(\iota_0, \ldots, \hat{\iota}_\lambda, \ldots, \iota_{i+1}) = (\delta d_*\xi)(\iota_0, \ldots, \iota_{i+1});$$

therefore

$$d'd'' + d''d' = \delta_{(j+1)}(-1)^i d_* + (-1)^{i+1} d_*\delta_{(j)} = (-1)^i \cdot (\delta_{(j+1)} d_* - d_*\delta_{(j)}) = 0.$$

Thus (C_{ij}, d', d'') is a double complex which we shall henceforth describe as the canonical double complex of $(X, \mathcal{S}, \mathfrak{U})$. We obtain the following diagram.

$$\begin{array}{ccccccccc}
 & 0 & & 0 & & 0 & & 0 & \\
 & \downarrow & & \downarrow & & \downarrow & & \downarrow & \\
0 \longrightarrow & \Gamma(X, \mathcal{S}) & \xrightarrow{\varepsilon_*} & \Gamma(X, \mathcal{S}_0) & \xrightarrow{d} & \Gamma(X, \mathcal{S}_1) & \xrightarrow{d} & \Gamma(X, \mathcal{S}_2) & \xrightarrow{d} \\
 & \downarrow{\scriptstyle\varepsilon} & & \downarrow{\scriptstyle\varepsilon} & & \downarrow{\scriptstyle\varepsilon} & & \downarrow{\scriptstyle\varepsilon} & \\
0 \longrightarrow & C^0(\mathfrak{U}, \mathcal{S}) & \xrightarrow{\varepsilon_*} & C^0(\mathfrak{U}, \mathcal{S}_0) & \xrightarrow{d''} & C^0(\mathfrak{U}, \mathcal{S}_1) & \xrightarrow{d''} & C^0(\mathfrak{U}, \mathcal{S}_2) & \xrightarrow{d''} \\
 & \downarrow{\scriptstyle\delta} & & \downarrow{\scriptstyle d'} & & \downarrow{\scriptstyle d'} & & \downarrow{\scriptstyle d'} & \\
0 \longrightarrow & C^1(\mathfrak{U}, \mathcal{S}) & \xrightarrow{\varepsilon_*} & C^1(\mathfrak{U}, \mathcal{S}_0) & \xrightarrow{d''} & C^1(\mathfrak{U}, \mathcal{S}_1) & \xrightarrow{d''} & C^1(\mathfrak{U}, \mathcal{S}_2) & \xrightarrow{d''} \\
 & \downarrow{\scriptstyle\delta} & & \downarrow{\scriptstyle d'} & & \downarrow{\scriptstyle d'} & & \downarrow{\scriptstyle d'} & \\
0 \longrightarrow & C^2(\mathfrak{U}, \mathcal{S}) & \xrightarrow{\varepsilon_*} & C^2(\mathfrak{U}, \mathcal{S}_0) & \xrightarrow{d''} & C^2(\mathfrak{U}, \mathcal{S}_1) & \xrightarrow{d''} & C^2(\mathfrak{U}, \mathcal{S}_2) & \xrightarrow{d''} \\
 & \downarrow{\scriptstyle\delta} & & \downarrow{\scriptstyle d'} & & \downarrow{\scriptstyle d'} & & \downarrow{\scriptstyle d'} &
\end{array}$$

Since all the hypotheses of Theorem 3.1 are fulfilled,

$$H^j(X, \mathcal{S}) \simeq H_{0j}, H^i(\mathfrak{U}, \mathcal{S}) \simeq H_{i0} \qquad \text{for all } i, j.$$

We can therefore use the homology groups H_{ij} of the canonical double complex to compute the flabby and Čech cohomology groups of X with coefficients in $\mathcal{S}$. Homomorphisms $\varphi_\ell : H^\ell(\mathfrak{U}, \mathcal{S}) \to H^\ell(X, \mathcal{S})$ will be constructed with the help of these double complexes.

Theorem 3.2. *Let (C_{ij}, d', d'') be a double complex.*

1. *Let the d'-sequences be exact at the locations (i, j) and $(i - 1, j)$. Then there are homomorphisms*

$$\varphi_{ij}: H_{ij} \to H_{i-1,j+1} \quad \text{for} \quad i \geqslant 1, \quad \text{with} \quad \varphi_{ij} \circ q_{ij} \circ d' = q_{i-1,j+1} \circ d''.$$

2. *Let the d''-sequences be exact at the locations $(i - 1, j + 1), (i - 1, j)$. Then there are homomorphisms*

$$\psi_{ij}: H_{i-1,j+1} \to H_{ij} \quad \text{for} \quad i \geqslant 1, \quad \text{with} \quad \psi_{ij} \circ q_{i-1,j+1} \circ d'' = q_{ij} \circ d'.$$

3. *If hypotheses* (1) *and* (2) *are satisfied simultaneously, then φ_{ij} is an isomorphism with $\varphi_{ij}^{-1} = \psi_{ij}$.*

PROOF

1. If $\xi_{ij} \in Z_{ij}$, then $d'\xi_{ij} = 0$. Therefore there is an $\eta_{i-1,j} \in C_{i-1,j}$ with $d'\eta_{i-1,j} = \xi_{ij}$. We set $\varphi_{ij}(q_{ij}(\xi_{ij})) := q_{i-1,j+1}(d''\eta_{i-1,j})$.

a. Let $\xi_{ij} = d'\eta = d'\eta^*$. Then $d'(\eta - \eta^*) = 0$; therefore there is a $\gamma \in C_{i-2,j}$ with $d'\gamma = \eta - \eta^*$ (for $i \geqslant 2$), and it follows that $d''\eta - d''\eta^* = d''d'\gamma \in B_{i-1,j+1}$.

Therefore

$$q_{i-1,j+1}(d''\eta) = q_{i-1,j+1}(d''\eta^*).$$

If $i = 1$, set $\gamma^* := \eta - \eta^*$. Then $d'\gamma^* = 0$. Therefore $d''\gamma^* \in B_{0,j+1}$, and furthermore $q_{0,j+1}(d''\eta) = q_{0,j+1}(d''\eta^*)$. The definition does not depend on the choice of $\eta_{i-1,j}$.

b. Let $\xi_{ij} \in B_{ij}$. If $i \geqslant 1$ and $j \geqslant 1$, then $\xi_{ij} = d'd''\gamma$ with $\gamma \in C_{i-1,j-1}$ and $d''(d''\gamma) = 0$. If $j = 0$, then $\xi_{ij} = d'\gamma^*$ with $d''\gamma^* = 0$. Therefore the definition depends only on the cohomology class of ξ_{ij}.

(c) $\quad d''(d''\eta_{i-1,j}) = 0, d'(d''\eta_{i-1,j}) = -d''(d'\eta_{i-1,j}) = d''(-\xi_{ij}) = 0.$

Therefore $d''\eta_{i-1,j}$ lies in $Z_{i-1,j+1}$.

Because of (a), (b), and (c), φ_{ij} actually defines a mapping from H_{ij} to $H_{i-1,j+1}$. It is clear that the map is a homomorphism.

2. The existence of ψ_{ij} follows exactly like that of φ_{ij}. If φ_{ij} and ψ_{ij} both exist, then

$$\varphi_{ij} \circ \psi_{ij} \circ q_{i-1,j+1} \circ d'' = \varphi_{ij} \circ q_{ij} \circ d' = q_{i-1,j+1} \circ d'',$$
$$\psi_{ij} \circ \varphi_{ij} \circ q_{ij} \circ d' = \psi_{ij} \circ q_{i-1,j+1} \circ d'' = q_{ij} \circ d'.$$

Hence it follows that $\varphi_{ij}^{-1} = \psi_{ij}$. □

Theorem 3.3. *Let X be a complex manifold, $\mathscr{S}$ a sheaf of R-modules over X, and $\mathfrak{U}$ an open covering of X. Then there is a (canonical) R-module homomorphism*

$$\varphi_\ell: H^\ell(\mathfrak{U}, \mathscr{S}) \to H^\ell(X, \mathscr{S}), \qquad \text{for} \qquad \ell \geqslant 1.$$

φ_1 is, in particular, injective.

PROOF

1. Let (C_{ij}, d', d'') be the canonical double complex of $(X, \mathscr{S}, \mathfrak{U})$. Then $H^j(X, \mathscr{S}) \simeq H_{0j}$, $H^i(\mathfrak{U}, \mathscr{S}) \simeq H_{i0}$, and we can define

$$\varphi_\ell := \varphi_{1,\ell-1} \circ \cdots \circ \varphi_{\ell-1,1} \circ \varphi_{\ell,0}$$

[Since all sheaves $\mathscr{S}_j, j \geqslant 0$ are flabby, we have $H^i(\mathfrak{U}, \mathscr{S}_j) = 0$ for $i \geqslant 1$, $j \geqslant 0$. Therefore the d'-sequences are exact!]

2. $\varphi_1 = \varphi_{10}: H_{10} \to H_{01}$ is given by $\varphi_{10} \circ q_{10} \circ d' = q_{01} \circ d''$. If $0 = \varphi_{10}(q_{10} \circ d'\eta) = q_{01} \circ d''\eta$, then $d''\eta$ lies in B_{01}, therefore there is an $\eta^* \in C_{00}$ with $d'\eta^* = 0$ and $d''\eta^* = d''\eta$. Then $d''(\eta - \eta^*) = 0$ and $d'(\eta - \eta^*) = d'\eta$, therefore $d'\eta \in B_{10}$; that is, $q_{10} \circ d'\eta = 0$. □

Def. 3.3. Let $\mathscr{S}$ be a sheaf of R-modules over X and $\mathfrak{U} = (U_\iota)_{\iota \in I}$ an open covering of X. $\mathfrak{U}$ is called a Leray covering of $\mathscr{S}$ if $H^\ell(U_{\iota_0 \cdots \iota_i}, \mathscr{S}) = 0$ for $\ell \geqslant 1$ and all $\iota_0, \ldots, \iota_i$.

Theorem 3.4. *If $\mathfrak{U}$ is a Leray covering of $\mathscr{S}$, then $\varphi_\ell: H^\ell(\mathfrak{U}, \mathscr{S}) \to H^\ell(X, \mathscr{S})$ is an isomorphism for every $\ell \geqslant 1$.*

PROOF. If $H^{\prime}(U_{\iota_0 \cdots \iota_i}, \mathscr{S}) = 0$, then by the definition of flabby cohomology the following sequence is exact:

$$\Gamma(U_{\iota_0 \cdots \iota_i}, \mathscr{S}_{j-1}) \xrightarrow{d_*} \Gamma(U_{\iota_0 \cdots \iota_i}, \mathscr{S}_j) \xrightarrow{d_*} \Gamma(U_{\iota_0 \cdots \iota_i}, \mathscr{S}_{j+1})$$

If $d_*\xi = 0$, then $0 = (d_*\xi)(\iota_0, \ldots, \iota_i) = d_*(\xi(\iota_0, \ldots, \iota_i))$ for all $(\iota_0, \ldots \iota_i)$. Therefore there are elements $\eta(\iota_0, \ldots, \iota_i)$ with $d^*(\eta(\iota_0, \ldots, \iota_i)) = \xi(\iota_0, \ldots, \iota_i)$. In each case it suffices to determine one ordering $\eta(\iota_0, \ldots, \iota_i)$ of the indices, since the values for other orderings are determined by the antisymmetric rule.

In this way a cochain η with $d_*\eta = \xi$ is determined.

The d''-sequences in the canonical double complex are therefore exact and the proposition follows. □

4. The Cohomology Sequence

Let X be a complex manifold, $\mathscr{S}^*$, $\mathscr{S}$, $\mathscr{S}^{**}$ sheaves of R-modules over X.

(A) Let $\varphi: \mathscr{S}^* \to \mathscr{S}$ be a homomorphism. Then $\mathfrak{W}(\varphi): \mathfrak{W}(\mathscr{S}^*) \to \mathfrak{W}(\mathscr{S})$ is a homomorphism between canonical flabby resolutions, defined by the mappings $W_i\varphi: \mathscr{S}_i^* \to \mathscr{S}_i$. These mappings induce mappings

$$(W_i\varphi)_*: \Gamma(X, \mathscr{S}_i^*) \to \Gamma(X, \mathscr{S}_i).$$

Theorem 4.1

1. If $\xi \in Z^i(X, \mathscr{S}^*)$, then $(W_i\varphi)_*\xi \in Z^i(X, \mathscr{S})$.
2. If $\xi \in B^i(X, \mathscr{S}^*)$, then $(W_i\varphi)_*\xi \in B^i(X, \mathscr{S})$.

PROOF. The following diagram is commutative:

$$\begin{array}{ccccc} \Gamma(X, \mathscr{S}_{i-1}^*) & \xrightarrow{d} & \Gamma(X, \mathscr{S}_i^*) & \xrightarrow{d} & \Gamma(X, \mathscr{S}_{i+1}^*) \\ \downarrow{\scriptstyle (W_{i-1}\varphi)_*} & & \downarrow{\scriptstyle (W_i\varphi)_*} & & \downarrow{\scriptstyle (W_{i+1}\varphi)_*} \\ \Gamma(X, \mathscr{S}_{i-1}) & \xrightarrow{d} & \Gamma(X, \mathscr{S}_i) & \xrightarrow{d} & \Gamma(X, \mathscr{S}_{i+1}) \end{array}$$

1. If $d\xi = 0$, then $d((W_i\varphi)_*\xi) = (W_{i+1}\varphi)_*(d\xi) = 0$.
2. If $\xi = d\eta$, then $(W_i\varphi)_*\xi = (W_i\varphi)_*d\eta = d((W_{i-1}\varphi)_*\eta)$. □

Corollary. *Let*

$$q_i^*: Z^i(X, \mathscr{S}^*) \to H^i(X, \mathscr{S}^*), \; q_i: Z^i(X, \mathscr{S}) \to H^i(X, \mathscr{S})$$

be the canonical residue class mappings. Then there exists a homomorphism $\bar{\varphi}: H^i(X, \mathscr{S}^*) \to H^i(X, \mathscr{S})$, *given by* $\bar{\varphi} \circ q_i^* = q_i \circ (W_i\varphi)_*$.

Theorem 4.2. $(H^i: \mathscr{S} \rightsquigarrow H^i(X, \mathscr{S}), \varphi \rightsquigarrow \bar{\varphi})$ *is a covariant functor, that is*:

1. $\overline{\mathrm{id}_{\mathscr{S}}} = \mathrm{id}_{H^i(X, \mathscr{S})}$.
2. $\overline{\psi \circ \varphi} = \bar{\psi} \circ \bar{\varphi}$.

The proof is trivial.

(B) Let $0 \to \mathscr{S}^* \xrightarrow{\varphi} \mathscr{S} \xrightarrow{\psi} \mathscr{S}^{**} \to 0$ be exact. Then we obtain the following commutative diagram with exact columns:

$$\begin{array}{ccccccccc}
 & 0 & & 0 & & 0 & & 0 & \\
 & \downarrow & & \downarrow & & \downarrow & & \downarrow & \\
\cdots \longrightarrow & \Gamma(X, \mathscr{S}^*_{i-1}) & \xrightarrow{d} & \Gamma(X, \mathscr{S}^*_i) & \xrightarrow{d} & \Gamma(X, \mathscr{S}^*_{i+1}) & \xrightarrow{d} & \Gamma(X, \mathscr{S}^*_{i+2}) & \longrightarrow \cdots \\
 & \downarrow (W_{i-1}\varphi)_* & & \downarrow (W_i\varphi)_* & & \downarrow (W_{i+1}\varphi)_* & & \downarrow (W_{i+2}\varphi)_* & \\
\cdots \longrightarrow & \Gamma(X, \mathscr{S}_{i-1}) & \xrightarrow{d} & \Gamma(X, \mathscr{S}_i) & \xrightarrow{d} & \Gamma(X, \mathscr{S}_{i+1}) & \xrightarrow{d} & \Gamma(X, \mathscr{S}_{i+2}) & \longrightarrow \cdots \\
 & \downarrow (W_{i-1}\psi)_* & & \downarrow (W_i\psi)_* & & \downarrow (W_{i+1}\psi)_* & & \downarrow (W_{i+2}\psi)_* & \\
\cdots \longrightarrow & \Gamma(X, \mathscr{S}^{**}_{i-1}) & \xrightarrow{d} & \Gamma(X, \mathscr{S}^{**}_i) & \xrightarrow{d} & \Gamma(X, \mathscr{S}^{**}_{i+1}) & \xrightarrow{d} & \Gamma(X, \mathscr{S}^{**}_{i+2}) & \longrightarrow \cdots \\
 & \downarrow & & \downarrow & & \downarrow & & \downarrow & \\
 & 0 & & 0 & & 0 & & 0 &
\end{array}$$

Theorem 4.3

1. *If* $\xi \in Z^i(X, \mathscr{S}^{**})$, *then there exists an* $\eta_0 \in \Gamma(X, \mathscr{S}_i)$ *and an* $\eta \in Z^{i+1}(X, \mathscr{S}^*)$ *with* $\xi = (W_i\psi)_*\eta_0$ *and* $d\eta_0 = (W_{i+1}\varphi)_*\eta$. η *is determined up to an element* $\eta^* \in B^{i+1}(X, \mathscr{S}^*)$.

2. *There exists a homomorphism, canonically induced by* (1),

$$\partial : H^i(X, \mathscr{S}^{**}) \to H^{i+1}(X, \mathscr{S}^*) \quad \text{with} \quad \bar{\varphi} \circ \partial = 0 \quad \text{and} \quad \partial \circ \bar{\psi} = 0.$$

PROOF

1. If $\xi \in Z^i(X, \mathscr{S}^{**})$, then $d\xi = 0$, and there exists an $\eta_0 \in \Gamma(X, \mathscr{S}_i)$ with $(W_i\psi)_*\eta_0 = \xi$. Clearly then $0 = d((W_i\psi)_*\eta_0) = (W_{i+1}\psi)_*d\eta_0$, that is, there exists an $\eta \in \Gamma(X, \mathscr{S}^*_{i+1})$ with $(W_{i+1}\varphi)_*\eta = d\eta_0$. The element η is a cycle, because $0 = dd\eta_0 = d((W_{i+1}\varphi)_*\eta) = (W_{i+2}\varphi)_*d\eta$ and therefore $d\eta = 0$. η is uniquely determined by η_0. If $\xi = (W_i\psi)_*\eta_0' = (W_i\psi)_*\eta_0''$, then there exists a $\rho \in \Gamma(X, \mathscr{S}^*_i)$ with $(W_i\varphi)_*\rho = \eta_0' - \eta_0''$ and we have

$$d\eta_0' - d\eta_0'' = d((W_i\varphi)_*\rho) = (W_{i+1}\varphi)_*d\rho;$$

therefore $\eta' - \eta'' = d\rho$.

2. A homomorphism $\tilde{\partial} : Z^i(X, \mathscr{S}^{**}) \to H^{i+1}(X, \mathscr{S}^*)$ is defined by $\tilde{\partial}(\xi) := q^*_{i+1}(\eta)$ such that

$$\bar{\varphi} \circ \tilde{\partial} \circ (W_i\psi)_*\eta_0 = \bar{\varphi} \circ q^*_{i+1}\,\eta = q_{i+1} \circ (W_{i+1}\varphi)_*\eta = q_{i+1}(d\eta_0) = 0.$$

If $\xi = d\xi^*$, then there is a $\sigma \in \Gamma(X, \mathscr{S}_{i-1})$ with $(W_{i-1}\psi)_*\sigma = \xi^*$; therefore $(W_i\psi)_*d\sigma = d((W_{i-1}\psi)_*\sigma) = \xi$. We can choose $\eta_0 = d\sigma$ and from the construction we obtain that $\tilde{\partial}(\xi) = 0$. Therefore $\tilde{\partial}$ induces a homomorphism

$$\partial : H^i(X, \mathscr{S}^{**}) \to H^{i+1}(X, \mathscr{S}^*) \quad \text{with} \quad \partial \circ q_i^{**} = \tilde{\partial}.$$

In particular (a) q_i^{**}, $(W_i\psi)_*$ are surjective and

$$\bar{\varphi} \circ \partial \circ q_i^{**} \circ (W_i\psi)_*(\eta_0) = \bar{\varphi} \circ \tilde{\partial} \circ (W_i\psi)_*(\eta_0) = 0.$$

(b) q_i is surjective, $(W_i\varphi)_*$ is injective, and for $\xi \in Z^i(X, \mathscr{S})$ we have

$$\partial \circ \bar{\psi} \circ q_i(\xi) = \tilde{\partial} \circ (W_i\psi)_*(\xi) = q_{i+1}^*(\eta) \qquad \text{with} \qquad (W_i\varphi)_*\eta = d\xi = 0$$

therefore $\eta = 0$. Hence $\bar{\varphi} \circ \partial = 0$ and $\partial \circ \bar{\psi} = 0$. □

Theorem 4.4. *Let* $0 \to \mathscr{S}^* \xrightarrow{\varphi} \mathscr{S} \xrightarrow{\psi} \mathscr{S}^{**} \to 0$ *be an exact sequence of sheaves of R-modules. Then the following long cohomology sequence is also exact*

$$0 \to \Gamma(X, \mathscr{S}^*) \xrightarrow{\varphi} \Gamma(X, \mathscr{S}) \xrightarrow{\psi} \Gamma(X, \mathscr{S}^{**}) \xrightarrow{\partial} H^1(X, \mathscr{S}^*) \to \cdots$$
$$\cdots \to H^{i-1}(X, \mathscr{S}^{**}) \xrightarrow{\partial} H^i(X, \mathscr{S}^*) \xrightarrow{\bar{\varphi}} H^i(X, \mathscr{S}) \xrightarrow{\bar{\psi}} H^i(X, \mathscr{S}^{**}) \to \cdots$$

PROOF

a. The sequence $0 \to \Gamma(X, \mathscr{S}^*) \to \Gamma(X, \mathscr{S}) \to \Gamma(X, \mathscr{S}^{**})$ is exact, since Γ is a left exact functor.

b. The cohomology sequence is exact at $H^i(X, \mathscr{S}^*)$, $i \geqslant 1$:

1. $\bar{\varphi} \circ \partial = 0$ by Theorem 4.3.
2. If $\xi \in Z^i(X, \mathscr{S}^*)$ and

$$0 = \bar{\varphi} \circ q_i^*(\xi) = q_i \circ (W_i\varphi)_*\xi,$$

then $(W_i\varphi)_*\xi = d\eta$ with $\eta \in \Gamma(X, \mathscr{S}_{i-1})$ and

$$d((W_{i-1}\psi)_*\eta) = (W_i\psi)_*d\eta = (W_i\psi)_*(W_i\varphi)_*\xi = 0.$$

$(W_{i-1}\psi)_*\eta$ thus lies in $Z^{i-1}(X, \mathscr{S}^{**})$ and

$$\partial \circ q_{i-1}^{**} \circ (W_{i-1}\psi)_*\eta = \tilde{\partial}(W_{i-1}\psi)_*\eta = q_i^*\xi.$$

Therefore $\operatorname{Ker} \bar{\varphi} \subset \operatorname{Im} \partial$.

c. Exactness at $H^i(X, \mathscr{S})$, $i \geqslant 1$:

1. By Theorem 4.2, $\bar{\psi} \circ \bar{\varphi} = 0$.
2. Let $\xi \in Z^i(X, \mathscr{S})$ and $0 = \bar{\psi} \circ q_i(\xi) = q_i^{**} \circ (W_i\psi)_*\xi$. Then

$$(W_i\psi)_*\xi = d\xi^* \qquad \text{with} \qquad \xi^* = (W_{i-1}\psi)_*\eta \in \Gamma(X, \mathscr{S}_{i-1}^{**})$$

and hence $d(\xi - d\eta) = 0$ and $(W_i\psi)_*(\xi - d\eta) = 0$. Thus there is a $\sigma \in \Gamma(X, \mathscr{S}_i^*)$ with $(W_i\varphi)_*\sigma = \xi - d\eta$. Clearly $d\sigma = 0$ also, and

$$\bar{\varphi} \circ q_i^*(\sigma) = q_i \circ (W_i\varphi)_*\sigma = q_i(\xi - d\eta) = q_i(\xi).$$

Therefore $\operatorname{Ker} \bar{\psi} \subset \operatorname{Im} \bar{\varphi}$.

d. Exactness at $H^i(X, \mathscr{S}^{**})$, $i \geqslant 1$:

1. $\partial \circ \bar{\psi} = 0$ by Theorem 4.3.
2. Let $d\xi = 0$ and

$$0 = \partial \circ q_i^{**}(\xi) = \tilde{\partial}\xi = q_{i+1}^*\eta \qquad \text{with} \qquad \xi = (W_i\psi)_*\eta_0$$

and

$$d\eta_0 = (W_{i+1}\varphi)_*\eta.$$

Then $\eta = d\sigma$, and

$$d(\eta_0 - (W_i\varphi)_*\sigma) = 0, (W_i\psi)_*(\eta_0 - (W_i\varphi)_*\sigma) = \xi;$$

therefore

$$\bar{\psi} \circ q_i(\eta_0 - (W_i\varphi)_*\sigma) = q_i^{**} \circ (W_i\psi)_*(\eta_0 - (W_i\varphi)_*\sigma) = q_i^{**}\xi.$$

Hence Ker $\partial \subset$ Im $\bar{\psi}$, and the proof is complete. □

(C) Let X be an n-dimensional complex manifold with structure sheaf $\mathcal{O}$. For every open set $U \subset X$ there is an associated multiplicative abelian group $M_U = \{f : f \text{ is holomorphic on } U \text{ and } f(x) \neq 0 \text{ for } x \in U\}$. M_U becomes a $\mathbb{Z}$-module (with $n \cdot f := f^n$), and together with the usual restriction mappings $r_V^U : M_U \to M_V$ yields a pre-sheaf of $\mathbb{Z}$-modules. The corresponding sheaf of $\mathbb{Z}$-modules $\mathcal{O}^*$ is called the *sheaf of germs of non-vanishing holomorphic functions.* We write the group operation in $\mathcal{O}^*$ and in the derived modules additively. If N_U is the additive abelian group of holomorphic functions, then there exists a $\mathbb{Z}$-module homomorphism $\exp_U : N_U \to M_U$ defined by $f \mapsto e^{2\pi i f}$. For $V \subset U$ the commutative law $\exp_V \circ r_V^U = r_V^U \circ \exp_U$ holds. This defines a sheaf homomorphism $\exp : \mathcal{O} \to \mathcal{O}^*$ with $\exp(rf) = r(e^{2\pi i f})$.

Theorem 4.5. *$0 \to \mathbb{Z} \to \mathcal{O} \overset{\exp}{\to} \mathcal{O}^* \to 0$ is an exact sequence of sheaves of $\mathbb{Z}$-modules (where $\mathbb{Z}$ also denotes the sheaf of germs of continuous $\mathbb{Z}$-valued functions).*

PROOF. Continuous $\mathbb{Z}$-valued functions are locally constant, in particular, locally holomorphic. Hence we can regard $\mathbb{Z}$ as a subsheaf of $\mathcal{O}$, and we need only show that Ker(exp) $= \mathbb{Z}$ and Im(exp) $= \mathcal{O}^*$.

1. Let $\sigma = (rf)(x) \in \mathcal{O}_x$, $f \in N_U$, $\exp(\sigma) = \mathbf{0}$. Then $\mathbf{0} = \exp(rf)(x) = (r(e^{2\pi i f}))(x)$. There exists a connected neighborhood $V(x) \subset U$ with $r(e^{2\pi i f})|V = \mathbf{0}$; that is, $e^{2\pi i f}|V = 1$. Then there is an $n \in \mathbb{Z}$ with $f|V = n$. Conversely if $\sigma \in \mathbb{Z}_x \subset \mathcal{O}_x$, it follows that $\exp(\sigma) = \mathbf{0}$.

2. Let $\rho = (rf)(x) \in \mathcal{O}_x^*$, $f \in M_U$, $x \in U$. Without loss of generality we may assume that U is an open set in $\mathbb{C}^n$, so that $\log(f)$ is holomorphically definable on U. Let

$$h := \frac{1}{2\pi i} \cdot \log(f), \sigma := (rh)(x) \in \mathcal{O}_x.$$

Then

$$\exp(\sigma) = \exp((rh)(x)) = (r(e^{2\pi i h}))(x) = (rf)(x) = \rho.$$ □

Theorem 4.6

1. *Let $f \in \Gamma(X, \mathcal{O}^*)$. Then there is an $h \in \Gamma(X, \mathcal{O})$ with $f = e^{2\pi i h}$ if and only if $\partial(f) = 0$.*
2. *If $H^\ell(X, \mathcal{O}) = 0$ for $\ell \geqslant 1$, then $H^\ell(X, \mathcal{O}^*) \simeq H^{\ell+1}(X, \mathbb{Z})$ for $\ell \geqslant 1$.*

PROOF. Look at the long exact cohomology sequence of the short exact sequence $0 \to \mathbb{Z} \to \mathcal{O} \to \mathcal{O}^* \to 0$. □

Def. 4.1. A system $(U_\iota, f_\iota)_{\iota \in I}$ is called the *Cousin II distribution on X* if

1. $\mathfrak{U} = (U_\iota)_{\iota \in I}$ is an open covering of X.
2. f_ι is holomorphic on U_ι and vanishes identically nowhere.
3. On $U_{\iota_0\iota_1}$ there is a nowhere vanishing holomorphic function $h_{\iota_0\iota_1}$ such that $f_{\iota_0} = h_{\iota_0\iota_1} \cdot f_{\iota_1}$ on $U_{\iota_0\iota_1}$.

A solution of this *Cousin II distribution* is a holomorphic function f on X such that $f_\iota = h_\iota : f$ with nowhere vanishing holomorphic functions h_ι on U_ι.

Remark. The functions $h_{\iota_0\iota_1}$ are uniquely determined by the distribution $(U_\iota, f_\iota)_{\iota \in I}$:

If $f_{\iota_0} = h_{\iota_0\iota_1} \cdot f_{\iota_1} = \tilde{h}_{\iota_0\iota_1} \cdot f_{\iota_1}$, then $0 = (h_{\iota_0\iota_1} - \tilde{h}_{\iota_0\iota_1}) \cdot f_{\iota_1}$.

If it were true that $(h_{\iota_0\iota_1} - \tilde{h}_{\iota_0\iota_1})(x_0) \neq 0$ for an $x_0 \in U_{\iota_0\iota_1}$, then it would also hold that $(h_{\iota_0\iota_1} - \tilde{h}_{\iota_0\iota_1})(x) \neq 0$ for $x \in V$, V an open neighborhood of x_0 in $U_{\iota_0\iota_1}$. Therefore $f_{\iota_1}|V = 0$ which is a contradiction.

Theorem 4.7. *Let $(U_\iota, f_\iota)_{\iota \in I}$ be a Cousin II distribution on X, $\mathfrak{U} = (U_\iota)_{\iota \in I}$. Then:*

1. *$h(\iota_0, \iota_1) := rh_{\iota_0\iota_1}$ defines an element $h \in Z^1(\mathfrak{U}, \mathcal{O}^*)$.*
2. *$(U_\iota, f_\iota)_{\iota \in I}$ is solvable if and only if h lies in $B^1(\mathfrak{U}, \mathcal{O}^*)$.*

PROOF.

1a. Because

$$f_{\iota_1} = h_{\iota_0\iota_1}^{-1} \cdot f_{\iota_0} = h_{\iota_1\iota_0} \cdot f_{\iota_0}$$

it follows that $h(\iota_1, \iota_0) = -h(\iota_0, \iota_1)$.

b. Because

$$h_{\iota_0\iota_1} \cdot h_{\iota_1\iota_2} \cdot f_{\iota_2} = h_{\iota_0\iota_1} \cdot f_{\iota_1} = f_{\iota_0} = h_{\iota_0\iota_2} \cdot f_{\iota_2}$$

it follows that $h(\iota_0, \iota_1) + h(\iota_1, \iota_2) = h(\iota_0, \iota_2)$.

2a. Let $(U_\iota, f_\iota)_{\iota \in I}$ be solvable. Then $f_\iota = h_\iota \cdot f$ with nowhere vanishing functions h_ι and $f_{\iota_0} = h_{\iota_0\iota_1} \cdot f_{\iota_1}$, therefore $h_{\iota_0} \cdot f = h_{\iota_0\iota_1} \cdot h_{\iota_1} \cdot f$. Let $\rho(\iota) := h_\iota$. Then

$$\rho(\iota_0) - \rho(\iota_1) = r(h_{\iota_0} \cdot h_{\iota_1}^{-1}) = r(h_{\iota_0\iota_1}) = h(\iota_0, \iota_1);$$

therefore $\delta\rho = h$.

b. If h lies in $B^1(\mathfrak{U}, \mathcal{O}^*)$, then for every $\iota \in I$ there exists a $\rho(\iota) \in \Gamma(U_\iota, \mathcal{O}^*)$ such that $\rho(\iota_0) - \rho(\iota_1) = h(\iota_0, \iota_1)$ on $U_{\iota_0\iota_1}$. Then $h_\iota := [\rho(\iota)]$ is a nowhere vanishing holomorphic function, and on $U_{\iota_0\iota_1}$ we have $h(\iota_0, \iota_1) = r(h_{\iota_0\iota_1}) = r(h_{\iota_0} \cdot h_{\iota_1}^{-1})$; therefore $h_{\iota_0} = h_{\iota_0\iota_1} \cdot h_{\iota_1}$.

Similarly we have $f_{\iota_0} = h_{\iota_0\iota_1} \cdot f_{\iota_1}$. Hence it follows that $f_{\iota_0} \cdot h_{\iota_0}^{-1} = f_{\iota_1} \cdot h_{\iota_1}^{-1}$.

Thus there is a holomorphic function f on X with $f_\iota = h_\iota \cdot f$ defined by $f|U_\iota := f_\iota \cdot h_\iota^{-1}$. □

Remark. The question of the solvability of a Cousin II distribution is a generalization of the Weierstrass problem.

Corollary. *If $H^1(\mathfrak{U}, \mathcal{O}^*) = 0$, then every Cousin II distribution belonging to the covering $\mathfrak{U}$ is solvable.*

Theorem 4.8. *Let X be an n-dimensional complex manifold with structure sheaf $\mathcal{O}$.*

1. *If $H^1(X, \mathcal{O}) = 0$, then every Cousin I distribution on X is solvable.*
2. *If $H^1(X, \mathcal{O}^*) = 0$, then every Cousin II distribution on X is solvable.*

PROOF. The canonical homomorphisms $H^1(\mathfrak{U}, \mathcal{O}) \to H^1(X, \mathcal{O})$ and $H^1(\mathfrak{U}, \mathcal{O}^*) \to H^1(X, \mathcal{O}^*)$ are injective for every covering $\mathfrak{U}$ (See Theorem 3.3). □

Def. 4.2. Let $h \in Z^1(\mathfrak{U}, \mathcal{O}^*)$ be the cocycle of a Cousin II distribution $(U_\iota, f_\iota)_{\iota \in I}$, $\underline{h}$ the corresponding cohomology class in $H^1(X, \mathcal{O}^*)$, and $\partial: H^1(X, \mathcal{O}^*) \to H^2(X, \mathbb{Z})$ the "boundary homomorphism" of the long exact cohomology sequence of $0 \to \mathbb{Z} \to \mathcal{O} \to \mathcal{O}^* \to 0$. Then $c(h) := \partial(\underline{h}) \in H^2(X, \mathbb{Z})$ is called the *Chern class* of h (resp. of $(U_\iota, f_\iota)_{\iota \in I}$).

Theorem 4.9. *If $H^\ell(X, \mathcal{O}) = 0$ for $\ell \geqslant 1$, then the Cousin II distribution $(U_\iota, f_\iota)_{\iota \in I}$ (with the corresponding cocycle h) is solvable if and only if $c(h) = 0$ (and that is a purely topological condition!).*

PROOF. By Theorem 4.6 $H^1(X, \mathcal{O}^*) \simeq H^2(X, \mathbb{Z})$, under ∂. h is thus solvable if and only if $\underline{h} = 0$, and that is the case if and only if $c(h) = \partial(\underline{h}) = 0$. □

EXAMPLE. There exist very simple domains of holomorphy on which not every Cousin II distribution is solvable. Suppose,

$$X := \{(z, w) \in \mathbb{C}^2 : \big||z| - 1\big| < \varepsilon, \big||w| - 1\big| < \varepsilon\}.$$

X is a Reinhardt domain and, as one can readily see, is logarithmically convex, therefore a domain of holomorphy.

The "center of X"

$$T := \{(z, w) \in \mathbb{C}^2 : |z| = 1, |w| = 1\} \subset X$$

is the real torus.

$$\mathfrak{g} := \{(z, w) \in \mathbb{C}^2 : w = z - 1\}$$

is a complex line, and therefore a real 2-dimensional plane.

For $(z, w) \in \mathfrak{g}$

$$|w|^2 = w \cdot \overline{w} = (z - 1)(\overline{z} - 1) = z\overline{z} + 1 - (z + \overline{z}) = |z|^2 + 1 - 2x \quad (\text{with} \quad z = x + iy)$$

If $|z| = 1$, then in particular we have $|w|^2 = 2 - 2x$, so $|w| = 1$ if and only if $x = 1/2$. Let

$$z_1 := \tfrac{1}{2}(1 + i\sqrt{3}), z_2 := \tfrac{1}{2}(1 - i\sqrt{3}), w_1 := z_1 - 1, w_2 := z_2 - 1.$$

Hence it follows that

$$T \cap \mathfrak{g} = \{(z_1, w_1), (z_2, w_2)\}$$

The mapping $\Phi: \mathfrak{g} \to \mathbb{C}$ with $\Phi(z, w) := z$ is topological with $\Phi^{-1}(z) = (z, z - 1)$. Let

$$R_\varepsilon := \{z \in \mathbb{C}: 1 - \varepsilon < |z| < 1 + \varepsilon\} = \{z \in \mathbb{C}: ||z| - 1| < \varepsilon\},$$
$$\tilde{R}_\varepsilon := \{z \in \mathbb{C}: z - 1 \in R_\varepsilon\} = \{z \in \mathbb{C}: ||z - 1| - 1| < \varepsilon\}.$$

R_ε, $\tilde{R}_\varepsilon$ are two congruent annuli displaced from one another with

$$R_\varepsilon \cap \tilde{R}_\varepsilon = \Phi(\mathfrak{g} \cap X) \supset \Phi(\mathfrak{g} \cap T) = \{z_1, z_2\}.$$

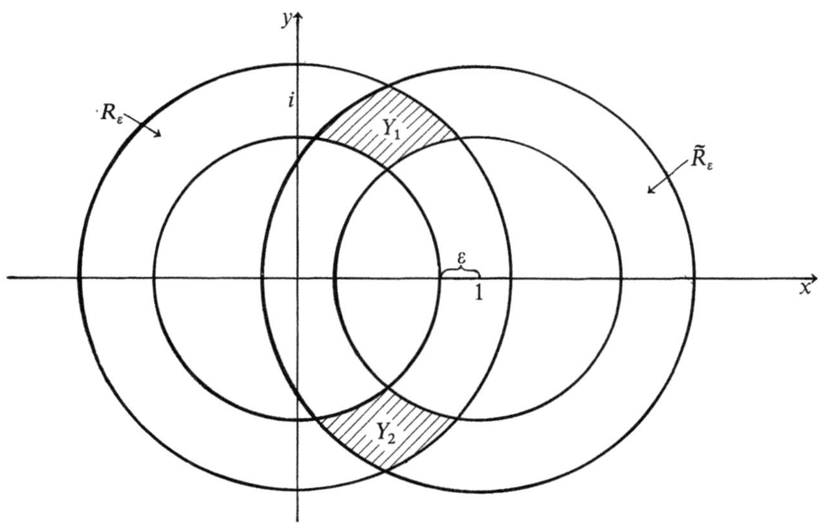

Figure 25. Illustration for the example.

If we choose ε sufficiently small, then $R_\varepsilon \cap \tilde{R}_\varepsilon$ decomposes into two connected components Y_1, Y_2.

Let $F_\lambda := \Phi^{-1}(Y_\lambda)$ for $\lambda = 1, 2$. Then $\mathfrak{g} \cap X = F_1 \cup F_2$ with $F_1 \cap F_2 = \emptyset$ and the sets F_λ are analytic in X. Let

$$U_1 := X - F_2, U_2 := X - F_1, g(z, w) := w - z + 1,$$

as well as

$$f_1 := g|U_1, f_2 := 1|U_2.$$

g has no zeroes in $U_{12} = X - (F_1 \cup F_2) = X - \mathfrak{g}$, and we have $f_1|U_{12} = g \cdot f_2|U_{12}$. Therefore $((U_1, f_1), (U_2, f_2))$ is a Cousin II distribution on X. We can introduce real coordinates on T:

$$(z, w) = (e^{i\varphi}, e^{i\theta}) \mapsto (\varphi, \theta).$$

Then $g|T = e^{i\theta} - e^{i\varphi} + 1 = (\cos\theta - \cos\varphi + 1) + i(\sin\theta - \sin\varphi)$, and the mapping.

$$\tau: U_1 \cap T \to \mathbb{R}^2 \qquad \text{with} \qquad \tau(\varphi, \theta) := (\cos\theta - \cos\varphi + 1, \sin\theta - \sin\varphi)$$

is real analytic and has exactly one zero $(\varphi_0, \theta_0) \approx (z_1, z_1 - 1)$.

For the functional determinant we have

$$\det J_\tau(\varphi_0, \theta_0) = \det\begin{pmatrix} \sin\varphi_0 & -\sin\theta_0 \\ -\cos\varphi_0 & \cos\theta_0 \end{pmatrix} = \det\begin{pmatrix} \sin\varphi_0 & -\sin\varphi_0 \\ -\cos\varphi_0 & \cos\varphi_0 - 1 \end{pmatrix}$$
$$= \det\begin{pmatrix} \sin\varphi_0 & 0 \\ -\cos\varphi_0 & -1 \end{pmatrix} = -\sin\varphi_0 = -\operatorname{Im} z_1 = -\tfrac{1}{2}\sqrt{3} \neq 0.$$

Hence we can find a neighborhood $V = V(\varphi_0, \theta_0) \subset U_1 \cap T$ which is mapped by τ biholomorphically onto a domain of $\mathbb{R}^2$. Let $V^* := V - \{(\varphi_0, \theta_0)\}$.

We can regard τ as a complex valued function. Then on V^* the differential form $\omega = d\tau/\tau$ is defined and clearly $d\omega = 0$.

We now choose an open subset $B \subset\subset V$ which relative to $\tau|V$ is the inverse image of a circular disc $\{z \in \mathbb{C} : |z| \leqslant s\}$. Let $H := \partial B$. Then

$$\int_H \omega = \int_{|\tau| = s} \frac{d\tau}{\tau} \neq 0.$$

Now suppose there is a solution f of the above Cousin II problem. Then $f|U_1 = g \cdot h$, with a nowhere vanishing holomorphic function h in U_1, and $f|T$ has a zero only at (φ_0, θ_0). Therefore $\tilde{\omega} := dh/h$ is a differential form on $U_1 \cap T$, $\alpha := df/f$ a differential form on $T - \{(\varphi_0, \theta_0)\}$ and $d\tilde{\omega} = 0$, $d\alpha = 0$ and $\alpha|V^* = \omega + \tilde{\omega}$. Thus it follows that

$$\int_H \tilde{\omega} = \int_{\partial B} \tilde{\omega} = \int_B d\tilde{\omega} = 0,$$
$$\int_H \alpha = -\int_{\partial(T-B)} \alpha = -\int_{T-B} d\alpha = 0,$$

but

$$\int_H \alpha = \int_H \omega + \int_H \tilde{\omega} = \int_H \omega \neq 0.$$

That is a contradiction. A solution f cannot exist. □

5. Main Theorem on Stein Manifolds

The two following theorems of Cartan–Serre are the basis for the theory of Stein manifolds. The proofs are difficult and cannot be included here.

Theorem 5.1 (Theorem A). *Let $(X, \mathcal{O})$ be a Stein manifold, $\mathcal{S}$ a coherent analytic sheaf over X. Then for every point $x_0 \in X$ there are finitely many global sections $s_1, \ldots, s_k \in \Gamma(X, \mathcal{S})$ which generate $\mathcal{S}_{x_0}$ over $\mathcal{O}_{x_0}$.*

Theorem 5.2 (Theorem B). *Let X be a Stein manifold, S a coherent analytic sheaf over X. Then $H^\ell(X, \mathcal{S}) = 0$ for $\ell \geqslant 1$. (For the definition of a Stein manifold, see Chapter V, Section 2.)*

Theorem 5.3. *Let X be a Stein manifold, and $0 \to \mathscr{S}^* \to \mathscr{S} \to \mathscr{S}^{**} \to 0$ an exact sequence of coherent analytic sheaves over X. Then*

$$0 \to \Gamma(X, \mathscr{S}^*) \to \Gamma(X, \mathscr{S}) \to \Gamma(X, \mathscr{S}^{**}) \to 0$$

is exact and

$$\Gamma(X, \mathscr{S}^{**}) \simeq \frac{\Gamma(X, \mathscr{S})}{\Gamma(X, \mathscr{S}^*)}.$$

PROOF. The cohomology sequence

$$0 \to \Gamma(X, \mathscr{S}^*) \to \Gamma(X, \mathscr{S}) \to \Gamma(X, \mathscr{S}^{**}) \to H^1(X, \mathscr{S}^*) \to \cdots$$

is exact and by Theorem B, $H^1(X, \mathscr{S}^*) = 0$. □

Theorem 5.4. *Let X be a complex manifold and U_1, $U_2 \subset X$ open Stein manifolds. Then $U: = U_1 \cap U_2$ is also a Stein manifold.*

PROOF

1. If $x_0 \in U$, then there are holomorphic functions $f_1, \ldots, f_\ell$ on U_1 such that x_0 is an isolated point in

$$\{x \in U_1 : f_1(x) = \cdots = f_\ell(x) = 0\}.$$

Then the functions $f_1|U, \ldots, f_\ell|U$ are holomorphic and x_0 is also isolated in

$$\{x \in U : f_1(x) = \cdots = f_\ell(x) = 0\}.$$

Therefore U is holomorphically separable.

2. Let $K \subset U$ be compact. Then K is also compact in U_i, and so, for the holomorphically convex hulls, $K \subset \hat{K}_i$, $\hat{K}_i$ compact. Clearly $\hat{K}$ is contained in $\hat{K}_1 \cap \hat{K}_2$. $U - \hat{K}$ is open; therefore $\hat{K}_1 \cap \hat{K}_2 - \hat{K} = \hat{K}_1 \cap \hat{K}_2 \cap (U - \hat{K})$ is open in $\hat{K}_1 \cap \hat{K}_2$. Since $\hat{K}_1 \cap \hat{K}_2$ is compact, it follows that $\hat{K}$ is compact. □

Def. 5.1. Let X be a complex manifold. An open covering $\mathfrak{U} = (U_\iota)_{\iota \in I}$ of X is called *Stein* if all the sets U_ι are Stein.

Theorem 5.5 (Leray). *Lex X be a complex manifold, $\mathscr{S}$ a coherent analytic sheaf on X, $\mathfrak{U}$ a Stein covering of X. Then $\mathfrak{U}$ is a Leray covering of X and for all ℓ, $H^\ell(\mathfrak{U}, \mathscr{S}) \simeq H^\ell(X, \mathscr{S})$.*

PROOF. If $\mathfrak{U} = (U_\iota)_{\iota \in I}$ is Stein, then by Theorem 5.4 all sets $U_{\iota_0 \cdots \iota_i}$ are Stein, and by Theorem B, $H^\ell(U_{\iota_0 \cdots \iota_i}, \mathscr{S}) = 0$ for $\ell \geqslant 1$. Therefore $\mathfrak{U}$ is a Leray covering and $\varphi_\ell : H^\ell(\mathfrak{U}, \mathscr{S}) \to H^\ell(X, \mathscr{S})$ is an isomorphism. □

Theorem 5.6. *If X is a complex manifold, then there are arbitrarily fine Stein coverings of X. If $\mathscr{S}$ is coherent analytic on X, then for every open covering $\mathfrak{U}$ of X there exists a refinement $\mathfrak{V}$ such that $H^\ell(\mathfrak{V}, \mathscr{S}) \simeq H^\ell(X, \mathscr{S})$ for all $\ell \geqslant 0$.*

Proof. Let $\mathcal{H}$ be the structure sheaf of X. If $x_0 \in X$, then there is an open neighborhood $U(x_0) \subset X$, a domain $G \subset \mathbb{C}^n$ and an isomorphism $\varphi:(U, \mathcal{H}) \to (G, \mathcal{O})$. If $V(x_0)$ is an arbitrary neighborhood, then there exists a polycylinder P with $\tilde{\varphi}^{-1}(P) \subset\subset V \cap U$. Then $\tilde{\varphi}^{-1}(P)$ is Stein. Therefore there exist arbitrarily small Stein neighborhoods and hence arbitrarily small Stein coverings. □

Theorem 5.7. *Let X be a Stein manifold, $\mathcal{S}$ a coherent analytic sheaf over X, $\mathfrak{U}$ an arbitrary open covering of X. Then $H^1(\mathfrak{U}, \mathcal{S}) = 0$. In particular, every Cousin I distribution over X is solvable.*

Proof. $H^1(X, \mathcal{S}) = 0$ and $\varphi_1 : H^1(\mathfrak{U}, \mathcal{S}) \to H^1(X, \mathcal{S})$ is injective. □

Theorem 5.8. *If X is Stein, then for all $\ell \geqslant 1$, $H^\ell(X, \mathcal{O}^*) \simeq H^{\ell+1}(X, \mathbb{Z})$.*

Proof. Theorem B and Theorem 4.6. □

Theorem 5.9. *Let X be Stein, $(U_\iota, f_\iota)_{\iota \in I}$ a Cousin II distribution on X, $h \in Z^1(\mathfrak{U}, \mathcal{O}^*)$ the corresponding cocycle. Then $(U_\iota, f_\iota)_{\iota \in I}$ is solvable if and only if $c(h) = 0$.*

Proof. Theorem B and Theorem 4.9. □

At the end of the last section we gave an example of a Stein manifold on which not every Cousin II problem is solvable. Let us assume the following two (topological) results without proof:

1. If X is a connected non-compact Riemann surface (X is then Stein by a theorem of Behnke–Stein), then $H^2(X, \mathbb{Z}) = 0$.
2. If X is a Stein manifold which is continuously contractible to a point, then $H^2(X, \mathbb{Z}) = 0$.

Theorem 5.10. *If X is a Stein manifold and $H^2(X, \mathbb{Z}) = 0$, then every Cousin II problem on X is solvable.*

Proof. Immediate corollary of Theorem 5.9. □

Therefore, every Cousin II problem on X is solvable if X is a non-compact connected Riemann surface or an arbitrary contractible Stein manifold. Specifically it follows that if $G \subset \mathbb{C}$ is a domain, then all Mittag–Leffler and Weierstrass problems on G are solvable. So far we have only used Theorem B. Interesting possibilities for applications of Theorem A are found primarily in the area of analytic subsets of Stein manifolds.

Def. 5.2. Let A be an analytic subset of a complex manifold X. A complex valued function f on A is called *holomorphic* if for every point $x_0 \in A$ there is a neighborhood $U(x_0) \subset X$ and a holomorphic function $\hat{f}$ on U with $\hat{f}|U \cap A = f|U \cap A$.

For analytic sets which are free of singularities (therefore submanifolds) this coincides with the old notion of holomorphy.

Theorem 5.11. *Let $(X, \mathcal{O})$ be a Stein manifold, $A \subset X$ an analytic subset and f a function holomorphic on A. Then there is a holomorphic function $\hat{f}$ on X with $\hat{f}|A = f$. (Global continuation!)*

PROOF. We assign to every point $x \in A$ a neighborhood $U_x \subset X$ and a holomorphic function $\tilde{f}_x$ such that $\tilde{f}_x|A \cap U_x = f|A \cap U_x$. To every point $x \in X - A$ let there be assigned a neighborhood $U_x \subset X$ with $U_x \cap A = \varnothing$ and the function $\tilde{f}_x := 0|U_x$. Let

$$\mathfrak{U} := (U_x)_{x \in X}, \ \eta(x) := \tilde{f}_x \in \Gamma(U_x, \mathcal{O}).$$

Then $\eta \in C^0(\mathfrak{U}, \mathcal{O})$ and $\xi := \delta\eta \in Z^1(\mathfrak{U}, \mathcal{O})$. Moreover, for all $x_0, x_1 \in X$

$$\xi(x_0, x_1)|A \cap U_{x_0x_1} = \tilde{f}_{x_0}|A \cap U_{x_0x_1} - \tilde{f}_{x_1}|A \cap U_{x_0x_1} = 0.$$

Therefore $\xi \in Z^1(\mathfrak{U}, \mathcal{I}(A))$, where we denote the ideal sheaf of A by $\mathcal{I}(A)$. By Theorem B, $H^1(X, \mathcal{I}(A)) = 0$ and hence also $H^1(\mathfrak{U}, \mathcal{I}(A)) = 0$. Therefore there is a $\rho \in C^0(\mathfrak{U}, \mathcal{I}(A))$ with $\delta\rho = \xi$, that is, $\delta(\eta - \rho) = 0$. There is a holomorphic function $\hat{f} \in \Gamma(X, \mathcal{O})$ defined by

$$\hat{f}|U_x := \eta(x) - \rho(x) = \tilde{f}_x - \rho(x)$$

and

$$\hat{f}|A \cap U_x = \tilde{f}_x|A \cap U_x = f|A \cap U_x.$$

That is, $\hat{f}|A = f$. □

Theorem 5.12. *Let $(X, \mathcal{O})$ be Stein, $X' \subset\subset X$ open, $\mathcal{S}$ a coherent analytic sheaf over X. Then there are sections $s_1, \ldots, s_\ell \in \Gamma(X, \mathcal{S})$ which at each point $x \in X'$ generate the stalk $\mathcal{S}_x$ over $\mathcal{O}_x$.*

PROOF

1. Let $x_0 \in \bar{X}'$. Then there exists an open neighborhood $U(x_0) \subset X$ and sections $t_1, \ldots, t_q \in \Gamma(U, \mathcal{S})$ such that for every point $x \in U$ the stalk $\mathcal{S}_x$ over $\mathcal{O}_x$ is generated by $t_1(x), \ldots, t_q(x)$. Now, by Theorem A there are global sections $s_1, \ldots, s_p \in \Gamma(X, \mathcal{S})$ and elements $a_{ij} \in \mathcal{O}_{x_0}$ such that

$$t_i(x_0) = \sum_{j=1}^{p} a_{ij}s_j(x_0) \qquad \text{for } i = 1, \ldots, q.$$

There exists an open neighborhood $V(x_0) \subset U$ and sections $\hat{a}_{ij} \in \Gamma(V, \mathcal{O})$ with $\hat{a}_{ij}(x_0) = a_{ij}$ for all i, j. Hence it follows that there exists an open neighborhood $W(x_0) \subset V$ with $t_i\Big|W = \left(\sum_{j=1}^{p} \hat{a}_{ij}s_j\right)\Big|W$ for $i = 1, \ldots, q$; that is, $s_1, \ldots, s_p$ generate each stalk $\mathcal{S}_x$, $x \in W$.

2. Since $\bar{X}'$ is compact, we can find finitely many points $x_1, \ldots, x_r \in \bar{X}'$, open neighborhoods $W_i(x_i)$ and global sections

$$s_1^{(i)}, \ldots, s_{p(i)}^{(i)}, \qquad i = 1, \ldots, r$$

such that $W_1 \cup \cdots \cup W_r$ covers $\bar{X}'$;

$$s_1^{(i)}, \ldots, s_{p(i)}^{(i)}, \qquad i = 1, \ldots, r,$$

generate $\mathscr{S}$ on W_i. Then

$$s_1^{(i)}, \ldots, s_{p(i)}^{(i)}, \qquad i = 1, \ldots, r,$$

generate the sheaf $\mathscr{S}$ on $\bar{X}'$. □

Theorem 5.13. *Let $(X, \mathcal{O})$ be Stein, $X' \subset\subset X$ open, $A \subset X$ analytic. Then there are holomorphic functions $f_1, \ldots, f_\ell$ on X such that*

$$A \cap X' = \{x \in X' : f_1(x) = \cdots = f_\ell(x) = 0\}.$$

Proof. Since $\mathscr{I}(A)$ is a coherent analytic sheaf on X, by Theorem 5.12 there exist global sections $f_1, \ldots, f_\ell \in \Gamma(X, \mathscr{I}(A)) \subset \Gamma(X, \mathcal{O})$ which generate each stalk of $\mathscr{I}(A)$ over X'. Clearly

$$A \cap X' \subset \{x \in X' : [f_1(x)] = \cdots = [f_\ell(x)] = 0\},$$

so we need only show the converse. (Recall that for an element $f \in \Gamma(X, \mathcal{O})$ the corresponding holomorphic function is denoted by $[f]$.)

If $x_0 \in X' - A$, then there are elements $a_\nu \in \mathcal{O}_{x_0}$ with $\sum_{\nu=1}^{\ell} a_\nu f_\nu(x_0) = 1 \in \mathcal{O}_{x_0}$. Then in a neighborhood $V(x_0) \subset X' - A$ the function 1 has the representation $1 = \sum_{\nu=1}^{\ell} \hat{a}_\nu [f_\nu]$, where the $\hat{a}_\nu$ are holomorphic functions in V. But then not all the $[f_\nu]$ can vanish at x_0.

Therefore

$$\{x \in X' : [f_1(x)] = \cdots = [f_\ell(x)] = 0\} \subset A \cap X'.$$ □

We record the following sharpened version of Theorem 5.13 without proof.

Theorem 5.14. *Let X be an n-dimensional Stein manifold, $A \subset X$ an analytic subset. Then there exist holomorphic functions $f_1, \ldots, f_{n+1}$ on X such that $A = \{x \in X : f_1(x) = \cdots = f_{n+1}(x) = 0\}$.*

We note that the theorem does not imply that $\mathscr{I}(A)$ is globally finitely generated. Indeed, there is an example due to Cartan which shows that this is not possible, in general.

CHAPTER VII
Real Methods

1. Tangential Vectors

In this section X is always an n-dimensional complex manifold.

Def. 1.1. Let $k \in \mathbb{N}_0$. A *k-times differentiable local function* at $x_0 \in X$ is a pair (U, f) such that:

1. U is an open neighborhood of x_0 in X;
2. f is real-valued function on U continuous at x_0; and
3. there exist a neighborhood $V(x_0) \subset U$ and a biholomorphic mapping $\psi: V \to G \subset \mathbb{C}^n$ such that $f \circ \psi^{-1}$ at $\psi(x_0)$ is k-times differentiable.

Complex valued local functions can be defined correspondingly.

Let the set of all k-times differentiable functions at x_0 be denoted by $\mathscr{D}^k_{x_0}$. Instead of (U, f) we usually write f.

Remark. Since the coordinate transformations are biholomorphic, so in particular k-times differentiable for every k, Definition 1.1 is independent of the choice of the coordinate system (V, ψ). The elements of $\mathscr{D}^k_{x_0}$ can be added and multiplied by real or complex scalars in the obvious fashion. (For example $(U, f) + (U', f') := (U \cap U', f + f')$.)

A well-known theorem says that if $f \in \mathscr{D}^1_{x_0}$, $g \in \mathscr{D}^0_{x_0}$ and $f(x_0) = g(x_0) = 0$, then $f \cdot g \in \mathscr{D}^1_{x_0}$ (see [21]).

Def. 1.2. A *(real) tangent vector at* x_0 is a mapping $D: \mathscr{D}^1_{x_0} \to \mathbb{R}$ such that:

1. D is $\mathbb{R}$-linear;
2. $D(1) = 0$; and
3. $D(f \cdot g) = 0$ for $f \in \mathscr{D}^1_{x_0}$ and $g \in \mathscr{D}^0_{x_0}$ such that $f(x_0) = g(x_0) = 0$.

We call (2) and (3) the *derivation properties*. The set of all tangent vectors at x_0 is denoted by T_{x_0}.

Remark. T_{x_0} forms a real vector space. The partial derivatives

$$\frac{\partial}{\partial x_1}, \ldots, \frac{\partial}{\partial x_n}, \frac{\partial}{\partial y_1}, \ldots, \frac{\partial}{\partial y_n}$$

which depend on the choice of the coordinate system form a basis for T_{x_0} (see [21]). Therefore $\dim_{\mathbb{R}} T_{x_0} = 2n$. For complex valued local functions $f = g + ih$ at x_0 and $D \in T_{x_0}$ we set $D(f) := D(g) + iD(h)$. D remains $\mathbb{R}$-linear!

Theorem 1.1. *If $c_1, \ldots, c_n$ are arbitrary complex numbers then there exists exactly one tangent vector D with $D(f) = \sum_{\nu=1}^{n} c_\nu \frac{\partial}{\partial x_\nu}(f)$ for each function f holomorphic at x_0. In particular, a given tangent vector D is already uniquely determined by its values on the holomorphic functions. In local coordinates D has the representation*

$$D = \sum_{\nu=1}^{n} \operatorname{Re}(D(z_\nu)) \frac{\partial}{\partial x_\nu} + \sum_{\nu=1}^{n} \operatorname{Im}(D(z_\nu)) \frac{\partial}{\partial y_\nu}$$

PROOF. If $c_\nu = a_\nu + ib_\nu$ for $\nu = 1, \ldots, n$, we set

$$D := \sum_{\nu=1}^{n} a_\nu \frac{\partial}{\partial x_\nu} + \sum_{\nu=1}^{n} b_\nu \frac{\partial}{\partial y_\nu}.$$

Then for each function f holomorphic at x_0 (because $f_{y_\nu} = if_{x_\nu}$)

$$D(f) = \sum_{\nu=1}^{n} a_\nu f_{x_\nu} + \sum_{\nu=1}^{n} b_\nu f_{y_\nu} = \sum_{\nu=1}^{n} (a_\nu + ib_\nu) f_{x_\nu} = \sum_{\nu=1}^{n} c_\nu \frac{\partial}{\partial x_\nu}(f).$$

Hence $c_\nu = D(z_\nu)$ for $\nu = 1, \ldots, n$. It is clear that D is uniquely determined by its values on the holomorphic functions as well as by the numbers $c_1, \ldots, c_n$. □

Theorem 1.2. *If $c \in \mathbb{C}$ and $D \in T_{x_0}$, then there exists exactly one tangent vector $c \cdot D \in T_{x_0}$ such that $(c \cdot D)(f) = c \cdot (D(f))$ for every function f holomorphic at x_0.*

PROOF. There exist complex numbers $c_1, \ldots, c_n$ such that

$$D(f) = \sum_{\nu=1}^{n} c_\nu \frac{\partial}{\partial x_\nu}(f)$$

for every function f holomorphic at x_0; and by Theorem 1.1 there is exactly one tangent vector D^* with $D^*(f) = \sum_{\nu=1}^{n} (cc_\nu) \frac{\partial}{\partial x_\nu}(f) = c \cdot (D(f))$ for holomorphic f. We set $c \cdot D = D^*$. □

Theorem 1.3. *Let*

$$i \cdot \frac{\partial}{\partial x_\nu} = \frac{\partial}{\partial y_\nu} \quad \text{and} \quad i \cdot \frac{\partial}{\partial y_\nu} = -\frac{\partial}{\partial x_\nu} \quad \text{for} \quad \nu = 1, \ldots, n.$$

Then T_{x_0} is an n-dimensional complex vector space with basis $\{\partial/\partial x_1, \ldots, \partial/\partial x_n\}$, and this new complex structure is compatible with the given real structure on T_{x_0}.

Proof. If f is holomorphic at x_0, then

$$\left(i \cdot \frac{\partial}{\partial x_\nu}\right)(f) = i\left(\frac{\partial}{\partial x_\nu}(f)\right) = i \cdot f_{z_\nu} = \frac{\partial}{\partial y_\nu}(f).$$

The axioms of a $\mathbb{C}$-vector space are clearly satisfied; in particular

$$i \cdot \frac{\partial}{\partial y_\nu} = i \cdot \left(i \cdot \frac{\partial}{\partial x_\nu}\right) = (i \cdot i) \cdot \frac{\partial}{\partial x_\nu} = -\frac{\partial}{\partial x_\nu}.$$

Therefore

$$\left\{\frac{\partial}{\partial x_1}, \ldots, \frac{\partial}{\partial x_n}\right\}$$

forms a system of generators of T_{x_0} over $\mathbb{C}$.

If $\sum_{\nu=1}^{n} c_\nu \cdot \frac{\partial}{\partial x_\nu} = 0$ with $c_\nu = a_\nu + ib_\nu$ for $\nu = 1, \ldots, n$, then

$$0 = \sum_{\nu=1}^{n} a_\nu \frac{\partial}{\partial x_\nu} + i \cdot \sum_{\nu=1}^{n} b_\nu \frac{\partial}{\partial x_\nu} = \sum_{\nu=1}^{n} a_\nu \frac{\partial}{\partial x_\nu} + \sum_{\nu=1}^{n} b_\nu \frac{\partial}{\partial y_\nu},$$

therefore $a_\nu = b_\nu = 0$ for $\nu = 1, \ldots, n$. That is, $\{\partial/\partial x_1, \ldots, \partial/\partial x_n\}$ is a basis for T_{x_0} over $\mathbb{C}$. □

Remark. A *complex tangent vector at* x_0 is a $\mathbb{C}$-linear mapping $D: \mathscr{D}^1_{x_0} \to \mathbb{C}$ with the derivation properties (2) and (3) of Def. 1.2. Let the set of all complex tangent vectors at x_0 be denoted by $T^c_{x_0}$. Then we set

$$T'_{x_0} := \{D \in T^c_{x_0} : D(\bar{f}) = 0 \quad \text{if } f \text{ is holomorphic at } x_0\},$$
$$T''_{x_0} := \{D \in T^c_{x_0} : D(f) = 0 \quad \text{if } f \text{ is holomorphic at } x_0\}.$$

We call the elements of T'_{x_0} *holomorphic tangent vectors*, the elements of T''_{x_0} *antiholomorphic tangent vectors*. The partial derivatives $\partial/\partial z_1, \ldots, \partial/\partial z_n$ resp. $\partial/\partial \bar{z}_1, \ldots, \partial/\partial \bar{z}_n$ form a basis for T'_{x_0} resp. T''_{x_0}, and $T^c_{x_0} = T'_{x_0} \oplus T''_{x_0}$.

We can now assign to every element $D \in T_{x_0}$ complex tangent vectors $D' \in T'_{x_0}$ and $D'' \in T''_{x_0}$ such that $D = D' + D''$. If

$$D = \sum_{\nu=1}^{n} a_\nu \frac{\partial}{\partial x_\nu} + \sum_{\nu=1}^{n} b_\nu \frac{\partial}{\partial y_\nu},$$

we set

$$D' := \frac{1}{2} \sum_{\nu=1}^{n} (a_\nu + ib_\nu) \frac{\partial}{\partial z_\nu}$$

$$D'' := \frac{1}{2} \sum_{\nu=1}^{n} (a_\nu - ib_\nu) \frac{\partial}{\partial \bar{z}_\nu}$$

Clearly $D'(f) + D''(f) = D(f)$ for every $f \in \mathscr{D}^1_{x_0}$. Hence we can write every real tangent vector $D \in T_{x_0}$ in the form

$$D = \sum_{\nu=1}^{n} c_\nu \frac{\partial}{\partial z_\nu} + \sum_{\nu=1}^{n} \bar{c}_\nu \frac{\partial}{\partial \bar{z}_\nu}.$$

If $c \in \mathbb{C}$, then

$$c \cdot D = \sum_{\nu=1}^{n} cc_\nu \frac{\partial}{\partial z_\nu} + \sum_{\nu=1}^{n} \overline{cc_\nu} \frac{\partial}{\partial \bar{z}_\nu}.$$

Def. 1.3. An *r-dimensional complex differential form* at x_0 is an alternating $\mathbb{R}$-multilinear mapping

$$\varphi: \underbrace{T_{x_0} \times \cdots \times T_{x_0}}_{r\text{-times}} \to \mathbb{C}$$

The set of all r-dimensional complex differential forms at x_0 is denoted by $F^{(r)}_{x_0}$.

Remarks

1. $F^{(r)}_{x_0}$ is a complex vector space. We can represent an element $\varphi \in F^{(r)}_{x_0}$ uniquely in the form $\varphi = \mathrm{Re}(\varphi) + i\,\mathrm{Im}(\varphi)$, where $\mathrm{Re}(\varphi)$ and $\mathrm{Im}(\varphi)$ are real-valued differential forms (cf. [22]). It follows directly that

$$\dim_{\mathbb{R}} F^{(r)}_{x_0} = \binom{2n}{r} + \binom{2n}{r},$$

so that

$$\dim_{\mathbb{C}} F^{(r)}_{x_0} = \binom{2n}{r}.$$

2. By convention $F^{(0)}_{x_0} = \mathbb{C}$. For $r = 1$ we obtain $F^{(1)}_{x_0} = T^*_{x_0} \oplus iT^*_{x_0}$, with $T^*_{x_0} = \mathrm{Hom}_{\mathbb{R}}(T_{x_0}, \mathbb{R})$. $F^{(1)}_{x_0}$ is the *complexification* of the real dual space of T_{x_0}.

3. We associate with each element $\varphi \in F^{(r)}_{x_0}$ a complex-conjugate element $\bar{\varphi} \in F^{(r)}_{x_0}$ by setting

$$\bar{\varphi}(\xi_1, \ldots, \xi_r) := \overline{\varphi(\xi_1, \ldots, \xi_r)}$$

We have

a. $\overline{\overline{\varphi}} = \varphi$.
b. $\overline{(\varphi + \psi)} = \overline{\varphi} + \overline{\psi}$, $\overline{c\varphi} = \overline{c} \cdot \overline{\varphi}$.
c. φ is real if and only if $\varphi = \overline{\varphi}$.

If we define the element $dz_\nu \in F_{x_0}^{(1)}$ by $dz_\nu(\xi) := \xi(z_\nu)$, then we obtain an additional element $d\overline{z}_\nu \in F_{x_0}^{(1)}$ from

$$d\overline{z}_\nu(\xi) := \overline{dz}_\nu(\xi) = \overline{dz_\nu(\xi)} = \overline{\xi(z_\nu)} = \xi(\overline{z}_\nu).$$

$\{dz_1, \ldots, dz_n, d\overline{z}_1, \ldots, d\overline{z}_n\}$ is a basis of $F_{x_0}^{(1)}$. In general $\overline{\varphi} = \operatorname{Re}(\varphi) - i\operatorname{Im}(\varphi)$; as a special case

$$dz_\nu = dx_\nu + i\,dy_\nu,\; d\overline{z}_\nu = dx_\nu - i\,dy_\nu.$$

4. Let $\varphi \in F_{x_0}^{(r)}$, $\psi \in F_{x_0}^{(s)}$. The wedge product $\varphi \wedge \psi \in F_{x_0}^{(r+s)}$ is defined as in [22]:

$$\varphi \wedge \psi(\xi_1, \ldots, \xi_r, \xi_{r+1}, \ldots, \xi_{r+s}) := \frac{1}{r!s!} \sum_{\sigma \in \mathfrak{S}_{r+s}} (\operatorname{sgn} \sigma)\varphi(\xi_{\sigma(1)}, \ldots, \xi_{\sigma(r)}) \cdot \psi(\xi_{\sigma(r+1)}, \ldots, \xi_{\sigma(r+s)}).$$

Then:

a. $\varphi \wedge \psi = (-1)^{r \cdot s}\psi \wedge \varphi$ (anticommutative property);
b. $(\varphi \wedge \psi) \wedge \omega = \varphi \wedge (\psi \wedge \omega)$ (associative property).

In particular

$$dz_\nu \wedge dz_\nu = 0 = d\overline{z}_\nu \wedge d\overline{z}_\nu.$$

With the multiplication "$\wedge$" $F_{x_0}^{\bullet} := \bigoplus_{r=0}^{\infty} F_{x_0}^{(r)}$ becomes a *graded associative (non-commutative) ring with* 1.

5. For $j = 1, \ldots, n$ let $dz_{n+j} := d\overline{z}_j$. Then $F_{x_0}^{(r)}$ is generated by the elements $dz_{\nu_1} \wedge \cdots \wedge dz_{\nu_r}$ with $1 \leqslant \nu_1 < \cdots < \nu_r \leqslant 2n$. The number of these elements is exactly $\binom{2n}{r}$; so they form a basis.

Theorem 1.4. *If $z_1, \ldots, z_n$ are coordinates of X near x_0 and if $\varphi \in F_{x_0}^{(r)}$, then there is a uniquely determined representation*

$$\varphi = \sum_{1 \leqslant i_1 < \cdots < i_r \leqslant 2n} a_{i_1 \cdots i_r}\, dz_{i_1} \wedge \cdots \wedge dz_{i_r}$$

(normal form of φ with respect to $z_1, \ldots, z_n$). In particular $\varphi = 0$ for $r > 2n$; therefore $F_{x_0}^{(r)} = 0$ for $r > 2n$.

Def. 1.4. Let $p, q \in \mathbb{N}_0$ and $p + q = r$. $\varphi \in F_{x_0}^{(r)}$ is called a *form of type* (p, q) if

$$\varphi(c\xi_1, \ldots, c\xi_r) = c^p \cdot \overline{c}^q \varphi(\xi_1, \ldots, \xi_r).$$

for all $c \in \mathbb{C}$.

Theorem 1.5. *If $\varphi \in F_{x_0}^{(r)}$, $\varphi \neq 0$ and φ is of type (p, q), then p and q are uniquely determined.*

Proof. Suppose φ is of type (p, q) and of type (p', q'). Since $\varphi \neq 0$ there exist tangent vectors $\xi_1, \ldots, \xi_r$ such that $\varphi(\xi_1, \ldots, \xi_r) \neq 0$. Then

$$\varphi(c\xi_1, \ldots, c\xi_r) = \begin{cases} c^p \bar{c}^q \varphi(\xi_1, \ldots, \xi_r) \\ c^{p'} \bar{c}^{q'} \varphi(\xi_1, \ldots, \xi_r) \end{cases}$$

Therefore $c^{p'}\bar{c}^{q'} = c^{p'}c^{q'}$ for each $c \in \mathbb{C}$. Let $c = e^{i\theta}$ with arbitrary $\theta \in \mathbb{R}$. Then $e^{i\theta(p-q)} = e^{i\theta(p'-q')}$. That can hold for all θ only when $p - q = p' - q'$. Since $p + q = p' + q' = r$ by assumption, it follows that $p = p'$, $q = q'$. □

Theorem 1.6.

1. *If φ is of type (p, q), then $\bar{\varphi}$ is of type (q, p).*
2. *If φ, ψ are of type (p, q), $c \in \mathbb{C}$, then $\varphi + \psi$ and $c \cdot \varphi$ are of type (p, q).*
3. *If φ is of type (p, q), ψ of type (p', q'), then $\varphi \wedge \psi$ is of type $(p + p', q + q')$.*

Proof

(1) $\bar{\varphi}(c\xi_1,\ldots,c\xi_r) = \overline{\varphi(c\xi_1,\ldots,c\xi_r)} = \overline{c^p\bar{c}^q\varphi(\xi_1,\ldots,\xi_r)} = \bar{c}^p c^q \bar{\varphi}(\xi_1, \ldots, \xi_r)$.

(2) Trivial.

(3) $\varphi(c\xi_1,\ldots,c\xi_r)\psi(c\xi_{r+1},\ldots,c\xi_{r+s}) = c^p\bar{c}^q c^{p'}\bar{c}^{q'}\varphi(\xi_1,\ldots,\xi_r)\psi(\xi_{r+1},\ldots,\xi_{r+s})$.

Therefore

$$\varphi \wedge \psi(c\xi_1, \ldots, c\xi_{r+s}) = \frac{1}{r!s!} c^{p+p'}\bar{c}^{q+q'} \sum_{\sigma \in \mathfrak{S}_{r+s}} (\operatorname{sgn} \sigma)\varphi(\xi_{\sigma(1)}, \ldots, \xi_{\sigma(r)})$$
$$\times \psi(\xi_{\sigma(r+1)}, \ldots, \xi_{\sigma(r+s)}) = c^{p+p'}\bar{c}^{q+q'} \cdot \varphi \wedge \psi(\xi_1, \ldots, \xi_{r+s}). \quad \square$$

Theorem 1.7. *If $\varphi \in F_{x_0}^{(r)}$, then φ has a uniquely determined representation*

$$\varphi = \sum_{p+q=r} \varphi^{(p,q)}$$

where $\varphi^{(p,q)} \in F_{x_0}^{(r)}$ are forms of the type (p, q).

Proof. Clearly dz_ν is of type $(1, 0)$, $d\bar{z}_\nu$ of type $(0, 1)$. Hence it follows that monomials $dz_{i_1} \wedge \cdots \wedge dz_{i_p} \wedge d\bar{z}_{j_1} \wedge \cdots \wedge d\bar{z}_{j_q}$ (with $1 \leqslant i_1 < \cdots < i_p \leqslant n$ and $1 \leqslant j_1 < \cdots < j_q \leqslant n$) are forms of type (p, q).

$$\varphi = \sum_{p+q=r} \varphi^{(p,q)}$$

with

$$\varphi^{(p,q)} := \sum_{\substack{1 \leqslant i_1 < \cdots < i_p \leqslant n \\ 1 \leqslant j_1 < \cdots < j_q \leqslant n}} a_{i_1 \cdots i_p, n+j_1, \cdots, n+j_q} dz_{i_1} \wedge \cdots \wedge dz_{i_p} \wedge d\bar{z}_{j_1} \wedge \cdots \wedge d\bar{z}_{j_q}$$

is therefore a representation of the desired sort. Let

$$\varphi = \sum_{p+q=r} \varphi^{(p,q)} = \sum_{p+q=r} \tilde{\varphi}^{(p,q)}.$$

Then

$$\sum_{p+q=r} \psi^{(p,q)} = 0 \qquad \text{for} \qquad \psi^{(p,q)} = \varphi^{(p,q)} - \tilde{\varphi}^{(p,q)}$$

It follows that

$$0 = \sum_{p+q=r} \psi^{(p,q)}(c\xi_1, \ldots, c\xi_r) = \sum_{p+q=r} c^p \bar{c}^q \psi^{(p,q)}(\xi_1, \ldots, \xi_r)$$

For fixed $(\xi_1, \ldots, \xi_r)$ we obtain a polynomial equation in the polynomial ring $\mathbb{C}[c, \bar{c}]$. Then the coefficients $\psi^{(p,q)}(\xi_1, \ldots, \xi_r)$ also vanish for all p, q. Since we can choose $\xi_1, \ldots, \xi_r$ arbitrarily, we have $\varphi^{(p,q)} = \tilde{\varphi}^{(p,q)}$ for all p, q. □

2. Differential Forms on Complex Manifolds

Def. 2.1. Let X be a complex manifold. An *ℓ-form on X* is a mapping

$$\varphi : X \to \dot{\bigcup_{x \in X}} F_x^{(\ell)}$$

with the property that $\varphi(x) \in F_x^{(\ell)}$ for every $x \in X$. If $z_1, \ldots, z_n$ are coordinates on an open subset $U \subset X$, then for $x \in U$

$$\varphi_x := \varphi(x) = \sum_{1 \leq \iota_1 < \cdots < \iota_\ell \leq 2n} a_{\iota_1 \ldots \iota_\ell}(x)\, dz_{\iota_1} \wedge \cdots \wedge dz_{\iota_\ell}.$$

$x \mapsto a_{\iota_1 \ldots \iota_\ell}(x)$ defines a complex valued function $a_{\iota_1 \ldots \iota_\ell}$ on U. We call φ *k-times differentiable at* $x_0 \in U$ if all functions $a_{\iota_1 \ldots \iota_\ell}$ are k-times differentiable at x_0. This definition is independent of the choice of coordinates. φ is called *k-times differentiable* (*on* X) if φ is k-times differentiable at every point of X.

Henceforth the set of all arbitrarily often differentiable ℓ-forms will be denoted by $A^{(\ell)} = A^{(\ell)}(X)$, by $A^{(p,q)}$ the set of all arbitrarily often differentiable forms of the type (p, q).

Def. 2.2. If f is an arbitrarily often differentiable function on X (therefore an element of $A^{(0)}$), then we define an element $df \in A^{(1)}$ by $(df)_x(\xi) := \xi(f)$ for $\xi \in T_x$ (*total differential of f*).

Remarks

1. For the basis elements dz_ν, $d\bar{z}_\nu$ the definition does not change anything.
2. In local coordinates

$$df = \sum_{\nu=1}^{n} f_{z_\nu}\, dz_\nu + \sum_{\nu=1}^{n} f_{\bar{z}_\nu}\, d\bar{z}_\nu.$$

PROOF. We write

$$df = \sum_{\nu=1}^{n} a_\nu\, dz_\nu + \sum_{\nu=1}^{n} b_\nu\, d\bar{z}_\nu.$$

If $\xi \in T_{x_0}$, then

$$\xi = \sum_{\nu=1}^{n} c_\nu \frac{\partial}{\partial z_\nu} + \sum_{\nu=1}^{n} \bar{c}_\nu \frac{\partial}{\partial \bar{z}_\nu},$$

and hence

$$df(\xi) = \xi(f) = \sum_{\nu=1}^{n} c_\nu f_{z_\nu} + \sum_{\nu=1}^{n} \bar{c}_\nu f_{\bar{z}_\nu}.$$

In particular it follows that

$$f_{z_\nu} = df\left(\frac{\partial}{\partial z_\nu}\right) = a_\nu, \qquad f_{\bar{z}_\nu} = df\left(\frac{\partial}{\partial \bar{z}_\nu}\right) = b_\nu. \qquad \square$$

We can also define a total differential $d: A^{(\ell)} \to A^{(\ell+1)}$ on manifolds. It has the following properties:

1. d is $\mathbb{C}$-linear.
2. $d(f) = df$ (in the sense of Def. 2.2) for $f \in A^{(0)}$.
3. $d(\varphi|U) = (d\varphi)|U$.
4. If $\varphi \in A^{(r)}$, $\psi \in A^{(s)}$, then $d(\varphi \wedge \psi) = d\varphi \wedge \psi + (-1)^r \varphi \wedge d\psi$.
5. If
$$\varphi|U = \sum_{1 \leq \iota_1 < \cdots < \iota_\ell \leq 2n} a_{\iota_1 \cdots \iota_\ell} dz_{\iota_1} \wedge \cdots \wedge dz_{\iota_\ell},$$
then
$$d\varphi|U = \sum_{1 \leq \iota_1 < \cdots < \iota_\ell \leq 2n} da_{\iota_1 \cdots \iota_\ell} \wedge dz_{\iota_1} \wedge \cdots \wedge dz_{\iota_\ell}.$$
6. $d \circ d = 0$.
7. d is a real operator; that is, $d\bar{\varphi} = \overline{d\varphi}$. In particular then $d\varphi = d(\operatorname{Re}\varphi) + id(\operatorname{Im}\varphi)$.

Theorem 2.1. *If* $\varphi \in A^{(p,q)}$, *then* $d\varphi = d'\varphi + d''\varphi$ *with* $d'\varphi \in A^{(p+1,q)}$ *and* $d''\varphi \in A^{(p,q+1)}$.

Proof. One usually abbreviates the normal form of $\varphi^{(p,q)}$ as

$$\varphi^{(p,q)} = \sum_{I,J} a_{I,J} d\mathfrak{z}_I \wedge d\bar{\mathfrak{z}}_J.$$

Then

$$\begin{aligned} d\varphi^{(p,q)} &= \sum_{I,J} da_{I,J} \wedge d\mathfrak{z}_I \wedge d\bar{\mathfrak{z}}_J \\ &= \sum_{I,J} \sum_{\nu=1}^{n} \frac{\partial a_{I,J}}{\partial z_\nu} dz_\nu \wedge d\mathfrak{z}_I \wedge d\bar{\mathfrak{z}}_J + \sum_{I,J} \sum_{\nu=1}^{n} \frac{\partial a_{I,J}}{\partial \bar{z}_\nu} d\bar{z}_\nu \wedge d\mathfrak{z}_I \wedge d\bar{\mathfrak{z}}_J \end{aligned}$$

which is a decomposition of $d\varphi^{(p,q)}$ into a form $d'\varphi^{(p,q)}$ of type $(p+1, q)$ and a form $d''\varphi^{(p,q)}$ of type $(p, q+1)$. $\square$

If $\varphi = \sum_{p+q=1} \varphi^{(p,q)}$ is an arbitrary ℓ-form, then we call $d'\varphi := \sum_{p+q=1} d'\varphi^{(p,q)}$ the *total derivative of* φ *with respect to* $\mathfrak{z}$ and $d''\varphi := \sum_{p+q=1} d''\varphi^{(p,q)}$ the *total derivative with respect to* $\bar{\mathfrak{z}}$. (In the English literature one generally writes ∂ instead of d' and $\bar{\partial}$ instead of d''.)

Theorem 2.2.

1. *d' and d'' are $\mathbb{C}$-linear operators with $d' + d'' = d$.*
2. *$d'd' = 0$, $d''d'' = 0$ and $d'd'' + d''d' = 0$.*
3. *d', d'' are not real. Moreover $\overline{d'\varphi} = d''\bar{\varphi}$ and $\overline{d''\varphi} = d'\bar{\varphi}$.*
4. *If φ is an ℓ-form, ψ arbitrary, then*

$$d'(\varphi \wedge \psi) = d'\varphi \wedge \psi + (-1)^\ell \varphi \wedge d'\psi,$$
$$d''(\varphi \wedge \psi) = d''\varphi \wedge \psi + (-1)^\ell \varphi \wedge d''\psi.$$

Proof. It suffices to prove this for pure forms:

(1) is trivial. For (2):

$$0 = dd\varphi = (d' + d'') \circ (d' + d'')\varphi = d'd'\varphi + d'd''\varphi + d''d'\varphi + d''d''\varphi.$$

If φ has type (p, q), then $d'd'\varphi$ has type $(p + 2, q)$, $(d'd''\varphi + d''d'\varphi)$ has type $(p + 1, q + 1)$, and $d''d''\varphi$ has type $(p, q + 2)$. Since the decomposition into forms of pure type is uniquely determined, the proposition follows.

For (3), since $\overline{d\varphi} = d\bar{\varphi}$ it follows that

$$\overline{d'\varphi} + \overline{d''\varphi} = d'\bar{\varphi} + d''\bar{\varphi}; \qquad \text{therefore} \qquad (\overline{d'\varphi} - d''\bar{\varphi}) + (\overline{d''\varphi} - d'\bar{\varphi}) = 0.$$

Hence $\overline{d'\varphi} - d''\bar{\varphi}$ has type $(q, p + 1)$ and $\overline{d''\varphi} - d'\bar{\varphi}$ has type $(q + 1, p)$. Therefore both terms must vanish.

For (4), both formulas follow from Rule (4) for the total derivative d by comparing types as in (2) and (3). □

Remark. A real differentiable function f is holomorphic if and only if $f_{\bar{z}_\nu} = 0$ for $\nu = 1, \ldots, n$, that is if $d''f = 0$. Correspondingly it follows for

$$\varphi = \varphi^{(p,0)} = \sum_{1 \leq i_1 < \cdots < i_p \leq n} a_{i_1 \ldots i_p}\, dz_{i_1} \wedge \cdots \wedge dz_{i_p}$$

that $d''\varphi = 0$ if and only if $a_{i_1 \cdots i_p}$ is always holomorphic. Hence we make the following definitions.

Def. 2.3. $\varphi \in A^{(\ell)}$ is called *holomorphic* if

1. φ is of type $(p, 0)$, and
2. $d''\varphi = 0$.

$\varphi \in A^{(\ell)}$ is called antiholomorphic if

1. φ is of type $(0, q)$, and
2. $d'\varphi = 0$.

Remark. Clearly φ is antiholomorphic if and only if $\bar{\varphi}$ is holomorphic.

3. Cauchy Integrals

The Poincaré Lemma from real analysis (see, for example, [22]) can be formulated as follows:

Let $B \subset \mathbb{C}^n$ be a star-shaped region (for example, a polycylinder), $\varphi \in A^{(\ell)}$, $\ell > 0$, $d\varphi|B = 0$. Then there exists a $\psi \in A^{(\ell-1)}$ with $d\psi = \varphi$.

We will below prove a similar theorem for the d'' operator. In order to do this, we must first generalize the Cauchy integral formula.

If $B \subset\subset \mathbb{C}$ is a region and f a complex valued, continuous, bounded function on B, then there exists a continuous function $\mathrm{Ch}_f^{(B)}$ on $\mathbb{C}$ defined by

$$\mathrm{Ch}_f^{(B)}(w) := \frac{1}{2\pi i}\int_B \frac{f(z)}{z-w}\, dz \wedge d\bar{z}.$$

Specifically, let $\Phi: [0, \infty) \times [0, 2\pi) \to \mathbb{C}$ be defined by $\Phi(r, \theta) := \mathrm{r}e^{i\theta} + w$, and let B^* be the region $\Phi^{-1}(B)$. Then

$$\left(\frac{f(z)}{z-w}\, dz \wedge d\bar{z}\right) \circ \Phi = \frac{f(\Phi(r,\theta))}{\Phi(r,\theta)}\, d\Phi \wedge d\bar{\Phi}$$
$$= 2i \cdot f(r \cdot e^{i\theta} + w) \cdot e^{-i\theta}\, dr \wedge d\theta$$

is a continuous, bounded differential form on B^*. Hence

$$\frac{f(z)}{z-w}\, dz \wedge d\bar{z}$$

is integrable over B, the integral is continuously dependent on w, and

$$\mathrm{Ch}_f^{(B)}(w) = \pm\frac{1}{\pi}\int_{B^*} f(re^{i\theta} + w)e^{-i\theta}\, dr \wedge d\theta.$$

If the real numbers $R, k > 0$ are chosen so that $|z_1 - z_2| \leqslant R$ for $z_1, z_2 \in B$ and $|f(z)| \leqslant k$ for $z \in B$, then we get the following estimate:

$$|\mathrm{Ch}_f^{(B)}(w)| \leqslant \frac{k}{\pi}\int_{B^*} dr \wedge d\theta \leqslant 2kR.$$

Now let $P \subset \mathbb{C}$ be a circular disk (therefore a polycylinder), and $T := \partial P$. If g is holomorphic on $\bar{P}$, then the Cauchy integral formula holds:

$$g(w) = \mathrm{ch}(g|T)(w) = \frac{1}{2\pi i}\int_T \frac{g(z)}{z-w}\, dz, \qquad \text{for } w \in P.$$

As a generalization we obtain

Theorem 3.1. *Let g be continuously differentiable on P, $f: = g_{\bar z}$ bounded. Then for $w \in P$*

$$g(w) = \mathrm{ch}(g|T)(w) + \mathrm{Ch}_f^{(P)}(w).$$

Proof. Let $w \in P$, H_r a small circular disk about w with $H_r \subset\subset P$, and $T_r: = \partial H_r$. If T and T_r are given the usual orientation, then it follows from Stokes' theorem (see [22]) that:

$$\begin{aligned}
\mathrm{Ch}_f^{(P)}(w) &= -\frac{1}{2\pi i}\int_{P-H_r} d\left(\frac{g(z)}{z-w}\,dz\right) + \frac{1}{2\pi i}\int_{H_r}\frac{f(z)}{z-w}\,dz \wedge d\bar z \\
&= -\frac{1}{2\pi i}\int_{\partial(P-H_r)}\frac{g(z)}{z-w}\,dz + \mathrm{Ch}_f^{(H_r)}(w) \\
&= -\frac{1}{2\pi i}\int_{T}\frac{g(z)}{z-w}\,dz + \frac{1}{2\pi i}\int_{T_r}\frac{g(z)}{z-w}\,dz + \mathrm{Ch}_f^{(H_r)}(w) \\
&= -\mathrm{ch}(g|T)(w) + \mathrm{ch}(g|T_r)(w) + \mathrm{Ch}_f^{(H_r)}(w).
\end{aligned}$$

Hence the function $\rho(r): = \mathrm{ch}(g|T_r)(w) + \mathrm{Ch}_f^{(H_r)}(w)$ has the constant value $\mathrm{ch}(g|T)(w) + \mathrm{Ch}_f^{(P)}(w)$, and it suffices to consider the limit for $r \to 0$:

$$\rho(r) = a(r) + b(r) + c(r)$$

with

$$a(r): = \frac{1}{2\pi i}\int_{T_r}\frac{g(w)}{z-w}\,dz = g(w)\cdot\frac{1}{2\pi i}\int_{T_r}\frac{dz}{z-w} = g(w),$$

$$b(r): = \frac{1}{2\pi i}\int_{T_r}\frac{g(z)-g(w)}{z-w}\,dz \quad \text{and} \quad c(r): = \mathrm{Ch}_f^{(H_r)}(w).$$

Since g is continuously differentiable as a function of w there exist functions Δ', Δ'' which depend continuously on w such that

$$g(z) = g(w) + (z-w)\cdot\Delta'(z) + (\bar z - \bar w)\cdot\Delta''(z).$$

If we choose r_0 and M such that $|\Delta'(z)|, |\Delta''(z)| < M$ for $z \in H_r$ and $r \leqslant r_0$, then we get

$$\left|\frac{g(z)-g(w)}{z-w}\right| \leqslant |\Delta'(z)| + |\Delta''(z)|\cdot\left|\frac{\bar z - \bar w}{z-w}\right| \leqslant 2M \quad \text{for} \quad z \in T_r \quad \text{and} \quad r < r_0;$$

therefore

$$|b(r)| \leqslant \frac{1}{2\pi}\int_{T_r}\left|\frac{g(z)-g(w)}{z-w}\right| dz \leqslant 2M\cdot r \quad \text{for} \quad r < r_0.$$

Hence

$$|b(r) + c(r)| \leqslant 2Mr + |\mathrm{Ch}_f^{(H_r)}(w)| \leqslant 2r\cdot(M + 2\cdot\sup|f(\mathrm{P})|),$$

and this expression becomes arbitrarily small. Hence it follows that $\rho(r) \equiv g(w)$. □

Theorem 3.2. *Let f be continuously differentiable on $\mathbb{C}$, $\operatorname{Supp}(f) \subset\subset \mathbb{C}$, $P \subset \mathbb{C}$ a circular disk with $\operatorname{Supp}(f) \subset P$.*

Then $g := \mathrm{Ch}_f^{(P)}$ is continuously differentiable on $\mathbb{C}$, with $g_{\bar z} = f$.

PROOF. Let

$$P_c := \{z \in \mathbb{C} : z + c \in P\},$$

$$\gamma(w, c) := \mathrm{Ch}_f^{(P)}(w + c) = \frac{1}{2\pi i} \int_P \frac{f(z)}{z - w - c}\, dz \wedge d\bar z$$

$$= \frac{1}{2\pi i} \int_{P_c} \frac{f(z + c)}{z - w}\, dz \wedge d\bar z.$$

Because

$$\operatorname{Supp}(f) \subset P, \qquad \gamma(w, c) = \frac{1}{2\pi i} \int_{\mathbb{C}} \frac{f(z + c)}{z - w}\, dz \wedge d\bar z.$$

By known theorems on parametric integrals (see [22]), γ is continuously differentiable with respect to c and $\bar c$. Since $\gamma(0, c) = g(c)$, g is differentiable. Applying formulas for the derivative of parametric integrals and the chain rule gives

$$g_z(c) = \frac{1}{2\pi i} \int_{\mathbb{C}} \frac{f_z(z + c)}{z}\, dz \wedge d\bar z = \frac{1}{2\pi i} \int_P \frac{f_z(z)}{z - c}\, dz \wedge d\bar z = \mathrm{Ch}_{f_z}^{(P)}(c),$$

$$g_{\bar z}(c) = \frac{1}{2\pi i} \int_P \frac{f_{\bar z}(z)}{z - c}\, dz \wedge d\bar z = \mathrm{Ch}_{f_{\bar z}}^{(P)}(c).$$

Since f vanishes on $T := \partial P$, it also follows from Theorem 3.1 that $g_{\bar z} = \mathrm{Ch}_{f_{\bar z}}^{(P)} = f - \mathrm{ch}(f|T) = f$. □

Theorem 3.3. *Let $B \subset\subset \mathbb{C}$ be a region, f continuously differentiable and bounded on B.*

Then $g := \mathrm{Ch}_f^{(B)}$ is continuously differentiable on B and $g_{\bar z} = f$.

PROOF. Let $w_0 \in B$ be given, H an open circular disk about w_0 with $H \subset\subset B$. We can then find an arbitrarily often differentiable function $\rho : \mathbb{C} \to \mathbb{R}$ for which

1. $0 \leqslant \rho \leqslant 1$,
2. $\rho|H = 1$,
3. $\operatorname{Supp}(\rho) \subset\subset B$.

Then let $f_1 := \rho \cdot f$, $f_2 := f - f_1$. Clearly

$$f_1 + f_2 = f \quad \text{and} \quad \mathrm{Ch}_{f_1}^{(B)} + \mathrm{Ch}_{f_2}^{(B)} = \mathrm{Ch}_f^{(B)}.$$

Moreover, $f_1|H = f|H$ and $f_2|H = 0$. f_1 is even continuously differentiable on all of $\mathbb{C}$ and if P is a circular disk with $B \subset P$, then $\mathrm{Ch}_{f_1}^{(B)} = \mathrm{Ch}_{f_1}^{(P)}$. Hence it follows from Theorem 3.2 that $\mathrm{Ch}_{f_1}^{(B)}$ is continuously differentiable on $\mathbb{C}$ and $(\mathrm{Ch}_{f_1}^{(B)})_{\bar z} = f_1$.

For $w \in H$ we also have

$$\mathrm{Ch}_{f_2}^{(B)}(w) = \frac{1}{2\pi i}\int_B \frac{f_2(z)}{z - w}\,dz \wedge d\bar{z} = \frac{1}{2\pi i}\int_{B-H} \frac{f_2(z)}{z - w}\,dz \wedge d\bar{z},$$

the integrand is continuous and bounded on $B - H$, as well as holomorphic with respect to w. From the theory of parametric integrals it follows that $\mathrm{Ch}_{f_2}^{(B)}|H$ is continuously differentiable and

$$(\mathrm{Ch}_{f_2}^{(B)}|H)_{\bar{w}} = 0$$

Therefore $g|H$ is continuously differentiable and $(g|H)_{\bar{z}} = f|H$. □

Remark. If $\hat{B} \subset \mathbb{C}$, $B^* \subset \mathbb{R}^n$ are regions, $B \subset\subset \hat{B}$ open and $f:\hat{B} \times B^* \to \mathbb{C}$ arbitrarily often differentiable, then it follows from the theory of parametric integrals that $\mathrm{Ch}_f^{(B)}$ with $\mathrm{Ch}_f^{(B)}(w, \mathfrak{x})\colon = \frac{1}{2\pi i}\int_B \frac{f(z, \mathfrak{x})}{z - w}\,dz \wedge d\bar{z}$ is *arbitrarily often* differentiable on $B \times B^*$, and

$$(\mathrm{Ch}_f^{(B)})_{x_\nu}(w, \mathfrak{x}) = \frac{1}{2\pi i}\int_B \frac{f_{x_\nu}(z, \mathfrak{x})}{z - w}\,dz \wedge d\bar{z},\ (\mathrm{Ch}_f^{(B)})_{\bar{w}} = f.$$

4. Dolbeault's Lemma

Theorem 4.1 (Dolbeault's lemma): *Let $K_\nu \subset \mathbb{C}$ be compact sets for $\nu = 1, \ldots, n$, U_ν open neighborhoods of K_ν, $K\colon = K_1 \times \cdots K_n$, $U\colon = U_1 \times \cdots \times U_n$.*

Moreover, let $\varphi = \varphi^{(0,q)} \in A^{(0,q)}(U)$ with $d''\varphi = 0$, $q > 0$. Then there exist an open set U' with $K \subset U' \subset U$ and a $\psi \in A^{(0,q-1)}(U')$ with $d''\psi = \varphi|U'$.

If φ is arbitrarily often differentiable as a function of real parameters, then ψ is also arbitrarily often differentiable as a function of these parameters.

Proof. By induction on n.

1. If $n = 1$, then also $q = 1$ and φ has the form $\varphi = a(z, \mathfrak{x})\,d\bar{z}$. Let $U' \subset\subset U$ be open with $K \subset U'$. Then $\mathrm{Ch}_a^{(U')}$ is arbitrarily often differentiable, and

$$d''(\mathrm{Ch}_a^{(U')}) = (\mathrm{Ch}_a^{(U')})_{\bar{z}}\,d\bar{z} = a d\bar{z} = \varphi$$

(see Theorem 3.3 and Remark).

2. Now suppose the theorem proved for the case $n - 1$, $n > 1$. The operators d''_* and $\partial/\partial\bar{z}_1$ are defined by

$$d''_*\left(\sum_J a_J d\bar{\mathfrak{z}}_J\right)\colon = \sum_J \sum_{\nu=2}^{n} \frac{\partial a_J}{\partial \bar{z}_\nu}\,d\bar{z}_\nu \wedge d\bar{\mathfrak{z}}_J \quad\text{and}\quad \frac{\partial}{\partial \bar{z}_1}\left(\sum_J a_J d\bar{\mathfrak{z}}_J\right)\colon = \sum_J \frac{\partial a_J}{\partial \bar{z}_1}\,d\bar{\mathfrak{z}}_J$$

so that

$$d''\varphi = d''_*\varphi + d\bar{z}_1 \wedge \frac{\partial \varphi}{\partial \bar{z}_1}.$$

If we write φ in the form $\varphi = d\bar{z}_1 \wedge \varphi_1 + \varphi_2$, where φ_1, φ_2 no longer contain $d\bar{z}_1$, then

$$0 = d''\varphi = d\bar{z}_1 \wedge \left(-d''_*\varphi_1 + \frac{\partial \varphi_2}{\partial \bar{z}_1}\right) + d''_*\varphi_2.$$

Since $d''_*\varphi_2$ contains no $d\bar{z}_1$, it follows that $d''_*\varphi_2 = 0$.

Now regard z_1 as an additional parameter and apply the induction hypothesis.

Let

$$K_*: = K_2 \times \cdots \times K_n, U_*: = U_2 \times \cdots \times U_n.$$

There is an open set U'_* with $K_* \subset U'_* \subset U_*$ and a $\psi = \psi^{(0,q-1)}$ in which z_1 appears as a parameter, such that

$$d''_*\psi | U_1 \times U'_* = \varphi_2 | U_1 \times U'_*.$$

On $U' : = U_1 \times U'_*$ we have

$$\varphi - d''\psi = \varphi - d''_*\psi - d\bar{z}_1 \wedge \frac{\partial \psi}{\partial \bar{z}_1} = d\bar{z}_1 \wedge \left(\varphi_1 - \frac{\partial \psi}{\partial \bar{z}_1}\right),$$

where $\varphi_1 - (\partial\psi/\partial\bar{z}_1)$ contains no $d\bar{z}_1$. On the other hand

$$0 = d''(\varphi - d''\psi) = d\bar{z}_1 \wedge d''_* \left(\varphi_1 - \frac{\partial \psi}{\partial \bar{z}_1}\right);$$

therefore

$$d''_*[\varphi_1 - (\partial\psi/\partial\bar{z}_1)] = 0.$$

For the case $q \geqslant 2$ by the induction hypothesis there are an open set U''_* with $K_* \subset U''_* \subset U'_*$ and a $\tilde{\psi} = \tilde{\psi}^{(0,q-1)}$ on U''_* such that

$$d''_*\tilde{\psi} = \left(\varphi_1 - \frac{\partial \psi}{\partial \bar{z}_1}\right) \Big| U''_*.$$

Hence on $U'' : = U_1 \times U''_*$

$$d''(d\bar{z}_1 \wedge \tilde{\psi}) = -d\bar{z}_1 \wedge d''_*\tilde{\psi} = -d\bar{z}_1 \wedge \left(\varphi_1 - \frac{\partial \psi}{\partial \bar{z}_1}\right) = d''\psi - \varphi,$$

therefore $\varphi = d''(\psi - d\bar{z}_1 \wedge \tilde{\psi})$.

For $q = 1$, $\varphi_1 - (\partial\psi/\partial\bar{z}_1)$ is a function $a = a(z_1, z_2, \ldots, z_n)$ which is holomorphic in $z_2, \ldots, z_n$. We regard $z_2, \ldots, z_n$ as additional parameters and determine a region U'_1 with $K \subset U'_1 \subset\subset U_1$ by (1), and a function $f = \mathrm{Ch}_a^{(U'_1)}$ with $d''f = a\, d\bar{z}_1 = \varphi - d''\psi$. Then $\varphi = d''(f + \psi)$. This completes the proof. □

We immediately obtain the following result for manifolds X.

If $\varphi \in A^{(0,q)}(U)$, $q \geqslant 1$, $U \subset X$ open and $d''\varphi = 0$, then for every $x \in U$ there exists an open neighborhood $V(x) \subset U$ and a $\psi = \psi^{(0,q-1)}$ on V with $d''\psi = \varphi | V$. (Let $K = \{x\}$.) We present the following theorem without proof. It provides us with some more precise information.

Theorem 4.2 (Lieb). *Let $G \subset \mathbb{C}^n$ be a domain with smooth boundary ∂G (see Def. 2.5 in Chapt. II; the function φ defining the boundary are arbitrarily often differentiable). Let the Levi form of the defining functions be positive definite everywhere. (In such a case one calls G strongly pseudo-convex.)*

Let $\omega = \sum_J a_J \, d\bar{\mathfrak{z}}_J$ be an arbitrarily often differentiable form of type $(0, q)$ on G with $d''\omega = 0$. Moreover let there be given a real constant M with

$$\|\omega\|: = \max_J \sup |a_J(G)| \leqslant M.$$

Then there exists a constant k independent of ω and a form ψ of type $(0, q-1)$ on G with $d''\psi = \omega$ and $\|\psi\| \leqslant k \cdot M$.

According to Siu and Range there exists a generalization of this theorem for domains with piece-wise smooth boundaries (See R. M. Range, and Y. -T. Siu: Uniform estimates for the $\bar{\partial}$-equation on intersections of strictly pseudo-convex domains. *Bull. Amer. Math. Soc.*, *78*(5):721–722, 1972).

5. Fine Sheaves (Theorems of Dolbeault and de Rham)

In this section X is always a paracompact complex manifold.

Def. 5.1. A *test function on* X is an arbitrarily often differentiable function $t: X \to \mathbb{R}$ with compact support.

Let the ring (without 1) of all test functions be denoted by T. Let $\mathscr{T}$ be the sheaf of germs of test functions on X.

Remark. Let $\mathfrak{U} = (U_\iota)_{\iota \in I}$ be an open covering of X. Since X is paracompact, there exists for $\mathfrak{U}$ a subordinate partition of unity, that is, a system $(t_\iota)_{\iota \in I}$ of test functions with the following properties:

1. $0 \leqslant t_\iota \leqslant 1$ for every $\iota \in I$;
2. $\operatorname{Supp}(t_\iota) \subset U_\iota$ for every $\iota \in I$;
3. the system of sets $\operatorname{Supp}(t_\iota)$ is locally finite;
4. $\sum_{\iota \in I} t_\iota = 1$ [by (3) the sum is finite at each point].

Def. 5.2. Let $\mathscr{S}$ be a sheaf of T-modules over X. $\mathscr{S}$ is called *fine* if for all $x \in X$, $\sigma \in \mathscr{S}_x$ and $t \in T$

1. $t \cdot \sigma = \mathbf{O}$ if $x \notin \operatorname{Supp}(t)$
2. $t \cdot \sigma = \sigma$ if $x \notin \operatorname{Supp}(1 - t)$.

Remarks

1. If $\mathscr{S}_1, \ldots, \mathscr{S}_\ell$ are fine sheaves, then $\mathscr{S}_1 \oplus \cdots \oplus \mathscr{S}_\ell$ is also fine.

2. The sheaf $\mathscr{A}^{p,q}$ of germs of (arbitrarily often differentiable) forms of type (p, q) defined by the pre-sheaf $\{A^{p,q}(U), r_V^U\}$ is clearly a fine sheaf.

The sheaf

$$\mathscr{A}^\ell := \bigoplus_{p+q=1} \mathscr{A}^{p,q}$$

is fine, by (1). Here

$$\Gamma(U, \mathscr{A}^\ell) = \bigoplus_{p+q=1} \Gamma(U, \mathscr{A}^{p,q}) = \bigoplus_{p+q=1} A^{p,q}(U) = A^\ell(U),$$

that is, $\mathscr{A}^\ell$ is the sheaf of germs of arbitrarily often differentiable ℓ-forms.

Theorem 5.1. *Let $\mathscr{S}$, $\mathscr{S}'$ be fine sheaves over X, $\varphi: \mathscr{S} \to \mathscr{S}'$ an epimorphism of sheaves of T-modules. Then $\varphi_*: \Gamma(X, \mathscr{S}) \to \Gamma(X, \mathscr{S}')$ is surjective.*

Proof

1. Let $s' \in \Gamma(X, \mathscr{S}')$, $x \in X$. Then there exist a $\sigma \in \mathscr{S}_x$ with $\varphi(\sigma) = s'(x)$, a neighborhood $W(x) \subset X$ and a section $s^* \in \Gamma(W, \mathscr{S})$ with $s^*(x) = \sigma$, so that $\varphi \circ s^*(x) = s'(x)$. We can find a neighborhood $U_x(x) \subset W$ with $\varphi \circ s^*|U_x = s'|U_x$. Let $s_{(x)} := s^*|U_x$.

2. $\mathfrak{U} = \{U_x : x \in X\}$ is an open covering of X. Let $(t_{(x)})_{x \in X}$ be a subordinate partition of unity. For $x \in X$ $t_{(x)} \cdot s_{(x)}$ is an element of $\Gamma(X, \mathscr{S})$. Since the system of sets $\operatorname{Supp}(t_{(x)})$ is locally finite, for fixed x_0 we have $t_{(x)} \cdot s_{(x)}(x_0) = \mathbf{0}$ for almost all $x \in X$. Therefore

$$s := \sum_{x \in X} t_{(x)} \cdot s_{(x)}$$

is also an element of $\Gamma(X, \mathscr{S})$ and

$$\begin{aligned}(\varphi \circ s)(x_0) &= \varphi\left(\sum_{x \in X} t_{(x)} \cdot s_{(x)}(x_0)\right) = \sum_{x \in X} t_{(x)} \cdot \varphi(s_{(x)}(x_0)) = \sum_{x \in X} t_{(x)} \cdot s'(x_0) \\ &= \sum_{x \in E} (t_{(x)} \cdot s'(x_0)) = \left(\sum_{x \in E} t_{(x)}\right) \cdot s'(x_0) = s'(x_0),\end{aligned}$$

where E is a finite set and $\sum_{x \in E} t_{(x)} \equiv 1$ near x_0. □

Theorem 5.2. *If $\mathscr{S}$ is fine, then $H^\ell(X, \mathscr{S}) = 0$ for $\ell \geqslant 1$.*

Proof. Let $0 \to \mathscr{S} \to \mathscr{S}_0 \to \mathscr{S}_1 \to \mathscr{S}_2 \to \cdots$ be the canonical flabby resolution of $\mathscr{S}$. $\mathscr{S}$ and all the $\mathscr{S}_\nu$ are sheaves of T-modules. By induction on ν we show that the $\mathscr{S}_\nu$ are all fine.

$\mathscr{S}_0 = W(\mathscr{S})$ is defined by the pre-sheaf $\{\hat{\Gamma}(U, \mathscr{S}), r_V^U\}$. $\hat{\Gamma}(U, \mathscr{S})$ is a T-module with $t \cdot s = 0$ if $\operatorname{Supp}(t) \cap U = \varnothing$ and $t \cdot s = s$ if $\operatorname{Supp}(1 - t) \cap U = \varnothing$. Therefore $\mathscr{S}_0$ is fine.

Now let $\mathscr{S}_0, \ldots, \mathscr{S}_\ell$ be fine and $\ell \geqslant 0$. The homomorphisms which appear take into account the T-module structure. Therefore the subsheaves

$$\mathscr{B}_i := \operatorname{Im}(\mathscr{S}_{i-1} \to \mathscr{S}_i)$$

are fine for $i = 0, \ldots, \ell$ and $\mathscr{S}_{-1} := \mathscr{S}$; hence $\mathscr{S}_{\ell+1} = W(\mathscr{S}_\ell/\mathscr{B}_\ell)$ is fine.

Since all the sheaves $\mathscr{K}_i := \operatorname{Ker}(\mathscr{S}_i \to \mathscr{S}_{i+1})$ are fine, we obtain epimorphisms of fine sheaves: $\mathscr{S}_{i-1} \twoheadrightarrow \mathscr{K}_i = \mathscr{B}_i$. By Theorem 5.1, $\Gamma(X, \mathscr{S}_{i-1}) \to$

$\Gamma(X, \mathscr{K}_i)$ is also surjective, and therefore

$$\operatorname{Im}(\Gamma(X, \mathscr{S}_{i-1}) \to \Gamma(X, \mathscr{S}_i)) = \operatorname{Ker}(\Gamma(X, \mathscr{S}_i) \to \Gamma(X, \mathscr{S}_{i+1})),$$

i.e., $H^i(X, \mathscr{S}) = 0$ for $i \geqslant 1$. □

Def. 5.3. The sheaf of germs of holomorphic $(p, 0)$-forms on X will be denoted by Ω^p. A holomorphic $(p, 0)$-form $\varphi = \varphi^{(p,0)}$ has a *local representation*

$$\varphi = \sum_{1 \leqslant i_1 < \cdots < i_p \leqslant n} a_{i_1 \cdots i_p}\, dz_{i_1} \wedge \cdots \wedge dz_{i_p},$$

with holomorphic coefficients $a_{i_1 \ldots i_p}$.

Thus the sheaf Ω^p is locally isomorphic to the (free) sheaf $\binom{n}{p} \cdot \mathcal{O}$. We also call Ω^p a locally free sheaf. In particular Ω^p is coherent.

There is a canonical injection $\varepsilon: \Omega^p \hookrightarrow \mathscr{A}^{p,0}$ and the differential

$$d'': A^{p,q}(U) \to A^{p,q+1}(U)$$

induces homomorphisms of sheaves of abelian groups:

$$d'': \mathscr{A}^{p,q} \to \mathscr{A}^{p,q+1}.$$

Theorem 5.3. *The following sheaf-sequence is exact:*

$$0 \to \Omega^p \overset{\varepsilon}{\hookrightarrow} \mathscr{A}^{p,0} \overset{d''}{\to} \mathscr{A}^{p,1} \overset{d''}{\to} \mathscr{A}^{p,2} \to \cdots$$

PROOF

1. It is clear that $d'' \circ \varepsilon = 0$ and $d'' \circ d'' = 0$.

2. Let $x \in X$, U be a coordinate neighborhood of x in X. An element $\varphi \in A^{p,q}(U)$ has the form

$$\varphi = \sum_{1 \leqslant i_1 < \cdots < i_p \leqslant n} dz_{i_1} \wedge \cdots \wedge dz_{i_p} \wedge \varphi_{i_1 \cdots i_p},$$

with $\varphi_{i_1 \cdots i_p} \in A^{0,q}(U)$. Therefore

$$d''\varphi = \sum_{1 \leqslant i_1 < \cdots < i_p \leqslant n} (-1)^p\, dz_{i_1} \wedge \cdots \wedge dz_{i_p} \wedge d''\varphi_{i_1 \cdots i_p},$$

and so $d''\varphi = 0$ implies that $d''\varphi_{i_1 \cdots i_p} = 0$ for all $i_1, \ldots, i_p$.

According to Dolbeault there are neighborhoods $U_{i_1 \cdots i_p}$ of x with $U_{i_1 \cdots i_p} \subset U$, as well as forms $\psi_{i_1 \cdots i_p}$ of type $(0, q-1)$ on $U_{i_1 \cdots i_p}$ such that $d''\psi_{i_1 \cdots i_p} = \varphi_{i_1 \cdots i_p} | U_{i_1 \cdots i_p}$. Let U' be the intersection of all sets $U_{i_1 \cdots i_p}$ and

$$\psi := \sum_{1 \leqslant i_1 < \cdots < i_p \leqslant n} (-1)^p\, dz_{i_1} \wedge \cdots \wedge dz_{i_p} \wedge \psi_{i_1 \cdots i_p} | U'.$$

Then $d''\psi = \varphi | U'$.

3. Let $\sigma \in \mathscr{A}^{p,0}$, $x \in U$, U an open neighborhood of x and $\varphi \in A^{p,0}(U)$ such that $\sigma = r\varphi(x)$. $\mathbf{0} = d''\sigma = r(d''\varphi)(x)$ if and only if $d''\varphi = 0$ near x,

if therefore φ is holomorphic near x. That means $\sigma \in \Omega^p$. The sequence is exact at $\mathscr{A}^{p,0}$.

4. Let $q \geqslant 1$, $\sigma \in \mathscr{A}_x^{p,q}$, U a neighborhood of x and $\varphi \in A^{p,q}(U)$ be such that $\sigma = r\varphi(x)$. $\mathbf{O} = d''\sigma = r(d''\varphi)(x)$ if and only if $d''\varphi = 0$ near x, and without loss of generality, on U.

By (2) this is equivalent to the existence of a neighborhood $U'(x) \subset U$ and a $\psi \in A^{p,q-1}(U')$ with $d''\psi = \varphi | U'$, and that is equivalent to $\sigma = r\varphi(x) = r(d''\psi)(x) = d''(r\psi)(x)$. Therefore the sequence is exact at $\mathscr{A}^{p,q}$. □

Def. 5.4. The induced sequence

$$\mathrm{O} \to \Gamma(X, \Omega^p) \xrightarrow{\varepsilon} \Gamma(X, \mathscr{A}^{p,0}) \xrightarrow{d''} \Gamma(X, \mathscr{A}^{p,1}) \to \cdots$$

is called the *Dolbeault sequence*. Clearly we have an augmented cochain complex (of $\mathbb{C}$-vector spaces). The associated cohomology groups

$$H^{p,q}(X) := \frac{\operatorname{Ker}(\Gamma(X, \mathscr{A}^{p,q}) \to \Gamma(X, \mathscr{A}^{p,q+1}))}{\operatorname{Im}(\Gamma(X, \mathscr{A}^{p,q-1}) \to \Gamma(X, \mathscr{A}^{p,q}))}$$

are called the *Dolbeault groups*.

Theorem 5.4 (Dolbeault)

$$H^{p,q}(X) \simeq H^q(X, \Omega^p) \qquad \text{for} \qquad q \in \mathbb{N}_0.$$

PROOF. Let $\mathrm{O} \to \mathscr{A}^{p,q} \to \mathscr{A}_0^q \to \mathscr{A}_1^q \to \mathscr{A}_2^q \to \cdots$ be the canonical flabby resolutions of the sheaves $\mathscr{A}^{p,q}$ (all $\mathscr{A}_\nu^q$ are fine!). Let $C_{\nu\mu} := \Gamma(X, \mathscr{A}_\mu^\nu)$ for $\nu, \mu \in \mathbb{N}_0$. Let $\delta' : C_{\nu\mu} \to C_{\nu+1,\mu}$ and $\delta'' : C_{\nu\mu} \to C_{\nu,\mu+1}$ be the homomorphisms induced by the flabby solution $\mathrm{O} \to \Omega^p \to \mathscr{S}_0 \to \mathscr{S}_1 \to \cdots$ and the Dolbeault sequence, with signs so that $(C_{\nu\mu}, \delta', \delta'')$ is a double complex. We obtain the following diagram:

$$\begin{array}{ccccccccccc}
 & & 0 & & 0 & & 0 & & 0 & & \\
 & & \downarrow & & \downarrow & & \downarrow & & \downarrow & & \\
0 & \longrightarrow & \Gamma(X, \Omega^p) & \longrightarrow & \Gamma(X, \mathscr{S}_0) & \longrightarrow & \Gamma(X, \mathscr{S}_1) & \longrightarrow & \Gamma(X, \mathscr{S}_2) & \longrightarrow & \cdots \\
 & & \downarrow & & \downarrow & & \downarrow & & \downarrow & & \\
0 & \longrightarrow & \Gamma(X, \mathscr{A}^{p,0}) & \longrightarrow & C_{00} & \xrightarrow{\delta''} & C_{01} & \xrightarrow{\delta''} & C_{02} & \xrightarrow{\delta''} & \cdots \\
 & & \downarrow & & \downarrow{\scriptstyle\delta'} & & \downarrow{\scriptstyle\delta'} & & \downarrow{\scriptstyle\delta'} & & \\
0 & \longrightarrow & \Gamma(X, \mathscr{A}^{p,1}) & \longrightarrow & C_{10} & \xrightarrow{\delta''} & C_{11} & \xrightarrow{\delta''} & C_{12} & \xrightarrow{\delta''} & \cdots \\
 & & \downarrow & & \downarrow{\scriptstyle\delta'} & & \downarrow{\scriptstyle\delta'} & & \downarrow{\scriptstyle\delta'} & & \\
0 & \longrightarrow & \Gamma(X, \mathscr{A}^{p,2}) & \longrightarrow & C_{20} & \xrightarrow{\delta''} & C_{21} & \xrightarrow{\delta''} & C_{22} & \xrightarrow{\delta''} & \cdots \\
 & & \downarrow & & \downarrow{\scriptstyle\delta'} & & \downarrow{\scriptstyle\delta'} & & \downarrow{\scriptstyle\delta'} & &
\end{array}$$

All the hypotheses of Theorem 3.1 of Chapter VI are satisfied, so

$$H^{p,q}(X) \simeq H_{q0}, H^q(X, \Omega^p) \simeq H_{0q}.$$

Since $H^i(X, \mathscr{A}^{p,q}) = 0$ (for $i \geqslant 1$) the δ''-sequences are exact. Since the sequence $0 \to \Omega^p \to \mathscr{A}^{p,0} \to \mathscr{A}^{p,1} \to \cdots$ is exact and $\mathfrak{W}$ is an exact functor, all sequences $0 \to \mathscr{S}_\nu \to \mathscr{A}_\nu^0 \to \mathscr{A}_\nu^1 \to \cdots$ are exact. Since all sheaves are flabby, the δ'-sequences are exact. Therefore $H_{0q} \simeq H_{q0}$, and the theorem is proved. □

Theorem 5.5. *Let X be a Stein manifold, $q \geqslant 1$. If φ is a form of the type (p, q) on X with $d''\varphi = 0$, then on X there exists a form ψ of the type $(p, q - 1)$, with $d''\psi = \varphi$.*

PROOF. By Theorem B $H^q(X, \Omega^p) = 0$ for $q \geqslant 1$; therefore $H^{p,q}(X) = 0$ for $q \geqslant 1$. □

Remarks. With the help of Poincaré's Lemma one shows that the sequence $0 \to \mathbb{C} \overset{\varepsilon}{\hookrightarrow} \mathscr{A}^0 \overset{d}{\to} \mathscr{A}^1 \overset{d}{\to} \mathscr{A}^2 \to \cdots$ is exact. The associated cohomology groups

$$H^r(X) := \operatorname{Ker}(\Gamma(X, \mathscr{A}^r) \to \Gamma(X, \mathscr{A}^{r+1}))/\operatorname{Im}(\Gamma(X, \mathscr{A}^{r-1}) \to \Gamma(X, \mathscr{A}^r))$$

are called the *de Rham groups*. As above one shows

Theorem 5.6. $H^r(X) \simeq H^r(X, \mathbb{C})$ *for* $r \geqslant 0$.

Since

$$\mathscr{A}^\ell = \bigoplus_{p+q=\ell} \mathscr{A}^{p,q}$$

we would expect that a connection between the topological cohomology groups $H^r(X, \mathbb{C})$ and the analytically defined cohomology groups $H^q(X, \Omega^p)$ exists. That is in fact the case. If, for example, X is a Kähler manifold (for example, a projective-algebraic manifold), then according to Kodaira,

$$H^r(X, \mathbb{C}) \simeq \bigoplus_{p+q=r} H^q(X, \Omega^p).$$

As a consequence one obtains:

$$B_1(X) = 2p = \text{even};$$

on X there exist p linearly independent differentials of the first kind, that is, elements of $\Gamma(X, \Omega^1)$.

List of Symbols

Bibliography

Textbooks

1. Abhyankar, S. S.: *Local Analytic Geometry*. New York: Academic Press, 1964.
2. Behnke, H., Thullen, P.: Theorie der Funktionen mehrerer komplexer Veränderlichen. *Ergebn. d. Math.*, Bd. 51, 2. erw. Auflage. Berlin–Heidelberg–New York: Springer 1970.
3. Cartan, H.: *Elementary Theory of Analytic Functions of One or Several Complex Variables*. New York: Addison & Wesley, 1963.
4. Fuks, B. A.: *Introduction to the Theory of Analytic Functions of Several Complex Variables*. Transl. of Math. Monogr., 8. Providence, Rhode Island: American Mathematical Society, 1963.
5. Fuks, B. A.: *Special Chapters in the Theory of Analytic Functions of Several Complex Variables*. Transl. of Math. Monogr., 14. Providence, Rhode Island: American Mathematical Society, 1965.
6. Grauert, H., Remmert, R.: *Analytische Stellenalgebren*. Grundl. d. math. Wiss., Bd. 176. Berlin–Heidelberg–New York: Springer, 1971.
7. Gunning, R. C., Rossi, H.: *Analytic Functions of Several Complex Variables*. Englewood Cliffs, N.J.: Prentice-Hall, 1965.
8. Hörmander, L.: *An Introduction to Complex Analysis in Several Variables*. Princeton, N.J.: Van Nostrand, 1966.
9. Vladimirov, V. S.: *Les Fonctions de Plusieurs Variables Complexes (et leur application á la théorie quantique des champs)*. Paris: Dunod, 1967.

Older (Classical) Books

10. Bochner, S., Martin, W. T.: *Several Complex Variables*. Princeton: Princeton University Press, 1948.
11. Osgood, W. F.: *Lehrbuch der Funktionentheorie*, 2. Bd., 1. Lieferung. Leipzig–Berlin: Teubner, 1924.

Lecture Notes

12. Bers, L.: *Introduction to Several Complex Variables*. New York: Courant Institute of Mathematical Sciences, 1964.
13. Cartan, H.: *Séminaire Ecole Normale Supérieure* 1951/52, 1953/54, 1960/61. Paris.
14. Hervé, M.: *Several Complex Variables*. Tata Institute of Fundamental Research Studies in Math., 1. London. Oxford University Press, 1963.
15. Malgrange, B.: *Lectures on the Theory of Functions of Several Complex Variables*. Bombay: Tata Institute of Fundamental Research, 1958.
16. Narasimhan, R.: *Introduction to the Theory of Analytic Spaces*. Lecture Notes in Mathematics, Vol. 25. Berlin–Heidelberg–New York: Springer, 1966.
17. Schwartz, L.: *Lectures on Complex Analytic Manifolds*. Bombay: Tata Institute of Fundamental Research, 1955.

More Advanced and Supplementary Books

18. *Colloquium über Kählersche Mannigfaltigkeiten*. Göttingen: Ausarbeitung des Mathemat. Inst., 1961.
19. Ehrenpreis, L.: *Fourier Analysis in Several Complex Variables*. Pure and Applied Mathematics, Vol. 17. New York: Wiley-Interscience Publishers, 1970.
20. Godement, R.: *Toplologie algébrique et théorie des faisceaux*. Paris: Hermann, 1964.
21. Grauert, H., Fischer, W.: *Differential- und Integralrechnung II*. Heidelberger Taschenbücher, 36. Berlin–Heidelberg–New York: Springer, 1968.
22. Grauert, H., Lieb, I.: *Differential- und Integralrechnung III*. Heidelberger Taschenbücher, Bd. 43. Berlin–Heidelberg–New York: Springer, 1968.
23. Hirzebruch, F., Scheja, G.: *Garben- und Cohomologietheorie*. Ausarb. math. und physik. Vorl., 20. Münster, Westf.: Aschendorffsche Verlagsbuchhandlung, 1957.
24. De Rham, G.: *Variétés différentiables*. Paris: Hermann, 1960.
25. Weil, A.: *Introduction à l'étude des variétés kählériennes*. Paris: Hermann, 1958.

Index

Q

R

S

T

U

W

Z

GPSR Compliance

The European Union's (EU) General Product Safety Regulation (GPSR) is a set of rules that requires consumer products to be safe and our obligations to ensure this.

If you have any concerns about our products, you can contact us on ProductSafety@springernature.com

In case Publisher is established outside the EU, the EU authorized representative is:

Springer Nature Customer Service Center GmbH
Europaplatz 3
69115 Heidelberg, Germany

Batch number: 10682323

Printed by Printforce, the Netherlands